The Cosmography of Paradise: The Other World from Ancient Mesopotamia to Medieval Europe

Warburg Institute Colloquia
Edited by Charles Burnett and Jill Kraye

27

The Cosmography of Paradise: The Other World from Ancient Mesopotamia to Medieval Europe

Edited by Alessandro Scafi

The Warburg Institute
London 2016

Published by
The Warburg Institute
School of Advanced Study
University of London
Woburn Square
London WC1H 0AB

ISBN 978–1–908590–50–3
ISSN 1352–9986

Typeset by Waveney Typesetters, Wymondham, Norfolk
Printed by Henry Ling, The Dorset Press, Dorchester, Dorset

Table of Contents

Contributors

Jan N. Bremmer is Emeritus Professor of Religious Studies at the University of Groningen. He has been a guest professor at many universities and institutions, including the Getty Research Institute (2006–2007), the Internationales Kolleg Morphomata (Cologne: 2010–2011), the Institute for the Study of Ancient World (New York: 2012–13) and the Max-Weber-Kolleg (Erfurt: 2013–14) as well as the Universities of Edinburgh (2007), Munich (2011–12) and Freiburg (2014). He has published widely, in particular on Greek, Roman, Early Christian and Contemporary Religion as well as on Social History and History of Scholarship. His most recent publications include *The Rise and Fall of the Afterlife* (2002); *Greek Religion and Culture, the Bible and the Ancient Near East* (2008); *The Rise of Christianity through the Eyes of Gibbon, Harnack and Rodney Stark* (2010); *Initiation into the Mysteries of the Ancient World* (2014); and, as editor, *The Strange World of Human Sacrifice* (2007); *The Gods of Ancient Greece* (2010); *The Pseudo-Clementines* (2010); *Perpetua's Passions* (2012); *The Materiality of Magic* (2015), *The Ascension of Isaiah* (2015) and *Thecla: Paul's Disciple and Saint in the East and West* (2015).

Markham J. Geller is Jewish Chronicle Professor at UCL, Department of Hebrew and Jewish Studies, but is on secondment to the Freie Universität Berlin as Gastprofessor für Wissensgeschichte. His most recent book is *Melothesia in Babylonia* (2015). His research interests include: Ancient Babylonian Medicine, as it appears in Akkadian on cuneiform tablets and in Aramaic of the Babylonian Talmud; History of ancient science (*Wissensgeschichte*); Ancient Magic from Mesopotamia, found in Sumerian, Akkadian and Aramaic texts.

Antonio C. D. Panaino is Professor of Iranian Studies at the University of Bologna. His scientific interests include: Avestan and generally Mazdean literatures; history of religions in Pre-Islamic Iran; the mutual influences between Byzantium and Sasanian; astronomy and astrology in the ancient world; ethnolinguistics of the Yaghnobi and of the Iranian area; preservation of the ethno-cultural heritage in areas of crisis and conflict; history of games in Antiquity. He is author of various books and more than two hundred scientific articles. His most recent publications include *The Lists of Names of Ahura Mazdā (Yašt I) and Vayu (Yašt XV)* (2002); *I Magi evangelici* (2004); *Rite, parole et pensée dans l'Avesta ancien et récent* (2004); *Chronologia Avestica: tra cronologia linguistica e storia religiosa* (2007); *Politica religiosa e regalità sacra nell'Iran preislamico* (2007); *I Magi e la loro stella* 2012); *Mortality and Immortality: Yama's / Yima's Choice and the Primordial Incest* (2013).

Annette Yoshiko Reed (BA McGill; MTS Harvard; MA, PhD Princeton) is Associate Professor in the Department of Religious Studies and Program in Jewish Studies at the

University of Pennsylvania, where she also serves as Faculty Master of Fisher Hassenfeld College House. Her research spans Second Temple Judaism, early Christianity, and Jewish/Christian relations in Late Antiquity. Her publications include *Fallen Angels and the History of Judaism and Christianity* (2005); *Heavenly Realms and Earthly Realities in Late Antique Religions* (ed. with R. Boustan, 2004); *The Ways that Never Parted: Jews and Christians in Late Antiquity and the Early Middle Ages* (ed. with A.H. Becker, 2003 and 2007), and *Jews, Christians, and the Roman Empire* (ed. with N. Dohrmann, 2013). She is currently working on two monographs: one on the origins of Jewish angelology and demonology, and the other on the Pseudo-Clementine *Homilies* and the history of 'Jewish Christianity'.

Einar Thomassen, born in Bergen, Norway, received his PhD from the University of St Andrews in 1982 and taught at the universities of Uppsala and Oslo before his appointment as Professor of Religious Studies at the University of Bergen in 1993. While his main work centres on ancient Gnosticism, his interests and publications range widely over the religions of Antiquity, the formative periods of Christianity and Islam, Sufism, philological method and comparative studies. His main publications are *Le Traité Tripartite* (1989); *The Letters of Ahmad b. Idris* (1993); *The Spiritual Seed: The Church of the "Valentinians"* (2006); *Canon and Canonicity* (2010). He has also published several text books and books for a larger audience in Norwegian and numerous articles on a variety of topics.

Yuri Stoyanov obtained his PhD in Combined Historical/Religious Studies from the University of London (The Warburg Institute). He is based at the Department of the Near and Middle East, School of Oriental and African Studies, University of London and is currently a Senior Fellow at the Albright Institute of Archaeological Research, Jerusalem and a scholar in residence at the Tantur Ecumenical Institute, Jerusalem. His past assignments include Oxford and Wingate fellowships, British Academy awards, etc.; he is on the editorial board of several international academic journals and was Director of the Kenyon Institute (formerly the British School of Archaeology) in Jerusalem. He has lectured and published widely on various facets of the interaction between heretical and heterodox currents in Zoroastrianism, Judaism, Christianity and Islam and their survivals into the modern era.

His publications include *The Hidden Tradition in Europe* (1994); *The Other God* (2000); *Defenders and Enemies of the True Cross* (2011); his edited books include *Christian Dualist Heresies in the Byzantine World c. 650–c. 1450* (1998).

Dimitris J. Kyrtatas is Professor of Ancient History. He was born in Athens and educated in Thessaloniki and London. From 1985 until 2001 he lectured at the University of Crete and in 2002 moved to the University of Thessaly. His main field of research is the social and religious history of the Greek world in the Roman period. He has published many books and articles on related topics. His publications include *The Social Structure of the Early Christian Communities* (1987); 'The origins of Christian Hell', *Numen* 56.2–3 (2009); 'Historical aspects in the formation of the New Testament Canon', in *Canon and Canonicity* (2010); 'Slavery and economy in the Greek world', in *Cambridge World History of Slavery* (2011); 'Early Christianity in Macedonia' in *Brill's Companion to Ancient Macedon* (2011); 'Living in tombs: The secret of an early Christian mystical experience', in *Mystery and Secrecy in the Nag Hammadi Collection and Other Ancient Literature* (2012).

Sergey Minov received his doctorate in 2013 from the Hebrew University of Jerusalem. He completed his thesis, entitled *Syriac Christian Identity in Late Sasanian Mesopotamia: The Cave of Treasures in Context*, at the Department of Comparative Religion. He devised and keeps updating a long-term digital project *The Comprehensive Bibliography on Syriac Christianity*, an on-line collection of bibliographical data covering all topics related to the history and culture of Syriac-speaking Christians [http://www.csc.org.il/db/db.aspx?db=SB]. At present he is a research fellow at the University of Oxford, participating in the *The Cult of Saints*, a major five-year project, based at the Faculty of History at the University of Oxford, which will investigate the origins and development of the cult of Christian saints up until the end of Late Antiquity. His research interests include the history and culture of Syriac Christianity in Late Antiquity and the Middle Ages, Jewish-Christian relations in the Near East, Jewish and Christian traditions of biblical exegesis and apocryphal literature. His publications include (with A. Kulik), *Biblical Pseudepigrapha in Slavonic Tradition* (2015); 'The Story of Solomon's Palace at Heliopolis', *Le Muséon* 123:1–2 (2010).

Danuta Shanzer, a classicist and medievalist, was educated at Bryn Mawr College and Oxford University. She has taught at Oxford, the University of Manchester, the University of California, Berkeley, Harvard University, Cornell University (where she directed the Medieval Studies Program for nine years), and the University of Illinois at Urbana-Champaign. In 2011 she took up a position as Professor of Late Antique and Medieval Latin Philology at the University of Vienna. She is a Fellow of the Medieval Academy of America and a Fellow of the Österreichische Akademie der Wissenschaften. She specializes in Latin literature and the social and religious history of Late Antiquity and the Early Middle Ages. Some favorite authors include Martianus Capella, Prudentius, Avitus of Vienne, and Gregory of Tours; special interests include epistolography and hagiography. Her work ranges from the philological and literary to the historical and theological, involves most of the barbarian successor kingdoms, and includes a long-term project on the development of various judicial ordeals in Late Antiquity and the early

Middle Ages. She serves on the editorial boards of numerous journals and is Latin Editor for the Dumbarton Oaks Medieval Library.

Veronica della Dora is Professor of Human Geography at Royal Holloway University of London. She previously served as a Senior Lecturer at the University of Bristol and has been a fellow at Dumbarton Oaks (Washington, DC), the Getty Research Institute (Los Angeles), and the American School of Classical Studies at Athens. Her research interests and publications span historical and cultural geography, the history of cartography and Byzantine studies with a specific focus on sacred space and landscape. She is the author of *Imagining Mount Athos: Visions of a Holy Place from Homer to World War II* (2011) and *Landscape, Nature and the Sacred in Byzantium* (2016).

Rudolf Simek (PhD Vienna, Mtheol Vienna) held academic positions in Edinburgh (1976–79) and Vienna (1980–1995) and is currently Professor of Medieval German and Scandinavian Literature in Bonn, Germany. He held long-term guest professorships in Tromsø and Sydney, and shorter ones in Catania, Cagliari, Rome, Durham, Reykjavík, Bergen, Aarhus, and Rzeszow. He has published several books on early Germanic Religion and Culture, on Vikings and Viking ships (most recently *Die Schiffe der Wikinger*, 2014) as well as a series of translations of Old Norse Sagas into German, and numerous books and articles on medieval science and literature. His research interests include early medieval religion, Viking and medieval Norse studies, as well as late medieval religious literature and the history of science in the Middle Ages, especially in the field of astronomy and cosmography. He served as an advisor and/or presenter for several TV-documentaries and one feature film. Among his most recent publications are *Artus-Lexikon* (2012), *Religion und Mythologie der Germanen* (2nd ed. 2014) and *Monster im Mittelalter* (2015).

Christian Lange is Professor of Arabic and Islamic Studies at Utrecht University. His research focuses on the history of premodern Islamic law and theology, in particular criminal law and doctrines and practices related to the Muslim afterlife. He is the author of *Justice, Punishment and the Medieval Muslim Imagination* (2008), a study of notions of justice, both in this world and the next, in late medieval Islam, as well as the co-editor of two volumes in medieval Islamic history. His monograph *Paradise and Hell in Islamic Traditions*, a cultural history of the otherworld from the beginnings of Islam to the eve of modernity, was published in 2015. In addition, Lange is the editor of collection of articles on the Islamic hell, entitled *Locating Hell in Islamic Traditions* (2015). From 2011–15, he was the Principal Investigator of the research project *The Here and the Hereafter in Islamic Traditions*, funded by the European Research Council and hosted at Utrecht University [http://hhit.wp.hum.uu.nl/].

Emilie Savage-Smith, FBA, is recently retired as Professor of the History of Islamic Science at the Oriental Institute, University of Oxford. She continues as Fellow Archivist of St Cross College and is the recent recipient of a Senior Investigator Award in Medical Humanities from the Wellcome Trust for the project *A Literary History of Medicine: The 'Best Accounts of the Classes of Physicians' by Ibn Abi Usaybiʿah (d. 1270).* Her most recent publications include *A New Catalogue of Arabic Manuscripts in the Bodleian Library, University of Oxford. Volume I: Medicine* (2012) and (with Yossef Rapoport) *An Eleventh-Century Egyptian Guide to the Universe: The 'Book of Curiosities', edited with an annotated translation* (2013).

Anders Hultgård is Professor Emeritus in history of religions at the University of Uppsala. He was a visiting professor at the University of California, Santa Barbara in spring 1990, at the École Pratique des Hautes Études in 1998 and 2001, and at the University Marc Bloch of Strasbourg in 2000 for half a year. He worked as a research fellow at the Centre for Advances Studies (CAS) in Oslo the academic year 2007/2008 and at the Swedish Collegium for Advanced Study the autumn 2010. He has been *Correspondant étranger de l'Académie des Inscriptions et Belles Lettres* in Paris since 2004.

His research interests include ancient Scandinavian religion, Zoroastrianism, and the religions of the Hellenistic-Roman world. Among his main publications are: *Apocalyptique iranienne et dualisme qoumrânien* (1995, with M. Philonenko and G. Widengren); 'Das Paradies: vom Park des Perserkönigs zum Ort der Seligen', in *La Cité de Dieu. Die Stadt Gottes* (2000); 'Óðinn, Valhǫll and the Einherjar. Eschatological Myth and Ideology in the Late Viking Period', in *Ideology and Power in the Viking and Middle Ages* (2011).

Acknowledgements

The colloquium *The Cosmography of Paradise: The Other World from Ancient Mesopotamia to Medieval Europe* took place in London on the 4th, 5th and 6th March 2009. It was organised by Alessandro Scafi (The Warburg Institute), in cooperation with Markham J. Geller (University College London). The event was supported by The Warburg Institute, The Institute of Jewish Studies, University College London, The British Institute for the Study of Iraq, the University of Thessaly (Volos), the University of Groningen, and by the School of Advanced Study.

My thanks to Jenny Boyle and Charles Burnett for their precious assistance in the editorial process and to Francis Herbert, who was of great help in ensuring that both the prose and all the footnote and bibliographical references of the essays contained in this volume were consistent and correct. In connection with the project as a whole, including the original colloquium, I would like to extend my thanks to a large number of people, especially: Adelina Angusheva-Tikhanova, Florentina Badalanova-Geller, Jens Braarvig, Corin Braga, Sebastian Brock, John Carey, David Chambers, Simon Franklin, Markham J. Geller, Boriana Hristova, Liz James, Karl-Gunnar A. Johansson, Petr Kiztler, Christian Lange, Christopher Ligota, Nanno Marinatos, Roland Mayer, Anissava Miltenova, Michael Paschalis, Elena Paskaleva, François Quiviger, Will Ryan, Gro Steinsland, Yuri Stoyanov, Pétér Toth, Martin West, Nicolas Wyatt, and Philip Young. My special thanks also go to the anonymous readers of the various essays.

The core group of contributors gave papers at the Warburg colloquium in March 2009. Unfortunately, it was not possible to include some of the original papers but I would like to thank all those who contributed to this endeavour. This publication includes contributions from scholars who did not participate in the colloquium. The aim has been to offer a rich comparative investigation of the relationship between ideas of paradise and cosmography in history, involving scholars of different fields and from different intellectual backgrounds.

Alessandro Scafi

Abbreviations

CCSL = *Corpus Christianorum Series Latina*.

CSEL = *Corpus Scriptorum Ecclesiasticorum Latinorum*.

PG = *Patrologia Graeca*.

PL = *Patrologia Latina*.

SC = *Sources Chrétiennes*.

Classical texts are cited by author and title, with references according to the standard division of the texts. Punctuation and spelling of Latin (especially with regard to the distinction between *v* and *u*, *&* and *et*; and the vowel *j*) have been normalized; accents have been introduced or changed according to modern usage; abbreviations have been expanded.

Introduction

Alessandro Scafi

Most of the essays presented in this volume were first delivered in a colloquium held at the Warburg Institute in March 2009. Just few days after the Warburg event on paradise the announcement was made, with perfect timing, that the paradise on earth of the Bible had been found. A bestselling thriller was published in London, *The Genesis Secret*, focused on a Neolithic archeological site in Kurdish Turkey. Its author reported how the site had been discovered on the rolling plains of eastern Turkey by an old, solitary Kurdish shepherd, who noticed carved oblong stones standing in the ground. British journalist Sean Thomas, under the pseudonym of Tom Knox, claimed that German archaeologist Klaus Schmidt, in charge of the excavations at the site of Göbelki Tepe, told him that those ruins were formerly a Temple in Eden, the world's most ancient building. Carbon-dating showed that this sort of Turkish Stonehenge was at least 12,000 years old, pre-pottery, pre-writing, pre-everything. The claim that the writers of the Bible, when talking of Eden, were in effect describing a corner of Kurdish Turkey was a stunning and seductive idea for a fiction book that was meant to sell well. Shortly afterwards, however, the German Archeological Institute in Instanbul issued a press release to denounce a distorsion of their scientific work. Klaus Schmidt never linked the site of Göbelki Tepe, however important, with the Garden of Eden, and his interview had been falsified for the sake of publicity.

Paradise is a problem, has always been and continues to be, also when claims of amazing finds concerning its site cannot withstand later and serious scrutiny. Paradise is a notion which is universal, to which people relate easily and debate surrounding the topic can help build bridges between the academic community and the general public. With this in mind we have gathered in the present volume the work of a number of scholars who have approached the theme of paradise in their different fields. We hope to offer an historical account on the early development – through time and over space – of the notion of paradise, an account dealing with diverse materials and issues yet structured around a common theme and based on sound research.

Paradise is not only paradise lost, as Proust once put it. Visions of an attainable paradise are common to all humanity. All religious traditions, including the Abrahamic faiths, tell us about a perfect happiness attainable 'elsewhere' and 'out of time'. A 'paradise' existed yesterday, during a marvellous Golden Age; a 'paradise' will return tomorrow in the glory of a divine eternity; or 'paradise' is already here, in the depth of our soul or in some remote place on earth that we cannot reach. When talking about paradise at the beginning of history, we approach the mystery of a paradise within and at the end; by approaching the past, we go down in the depth and walk back into the future.

The Cosmography of Paradise: The Other World from Ancient Mesopotamia to Medieval Europe, Warburg Institute Colloquia 27, 2016

Where and indeed what is paradise? Is it a silent, peaceful island, or a realm beyond the skies? Is it a mountain with seven layers? A garden with shady trees? Or is it a heavenly city? A state of mind? How do we get to paradise? How do we know about paradise? More prosaicly, perhaps, to what extent does our vision of the other world depend on our experience of this world? Do visions of paradise reflect this world or some other imaginary ideas? Are different perspectives in ancient and medieval thought independent or interrelated? The contributions offered here by scholars coming from different fields but discussing the relevance of the same theme within their own field of work attempt different answers to these questions.

The thought common to all cultures is that paradise is an infinite, perfect state, beyond time and space. Yet, in each culture, through time and over space, the idea of paradise is shaped by that culture's concept of a finite, imperfect world. Hence the idea of the Warburg conference and now of this book, which aims to focus on how the notion of paradise within a diversity of cultures developed and was shared and transmitted from the ancient Near East to Judaism, Christianity and Islam, and, therefore, to the European Middle Ages. Paradise is a beautiful place, a charming notion, a beguiling word. The word 'paradise' was spelt in Persian, Median, Greek, Hebrew, Latin, and throughout the ages, across cultures, from place to place, each act of translation and transmission has been at the same time an act of redefinition.

The Cosmography of Paradise: The Other World from Ancient Mesopotamia to Medieval Europe considers the general theme of paradise from various comparative perspectives. The focus has been on the way the relationship between 'the other world' and the structure of the whole cosmos has been viewed in different ages and traditions around the Mediterranean basin, spanning from the ancient Near East to medieval Europe.

The notion of paradise has always attracted a great deal of scholarly attention. The tendency, however, has been to study the notion within a specific ancient or medieval tradition and through the lens of a particular theme. Those who look for a broader and comparative approach find only general surveys of the idea, usually focused on the Western tradition and including only preliminary references to Antiquity: such as McDannell and Lang's *Heaven: A History* (1988), Delumeau's *Une histoire du paradis* (1992), Bernheim and Stavridès's *Paradis, Paradis* (1991) and Burton Russell, *A History of Heaven* (1997). More focused comparative studies exploring the vision of paradise, however, have also appeared: *Paradise Interpreted*, eds Luttikhuizen et al. (1999), for example, discusses the reception of the biblical accounts in early Jewish writings, early Christian apocryphal and Gnostic literature, early and medieval mainstream Christianity; and *Paradise in Antiquity,* eds Bockmuehl and Stroumsa (2010), explores Jewish and Christian views in Antiquity. Scholarly enquiries have also been produced on the transmission of geographical and cosmographical ideas in Late Antiquity (such as Inglebert, *Interpretatio Christiana: Les Mutations des savoirs*, 2001).

Against this background, the aim of this volume is to investigate further the emergence and development of the notion of paradise from the ancient Near East to the European Middle Ages by focusing on the various ways the relationship between paradise and the general features of the universe has been viewed. This approach has not yet been attempted, but seems to be essential for elucidating further the notion of paradise, defined as a perfect state beyond time and space, in so far as the historical formation of such an idea relied heavily upon a variety of temporally and culturally conditioned concepts of the physical cosmos as a finite and imperfect realm. The conception of the wider cosmos is indeed relevant for a vision of perfection that was projected in another time and space. It seems also important to maintain a rigorous historical approach in order to study the transmission of ideas on paradise from the Near East and Antiquity to the Christian Middle Ages. It is precisely the emphasis on cosmography that allows the discussion of several traditions: Sumerian, ancient Iranian, Greek, Jewish, early Christian, Gnostic, Byzantine, Islamic, Scandinavian, and Latin Western.

The reader will find a discussion of the Greek tradition about paradisiacal lands in Jan N. Bremmer's essay. As Nanno Marinatos explained in her paper delivered at the 2009 Warburg Colloquium, in ancient Greek thought there was no exact equivalent for the Judeo-Christian notion of a carefree and timeless paradise. In the literature of ancient Greece, however, lands of the blessed and ages of immeasurable happiness abound. Bremmer examines references found in Pindar, Homer and Hesiod as well as Plato and the Orphic tradition about blessed lands of plenty. As Michael Paschalis pointed out in his paper delivered at the 2009 Warburg Colloquium, Roman state religion was not concerned with the reward, salvation or punishment of an individual after death, but dealt primarily with the world of the living. The myth of a lost Golden Age was associated with an idealized Roman past and located in early Italy, with the idea of its 'return' promoted in the Augustan Age and later periods. In Virgil's *Fourth Eclogue*, for instance, the poet saluted the beginning of a new era of plenty, peace and justice that was soon to come about. There were Roman poets, philosophers, and scholars who incorporated narratives of afterlife into their works, for instance Cicero's *Dream of Scipio*, which concludes his treatise *On the Commonwealth*, and Book Six of Virgil's *Aeneid*. Both accounts characteristically focus on the fate of the blessed souls which they place in different abodes: the former in the Milky Way and the latter in the Elysian Fields. But these visions were not so much concerned with the fate of man in the afterlife as with Romans and Rome.

In the opening essay, **Jan N. Bremmer** keeps a steady eye on the cosmography of paradise, but also on its contents in order to keep track of continuities and developments. He focuses on the major trajectories of the connection between a garden or *locus amoenus* and the afterlife, starting with Homer's Greek Elysian Fields (about 700 BC) and ending with the Christian vision of paradise in the *Apocalypse of Paul* (about AD 400). Bremmer points out that 'paradise' is originally a Median and Persian word that was adopted by Jews in a Greek-speaking milieu in Egypt. Within a few centuries that environment was conquered by the Romans and eventually became Christian. Not only did all these cultures and religions influence our terminology and representations of paradise, but, in the course of time, paradise even became associated or identified with the Heavenly Jerusalem.

Markham J. Geller explores Sumerian and Babylonian views of the Realm of the Dead, showing their differences. Sumerians adopted a flat-earth cosmology in which the afterlife was found in the 'mountain' or mountains while Semitic peoples ascribed it to an underworld or netherworld which became the dominant cosmological model (Hades, Sheol, Hell, etc.). Egyptian cosmology is less clear but seems to be similar to a Sumerian flat-earth cosmology, at least in earlier periods. The real question, Geller notes, centred around the sun: did it orbit the earth or not?

Antonio C. D. Panaino focuses on ideas about paradise in the Iranian pre-Islamic world. Our modern designations of 'paradise' depends on a Greek word, _parádeisos_, which, in its turn – Panaino reminds us – was clearly a loanword from an Iranian stem, probably Median, such as *_paridaiza-_, i.e. 'a (garden) surrounded by walls', attested only once in Avestan (_Widēwdād_ 3, 18) as _pairi.daēza-_, and once in Later Old Persian as _paradayadām_ ([par^{a}day^{a}dām] instead of *_paridaida-_, but present also in Christian Sogdian as _prdyz_, and in Modern Persian as _pālēz_. Panaino then discusses Zoroastrian doctrines regarding the fate of the individual after death and visions of the final place of resting and of beatification where the souls of the pious Mazdeans were supposed to go in the afterlife, a place thought to be located somewhere above the fourth heaven. In this essay also the Western reception of the Iranian model is explored.

Annette Yoshiko Reed focuses on the earliest extensive Jewish cosmographies – namely, the Enochic _Astronomical Book_ and the _Book of the Watchers_ (_c._ third century BC). She suggests that these sources preserve a poignant transitional moment in Jewish reflection both on the structure of the cosmos and on the fate of humankind. By means of detailed accounts of the travels of the antediluvian sage Enoch to heaven and the ends of the earth, these sources address a number of questions not explicitly explained in earlier biblical literature – including the boundaries between the domains of the spiritual and the fleshly, the locations of Eden and the Tree of Life, and the post-mortem and eschatological fates of the wicked and the righteous.

Einar Thomassen explains that, even though Gnostic texts are remarkably vague about the topographical features of the place where the redeemed will rest, several sources describe the journey towards that place. In these descriptions a certain number of traditional themes are discernable, among them guardians that interrogate the soul and passwords that must be spoken. Thomassen surveys this material, discussing its wider religious and historical background and defines the nature of the Gnostic reinterpretation of these traditional themes.

Yuri Stoyanov charts and analyzes the most significant and indicative adoptions and re-interpretations of traditional notions and narratives associated with biblical paradise in medieval Christian dualism (namely, Paulicianism, Bogomilism and related groups in Eastern Christendom and the radical and moderate dualist versions of Catharism in Western Christendom). His survey of medieval dualist conceptions of biblical paradise has a preliminary character but demonstrates again that the extant doctrinal evidence for Bogomilism and Catharism is complex and polyvalent and in the case of its paradise-related narrative and notions certainly does not represent or reflect medieval heresiological constructs built on earlier heresiological texts and stereotypes. The application of models and approaches drawn from contemporary anthropological and sociological theory to medieval Christian dualism need to be preceded by a proper text-critical treatment of the relevant primary sources which in the case of its treatments of the paradise tradition display unambiguously the patterns and traces of Christian dualist literary reworking and theological re-interpretation in sectarian milieus conversant with medieval redactions of some early Jewish and Christian pseudepigrapha.

Dimitris J. Kyrtatas explores early Christian discussions about the location and nature of paradise, in particular in the communities living in fourth-century Egypt. He points out that discussions about paradise were well on their way already in the second century. In the third century, debates and speculations led various groups to heated controversies and even open conflict. Kyrtatas notes that important material has been preserved by the communities living in fourth-century Egypt, where some Christians appear to have been even searching for paradise in the desert. The stories told among the earliest ascetics and hermits shed light on the overall problem of envisaging the destiny of the righteous dead and of grasping its significance for the living.

Sergey Minov discusses several distinctive traditions on the Gardenof Eden that were current among Syriac-speaking Christians during Late Antiquity, focusing on those aspects of paradisiacal imagery that are unique to Syriac Christianity and set it apart from the rest of ancient Christian traditions of interpretation of the Garden of Eden. Minov analyzes three different aspects of the image of paradise in Syriac tradition: the notion of paradise as a sacred, Templelike, place; the vision of paradise as a cosmic mountain; the use of feminine imagery with regard to the Garden of Eden. This survey allows Minov the opportunity to bring into focus different cultural forces that were at work during the formative period in the history of Syriac Christianity.

Danuta Shanzer discusses some of the intellectual problems involved in including all five senses in visions of the afterlife paradise in late antique Christian tradition. She explains how the senses were organized into a variety of different hierarchies, but taste invariably ended up low on the pecking order. Despite a doctrine of spiritual senses, taste still continued to be considered problematic, and 'tasting paradise' was seen as a momentary visionary action of the living rather than an activity of the blessed dead, as it emerges in martyrs' passions. Shanzer wonders why literary visions of paradise were shy of imagery of taste and scenes of eating and puts forward new interpretations of passages from the *Passio Perpetuae* and Augustine's *Confessions*, to shed light on the function of one particular (somewhat neglected) food in paradise.

Veronica Della Dora focuses on different 'mappings' of the Aristotelian cosmos on Byzantine and post-Byzantine icons. Her paper explores the tension between theological heaven (as a condition of the soul) and 'geographical' heaven (as a dimension encompassing the cosmos) in the Orthodox Christian tradition and the way in which this tension has been represented in Byzantine iconography and cosmography. While icons featuring theophanic moments in the life of Christ offer maps of the cosmos through changing iconographic symbols, della Dora argues that as objects embedded in the church and as part of liturgical performance, they also help 'enact' (theological) heaven in a strong ontological sense – beyond representation.

Rudolf Simek discusses geographical ideas about the location of paradise and surveyes medieval cartographical material. He notes that the extensive genre of visionary literature in medieval Europe offers, surprisingly enough, only rare glimpses of paradise, although purgatory and hell are described in much detail. However, as a place in scholarly cosmography, paradise is mentioned throughout encyclopedic and geographic literature and regularly depicted on medieval world maps. Simek reminds us how paradise was located and even described as an *earthly* region, placed in the east and inaccessible to living men. In this way paradise on earth appears in stark contrast to eschatological iconography, where paradise is shown as a rather symbolic *heavenly* place as the future abode of the saved after the day of judgement.

Christian Lange questions the way Western scholars of Islamic eschatology and cosmology have often stressed that Muslim theologians generally conceive of paradise as being far removed from this world, so remote in fact that humans can under no circumstances aspire to enter it during their lifetime. This idea, he remarks, resonates with the common notion that Islam stresses the remoteness and otherness of God, the lord of paradise. Lange argues, however, that such views underscoring the transcendentalist nature of Muslim conceptions of paradise, should be nuanced in the light of important strands of immanentist thinking about paradise in Islam. The Qurʾān, the Sunna of the Prophet (recorded in the *ḥadīth*), but also the manifold expressions of sacred geography in the Islamic world provide evidence pointing in this direction. Even in Muslim systematic theology, or *kalām*, there was a broad consensus that the world of the here-and-now and paradise co-exist closely in time and in space. Adam's primordial paradise was regularly declared to have been on earth, and to be identical with the eschatological garden to which humankind will return at the end of times.

Emilie Savage-Smith explains that paradise, as envisioned in Islamic theology, is an afterworld with a specific structure and concrete characteristics, many reflecting earthly physical phenomena. She explores the role of paradise in medieval Islamic cosmology, and discusses some late-medieval or early modern attempts to illustrate the Islamic paradise. She concludes her contribution with speculations as to why paradise is not illustrated on medieval Islamic maps when it is so commonly represented on medieval European ones.

Anders Hultgård points to the problematic nature of using the concept of 'paradise' (common to Jewish, Christian, Islamic traditions, as well as Zoroastrianism from where the term itself ultimately derives) in the study of other religious traditions. 'Paradise', he argues, implies specific connotations that may not apply to traditions having a different world-view. In spite of this difficulty, he has explored afterlife ideas in ancient Scandinavian religion that broadly corresponded to the common term 'paradise' and contrasted them with Christian eschatological ideas in the form they were presented to Scandinavians in the transition period, i.e. from the tenth to the twelfth centuries. Little is known about the afterlife ideas of the ancient Scandinavians whereas Christian medieval eschatology is well documented in Old Norse vernacular texts from the twelfth and thirteenth centuries. A 'warrior paradise' is Valhalla, where Odin welcomes fallen heroes. Individuals belonging to other groups in society were believed to join other deities. Hultgård points to the eddic poem Grímnismál to get some clues on the pre-Christian cosmography. The emergence of the Christian paradise can be seen in early vernacular texts (preaching and praise poetry) and in runic inscriptions which show clear affinities in terminology and ideas.

All these contributions seem to have in common one theme, i.e. the idea of a blessed condition paradoxicaly both separated from and mysteriously connected to the known and familiar world of everyday life. In a variety of traditions and historical contexts we discover the recurrent notion that such ideal and blessed places are far away from the normal human environment, yet they are still, somewhere, sometime, somehow, within the human world, here and now.

The quest for paradise is something that all peoples share, and it is seems to be all the more urgent today, when we risk losing sight of the promise of a better world. Scholarly research into the past finds its fulfilment when put at the service of the challenges of the present, and surely, we understand each other better when we know ourselves, and each other's history, better. Paradise is a universal notion which can help foster the understanding of a common heritage and an appreciation of the specific differences within that common heritage.

The Birth of Paradise: To Early Christianity via Greece, Persia and Israel

Jan N. Bremmer

In the last fifteen years we have had a splendid article by Anders Hultgård on the early history of paradise, Alessandro Scafi has mapped the history of later interpretations and imaginations in a wonderfully illustrated book, and the proceedings of three conferences have further illuminated all kinds of aspects of its narration and reception.[1] So what is there still to say after these eminent scholars? In fact, who would even dare to disagree with what must now be considered not only the *vox populi* but even the *vox dei* in modern definitions: *Wikipedia*. According to this omniscient institution, '"paradise" is an idealized place in which existence is positive, harmonious and timeless. It is conceptually a counter-image of the miseries of human civilization, and in paradise there is only peace, prosperity, and happiness. Paradise is a place of contentment, but it is not necessarily a land of luxury and idleness'.

Even if we seem to detect a somewhat Puritan protest against luxury and idleness in this definition and cannot fail to note a clearly secularized tendency – there is no reference to God or the afterlife – the definition is perfectly acceptable as a reflection of everyday usage of the term. It does not, though, have anything to say about the origin and development of paradise. And precisely those aspects of the concept of paradise are highly interesting. Paradise is an originally Median word that was adopted by Jews in a Greek-speaking milieu, which became within a few centuries dominated by the Romans and eventually converted to Christianity. Not only did all these cultures and religions influence our terminology and representations of paradise, but, in the course of time, paradise became even associated or identified with the Heavenly Jerusalem. It will be clear that we cannot discuss in detail all these aspects. It will be my aim in this

Author's Note: I am most grateful to Ton Hilhorst for assistance, to Eibert Tigchelaar for comments and to Nick Wyatt for kindly and skilfully correcting my English. Eline Veldt kindly helped with the preparation of the manuscript. If there is no translator given, it is my own translation.

Abbreviations used in this essay:

FGrH = F. Jacoby, *Die Fragmente der griechischen Historiker*, Berlin and Leiden, 1923–1958.

Fowler = R. L. Fowler, *Early Greek Mythography*, vol. 1, Oxford, 2000.

fr. = fragment, followed by the number and the name(s) of the editor(s) of the authoritative edition of the relevant author.

SEG = *Supplementum Epigraphicum Graecum*, Leiden, 1923–.

[1] A. Hultgård, 'Das Paradies: vom Park des Perserkönigs zum Ort der Seligen', in *La Cité de Dieu = Die Stadt Gottes*, eds M. Hengel et al., Tübingen, 2000, pp. 1–43; A. Scafi, *Mapping Paradise: A History of Heaven on Earth*, London and Chicago IL, 2006; *Beyond Eden: The Biblical Story of Paradise (Genesis 2–3) and its Reception History*, eds K. Schmid and C. Riedweg, Tübingen, 2008; *Paradise in Antiquity: Jewish and Christian Views*, eds M. Bockmuehl and G. Stroumsa, Cambridge, 2010; *Paradeisos. Genèse et métamorphose de la notion de paradis dans l'Antiquité*, ed. É. Morvillez, Paris, 2014.

contribution to focus on the major trajectories of the connection between a garden or *locus amoenus* and the afterlife.[2] I will thereby keep a steady eye on the cosmography of paradise, but also on its contents, in order to keep track of continuities and developments. It is not an easy journey, but if that had been the case, paradise would never have exerted the fascination it has exerted until our own time.

Now, if a classicist looks at the main modern studies of paradise one omission cannot fail to strike his eye. None of them pays much attention to ancient Greece or even wonders what the impact was of ancient Greece on early representations of paradise. Yet in the last centuries BC, formative Judaism lived in a culture that was strongly influenced by Greek traditions. Similarly, the early Christians either wrote Greek, like the apostle Paul, or lived in cities where Greek was the main language, like the authors of the Pseudo-Pauline Letters and most of the early Church Fathers. It seems, then, not unreasonable to take our point of departure in Homeric Greece, where we find a paradise-like place for the very first time in Western tradition, and look at the developments in the course of the subsequent millennium from that particular vantage point, and that is what we will do: the splendid book of Alessandro Scafi but also the important studies by Jean Delumeau, Christoph Auffarth and Peter Dinzelbacher mean that we need not go as far as the Middle Ages.[3] Yet sometimes we shall also look back in time, as there is no reason to think that Homer had invented something like paradise. On the contrary, he may well be just one moment in a long chain of representations.

1. Greece

So let us start with the great but elusive poet, Homer. When in the *Odyssey* Telemachus visits Menelaus in Sparta, his host tells him that Proteus had forecast to him that he would not die at home. Instead, as he said,

> the immortals will send you to the Elysian Plain and the boundaries of the earth, where fair Rhadamanthys [resides]. There life is very easy for mortals: no snow nor storm nor ever rain, but Okeanos always sends up winds of fresh-blowing Zephyr (the west wind) to refresh people (*Odyssey*, 4.563–8).

These few lines are no more than a vignette, but they raise a number of questions. First, we cannot but be struck by the manner in which this description is introduced. This is not first-hand information, but we, the audience, hear it through Menelaus who has it

[2] The classic analysis of the *locus amoenus* is E. R. Curtius, *Europäische Literatur und lateinisches Mittelalter*, Bern, 1948, pp. 219–23, but he overlooked the important discussion by G. Pasquali, *Orazio lirico*, Florence, 1920, pp. 521–65, cf. A. La Penna, 'Il *locus amoenus*, Pasquali e E. R. Curtius', *Maia*, 63, 2011, pp. 160–61.

[3] J. Delumeau, *Une Histoire du paradis*, 2 vols, Paris, 1992–95; also in English as *History of Paradise: The Garden of Eden in Myth and Tradition*, transl. M. O'Connell, New York, 1995; C. Auffarth, *Irdische Wege und himmlischer Lohn*, Göttingen, 2002, pp. 36–72; P. Dinzelbacher, *Von der Welt durch die Hölle zum Paradies: das mittelalterliche Jenseits*, Paderborn, 2007. See now also several studies in *Mondes clos, cultures et jardins = Asdiwal*, Supplement 1, eds D. Barbu et al, Gollion, 2013.

from Proteus as a promise.[4] In other words, this is not the same as a systematic description of an authority on afterlife belief. The reason why Menelaus is singled out is, as Proteus adds, that he is Helen's husband and thus a son-in-law of Zeus (4.569). As we are in Sparta at that moment, it may not be amiss to point out that Menelaus was actually worshipped as a god in Sparta in the Archaic Age. Inscriptions from the sixth century BC leave no doubt about his status, and the same is true regarding Helen: the couple were worshipped as gods in Sparta, and this may well have helped Menelaus to get to Elysium.[5] But what does that name mean?

Walter Burkert's explanation of Elysium as a place struck by lightning has long held the field, even though there is nowhere any connection between lightning and Elysium.[6] But in the late 1990s the Leiden Indo-Europeanist Rob Beekes demonstrated that Elysium is not a Greek word, but of a still indeterminate etymology.[7] Given the influence of Phoenician mythology on Greece in the Archaic Age, we may wonder if there is perhaps not an influence from that direction. Elishah is one of the sons of Javan, 'the Ionian',[8] in Genesis 10:4 (and I Chronicles 1:7). The names of his brothers reflect the names of other far-off lands, such as Tarshish and Kittim, and they are said to be the ancestors of the 'coastland peoples' (Genesis 10:5); in fact, the 'coasts of Elishah' are mentioned as the place for obtaining blue and purple awnings in Ezekiel (27:7).[9] Perhaps there is a link in the background with Cyprus, which in the Late Bronze Age was known in Egyptian and Ugaritic sources as Alashia. At present, though, we do not have sufficient data to decide the matter.

What else are we told about Elysium? It is apparently the home of Rhadamanthys, who is also said elsewhere in the *Odyssey* (7.323) to have visited the ends of the earth to have a good look at the criminal Tityos. In the *Iliad* (14.321–2) Rhadamanthys is the brother of Minos, the well-known king of Crete. Stephanie West even suggests that 'Rhadamanthys is closely associated with Crete',[10] but there is a much stronger connection of him with Boeotia, which we find already in Hesiod, where Rhadamanthys is the son of Theban Europe who had to swim to Crete to give birth to Minos, Rhadamanthys and

[4] This is aptly pointed out, in a long analysis of the passage, by C. Sourvinou-Inwood, *'Reading' Greek Death*, Oxford, 1995, pp. 17–107 (53).

[5] Menelaus: *SEG* 26.458–59, 35.321. Helen: *SEG* 26.458, cf. N. Laneres, 'L'Harpax de Therapne ou le *digamma* d'Hélène', in *Phônês charaktêr ethnikos*, ed. M. Hatzopoulos, Athens, 2007, pp. 237–69.

[6] W. Burkert, 'Elysion', *Glotta*, 39, 1960–61, pp. 208–13, repr. in W. Burkert, *Kleine Schriften IV*, Göttingen, 2011, pp. 129–34.

[7] R. Beekes, '*Hades* and *Elysion*', in *Mír curad: Studies in Honor of Calvert Watkins*, ed. J. Jasanoff, Innsbruck, 1998, pp. 17–28 (19–23).

[8] For early Near Eastern notices of the Ionians, see the bibliography in J. N. Bremmer, *Greek Religion and Culture, the Bible and the Ancient Near East*, Leiden, 2008, p. ix.

[9] E. Lipínski, *Dictionnaire de la Civilisation Phénicienne et Punique*, Turnhout, 1992, s.v. Élisha; C. López-Ruiz, *When the Gods Were Born: Greek Cosmogonies and the Near East*, Cambridge MA and London, 2010, p. 235 n. 19.

[10] S. West, in A. Heubeck, S. West and J. B. Hainsworth, *A Commentary on Homer's Odyssey*, I, Oxford, 1988, p. 227.

Sarpedon,[11] a combination that points to a later mythographical construction, as Sarpedon had nothing to do with Crete or Boeotia, but everything to do with Lycia.[12] Rhadamanthys was said to have educated Theban Heracles,[13] and to have lived with Alcmene, whom he had married, in Boeotia.[14] In fact, all we know points more to a Boeotian origin than a Cretan one, which is not mentioned again before 400 BC.[15] Even though ancient mythography has virtually nothing to say about Rhadamanthys's Cretan *Sitz im Leben*, the passage has spawned a whole literature about the Minoan origin of Elysium. The fact that we can hardly project these few lines hundreds of years back in time, or that they do not furnish any detail that is clearly reflected in Minoan iconography, is conveniently forgotten in these cases.[16]

When we stick to what the passage really says, we have to do with a very pleasant place where there is always a cooling western wind, one more indication that Homer originated from a coastal area in western Anatolia. Just such a wind prevails in Alcinoos's orchard (*Odyssey*, 7.119), another earthly paradise, as we shall see shortly. In fact, the latter comes closer to our idea of a paradise as it contains trees that never lose their fruit; nor is there winter or summer, but always an even tempered climate. Naturally such ideal places are far away from the normal human environment. Yet they are here still within the human world, as the river Okeanos, which encircled the world, separated the living from the dead.[17]

We shall wait a moment before posing the question, how old this idea is, and first look at a more or less contemporaneous paradise. When in his *Works and Days* Hesiod relates the Myth of the Five Races, he mentions that death took away some of the heroes of the fourth race,

> But upon others Zeus the father, Kronos's son, bestowed means of living and habitations far from human beings and settled them at the boundaries of the earth [as in the *Odyssey*]; and these dwell with a spirit free of care on the Islands of the Blessed beside deep-eddying Okeanos[18] –

[11] Hesiod fr. 140–1, eds Merkelbach and West; Pherecydes *FGrH* 3 F 84 = fr. 84, ed. Fowler; Euripides fr. 752g.26, ed. Kannicht; Aristophanes *FGrH* 379 F 8 = fr. 8 Fowler; Pliny, *Natural History*, 25.21.

[12] *Iliad*, XVI.419–683, with Richard Janko *ad loc.* The passage is used by Aeschylus fr. 99, ed. Radt.

[13] For Heracles's Theban origin, see A. Schachter, *Cults of Boiotia*, II, London, 1986, pp. 1–37; add the new evidence in E. Mackil, 'A Boiotian Proxeny Decree and Relief in the Museum of Fine Arts, Boston and Boiotian-Lakonian Relations in the 360s', *Chiron*, 38, 2008, pp. 157–94 (166 n. 25). For Rhadamanthys's education of Heracles, see my 'L'Initiation d'Héraclès, in *Du récit au rituel par la forme poétique et esthétique: pragmatique cultuelle en Grèce ancienne*, eds C.Calame and P. Ellinger, Paris, 2015, forthcoming.

[14] Apollodorus 2.70; Plutarch, *Moralia*, 577E and *Life of Lysander*, 28.9; E. Scheer, *Lycophronis Alexandra*, 2 vols, Berlin, 1908, II, p. 38 (scholion on line 50 of Lycophron).

[15] Epimenides *FGrH* 457 F 19 (78ff) = fr. **3 Fowler; Ephoros *FGrH* 70 F 23.

[16] As is rightly argued by Sourvinou-Inwood, *'Reading' Greek Death* (n. 4 above), pp. 32–49 and Beekes, '*Hades* and *Elysion*' (n. 7 above), p. 23. P. Warren, '*Êlusion Minôikon*', in *Stephanos aristeios*, eds F. Lang et al, Vienna, 2007, pp. 261–70, reverts to the older position.

[17] *Odyssey*, 10.508–12, 11.158–9, 24.11–14.

[18] The combination of 'Isles of the Blessed' and Okeanos also appears on a papyrus, cf. *The Oxyrhynchus Papyri, Part XXX*, Oxford, 1964, no. 2510.2–3, but this probably is, as M. L. West, 'Review: New Fragments of Greek Poetry', *Classical Review*, 16, 1966, pp. 21–4 (22), suggests, a late piece, despite being labelled a *fragmentum dubium* of an *Ilias Parva* (fr. °32) by Bernabé.

> happy heroes, for whom the fertile field bears delicious fruit flourishing three times a year (*Works and Days*, 167–73).

We see here a rather similar idea to that in Homer: a few chosen, surviving heroes are allowed to live far away from humanity at the edge of the world,[19] where the soil is extremely fertile – three being the symbol of fullness. We are clearly in an agricultural world, but that is to be expected in a work on agriculture such as the *Works and Days* is.

How old, though, is the idea of such a Club Méditerranée for the elect? The late Christiane Sourvinou-Inwood (1945–2007) strongly argued that this idea of a paradise as a place for a chosen few was relatively new in the time of Homer. This may be true, even though the absence of evidence for earlier beliefs does not necessarily make it so. Yet there is one aspect of the Homeric description that suggests that the poet drew on an older tradition. I can best illustrate this by pointing to another description of a place for the happy few, Olympus. Here the gods meet in a place that 'is not shaken by winds, nor drenched by rain nor covered by snow, but the sky is completely without clouds' (*Odyssey*, 6.43–45). In other words, Homer draws on a tradition of a select place that is characterised by negatives.

This tradition is clearly older than ancient Greece, as we find comparable descriptions of the otherworld in Celtic and Germanic traditions. A very nice example is to be found in Chrétien de Troyes's *Erec* where the Isle of Glass is described as: 'in this isle there is no thunder, no lightning strikes, nor storm; no toad or serpent there lives; it is not too hot, nor is there winter'.[20] The same way of describing can also serve to characterize the *Urzeit* in ancient Iranian literature where at the time of 'swift Yima' (the first king), 'there was no cold nor heat; there was no old age, nor death, nor envy created by the *daevas*' (*Yasna*, 9.4–5, transl. Lincoln). These examples, which could easily be multiplied,[21] demonstrate that Homer drew on an Indo-European tradition of an otherworld that was normally couched in negative terms. Originally, this otherworld, I suggest, was the world of gods and kings, as the Indo-Europeans used to imagine their underworld as a dark place.[22] In that respect, then, Christiane Sourvinou-Inwood must be right that the transfer of living persons at the end of their lives to such special places was a later development. How late, however, we simply cannot say.

Our next poet is Pindar, undoubtedly the greatest Greek poet in the first centuries after Homer and Hesiod. In his somewhat esoteric *Second Olympian Ode*, he says:

[19] The same idea is also found in an, unfortunately, only fragmentarily surviving passage of Hesiod's *Catalogue* (fr. 204.102–4, eds Merkelbach and West).

[20] Chrétien de Troyes, *Érec et Énide*, v. 1944–7: 'En cele isle n'ot l'en tonoirre, / Ne n'I chiet foudre ne tempeste, / ne bos ne serpenz n'I areste, / N'I fait trop chaut, ne n'I yverne.'

[21] M. Davies, 'Description by Negation: History of a Thought Pattern in Ancient Accounts of Blissful Life', *Prometheus*, 13, 1987, pp. 265–84; B. Lincoln, *Death, War, and Sacrifice*, Chicago IL, 1991, pp. 23–31 ('On the imagery of Paradise', 1980[1]).

[22] M. L. West, *Indo-European Poetry and Myth*, Oxford, 2007, p. 388.

> But those who had the courage to have lived three times in either realm, while keeping their souls free from all unjust deeds, travel the road of Zeus to the tower of Kronos, where Okeanos's breezes blow around the Isle of the Blessed, and flowers of gold are ablaze, some from radiant trees on land, while the water nurtures others; with these they entwined their arms with garlands and [made] crowns, because of the just counsels of Rhadamanthys, whom the great father keeps ever seated at his side, the husband of Rhea, she who has the highest throne of all. Peleus and Kadmos are numbered among them, and Achilles too ... .[23]

The passage is a subtle combination of traditional motifs and poetic innovations. We still have a place near the Ocean, although it is a single island, not a group as in Hesiod. There is also lush vegetation, and Rhadamanthys and other heroes are there. But it is clear that we have moved up in time, as Pindar now allows this paradisiacal island to be entered by mortals too, that is, if they have passed the test of having lived piously three times in the upper- and underworld. It is impossible to trace here the origins of these individual motifs: it is enough to observe the influence of new, alternative ideas here, a kind of Classical Greek New Age thinking, although we normally call it Pythagoreanism and Orphism. These two movements, however difficult to define, tried to adapt Greek religion to the more individualizing and reflective mood of the fifth century by, amongst other things, paying special attention to the afterlife. An important source must have been an Orphic epic *Katabasis*, a *Descent into the Underworld*, which was used by Virgil in the famous sixth book of the *Aeneid*.[24] It described the fate of both the sinners, who had to wallow in the mud, and the just, who would dance under an eternal sun on meadows smelling of roses.[25] The use of Orphic eschatological material by Pindar in general, but also the reference in our passage to wreaths and crowns, suggests that he knew an early version of this poem that soon seems to have circulated in various versions. The Orphic material was also used by Plato, who in his *Gorgias* refers to the Isles of the Blessed (524a), but is rather reticent about the fate of the just and has more to say about that of the damned, a characteristic that we will also find in many early Christian authors.

So what can we conclude from our Greek material? The Greeks knew of a special area for the upper-class elect which they called the Elysian Plain or Isle(s) of the Blessed and which they situated in the west at the edge of the Ocean. It was an area with a nice climate and lush vegetation, which in the fifth century was apparently also opened up to

[23] Pindar, *Olympian Odes*, 2.68–79, ed. and transl. William H. Race (Cambridge MS and London: Harvard University Press, 1997), pp. 70–71 (I have slightly adapted Race's translation):

ὅσοι δ᾽ ἐτόλμασαν ἐστρίς / ἑκατέρωθι μείναντες ἀπὸ πάμπαν ἀδίκων ἔχειν / ψυχάν, ἔτειλαν Διὸς ὁδὸν παρὰ Κρόνου τύρσιν: ἔνθα μακάρων / νᾶσον ὠκεανίδες / αὖραι περιπνέοισιν, ἄνθεμα δὲ χρυσοῦ φλέγει, / τὰ μὲν χερσόθεν ἀπ᾽ ἀγλαῶν δενδρέων, ὕδωρ δ᾽ ἄλλα φέρβει, / ὅρμοισι τῶν χέρας ἀναπλέκοντι καὶ στεφάνους / βουλαῖς ἐν ὀρθαῖσι Ραδαμάνθυος, / ὃν πατὴρ ἔχει μέγας ἑτοῖμον αὐτῷ πάρεδρον, / πόσις ὁ πάντων Ρέας ὑπέρτατον ἐχοίσας θρόνον. / Πηλεύς τε καὶ Κάδμος ἐν τοῖσιν ἀλέγονται: / Αχιλλέα τ᾽ ἔνεικ ...

[24] J. N. Bremmer, *Initiation into the Mysteries of the Ancient World*, Berlin and Boston, 2014, pp. 180–204.

[25] This is perhaps an elaboration of an earlier *Jenseits* motif, namely that of the presence of meadows in the afterlife: cf. West, *Indo-European Poetry and Myth* (n. 22 above), p. 393.

non-heroic persons. What we do not hear of yet are women or animals in paradise. In early Greece, it still is very much a human and male affair.[26] Yet it should be stressed that this paradisiacal picture of afterlife never became generally accepted among the Greeks, and, like its counterpart Hades, it is virtually unrepresented iconographically.[27] In fact we do not even hear the name Elysium again until Apollonius of Rhodes (*Argonautica*, 4.811) in the third century BC mentions it once more, if only fleetingly.[28]

2. Persia

Yet the third century BC will be a watershed in the origin of paradise, though to see this properly we first have to go back in time. As is well known, the nation that we nowadays call Iran once consisted of two tribes, the Medes and the Persians. The language of the former, not the latter as is often thought, has given us the term 'paradise', from their **paridaiza*, 'enclosure'. This influence of the Medes, rather than the Persians, may surprise, but we should not forget that both the Greeks and the Jews mentioned the Iranians first as Medes and only later as Persians or Medes and Persians. In Persian and Elamite usage, the term 'paradise', as could perhaps be expected from such a general term as enclosure, could be a storage place, a vineyard, forest or nursery of trees, amongst other things.[29]

Although we do not have any contemporary Persian texts that mention a 'paradise' in the Persian empire, the Persians must have called certain wild parks *paridaiza*, as that is the term used by the Greeks for a specific Persian institution. From an Athenian vase and authors like Xenophon, who had first-hand experience of the Persian Empire, we can gather that the Persian elite, that means kings, satraps and other aristocrats, had wild parks connected to their palaces, in which they used to go hunting and which the Greeks called 'paradises'. To sustain a good amount of game, these wild parks naturally had water and trees,[30] but also many other plants and meadows, as the Persians also used these parks for recreation and erected pavilions in them, the parks often being close to palatial residences.

Whereas hunting played a large role in the lives of the Persian aristocrats, this was much less the case with their successors in power, the Macedonian nobles who had accompanied Alexander the Great in his conquests. We stop hearing almost immediately of the connection between hunting and *paradeisos* after the fall of the Persian Empire at the end

[26] For Elysium and the Isles of the Blessed, see most recently H. Thesleff, 'Notes on the Paradise myth in Ancient Greece', *Temenos*, 22, 1986, pp. 129–39; M. Gelinne, 'Les Champs Élysées et les Îles des Bienheureux chez Homère, Hésiode et Pindare', *Les Études Classiques*, 56, 1988, pp. 225–40; S. Mace, 'Utopian and Erotic Fusion in a New Elegy by Simonides (fr. 22 West^2)', *Zeitschrift für Papyrologie und Epigraphik*, 113, 1996, pp. 233–47.

[27] A. Scholl, 'Hades und Elysion: Bilder des Jenseits in der Grabkunst des klassischen Athens', *Jahrbuch des Deutschen Archäologischen Instituts*, 122, 2007, pp. 51–79.

[28] Note that when Strabo and Plutarch, our next sources, refer to it and locate it in what was the extreme west at the time, Spain, they explicitly connect it with the passage of the *Odyssey*: Strabo 1.1.4, 1.2.31, 3.2.13; Plutarch, *Moralia*, 942F, *Life of Sertorius*, 8.5.

[29] See Bremmer, *Greek Religion and Culture* (n. 8 above), pp. 36–9.

[30] For a fascinating analysis of such trees as citron, Persian walnut, birch and cedar in a Jerusalem 'paradise', see D. Langgut et al., 'Fossil Pollen Reveals the Secrets of Royal Persian Garden at Ramat Rahel (Jerusalem)', *Palynology* 37, 2013, pp. 115–29.

of the fourth century BC. What we keep seeing, though, is the presence of royal *paradeisoi*, such as those of Solomon (*Antiquities of the Jews*, 8.186) and David (*Antiquities of the Jews*, 7.347) as mentioned by Flavius Josephus. These are clearly retrojections from his own time, but the ones of John Hyrcanus (135–104 BC), which were situated 17 km west of modern Amman (*Antiquities of the Jews*, 12.233), were probably within living memory. They will have looked like other royal *paradeisoi* in Hellenistic times, as they feature, if somewhat dimly, in various descriptions: royal parks with many trees, suitable for walking, less wild than their Persian predecessors but more wooded than their Roman descendants.[31]

Such parks of course fit the time of the Septuagint, which began to be compiled in Alexandria in the second quarter of the third century BC.[32] Our knowledge of early Alexandria is sketchy, but it has increasingly been recognized that the royal palace of Ptolemy II (282–246 BC) was inspired by the Persian palaces with their *paradeisoi*; his *paradeisos* actually seems to be reflected in the description of King Aeëtes's palace in Colchis by the already mentioned Apollonius of Rhodes (*Argonautica*, 3.219–29).[33] It must have been the presence of the royal *paradeisos* in their own city which gave the idea to the translators of the Torah to render the Hebrew *gan Eden*,[34] 'the Garden of Eden', with the Persian-Greek term *paradeisos*: Yahweh could hardly have had a smallish garden, a *kêpos*, for his status required the best contemporary garden there was, that of the king. There is also a clear indication of the association of Yahweh's *paradeisos* with the world of the Ptolemies, viz. the Septuagint translation of *Gan Eden* in Genesis (3:23) as *paradeisos tês tryphês*.[35] *Tryphê* was a term much used by the Ptolemaic monarchy to characterize its leisurely life with its prosperity and magnificence; in fact, three of its kings were surnamed Tryphon and various princesses Tryphaena.[36] Behind the primeval *paradeisos* of the heavenly king in the Septuagint version of Genesis, then, there looms the cultivated *paradeisos* of the all too earthly king of contemporary Egypt.

3. *Jews*

The Garden of Eden, as described by Genesis 2–3, does indeed share some characteristics with the Persian *paradeisoi*, even though it does not immediately impress the reader by a detailed account. On the contrary, the author of Genesis was fairly reticent in his

[31] W. Sonne, 'Hellenistische Herrschaftsgärten', in *Basileia. Die Paläste der hellenistischen Könige*, eds W. Hoepfner and G. Brands, Mainz, 1996, pp. 136–43.

[32] T. Rajak, *Translation & Survival. The Greek Bible of the Ancient Jewish Diaspora*, Oxford, 2011².

[33] G. Grimm, 'City Planning?,' in *Alexandria and Alexandrianism*, eds P. Green et al, Malibu CA, 1996, pp. 55–74; Sonne, 'Hellenistische Herrschaftsgärten' (n. 30 above), pp. 139–41; I. Nielsen, *Hellenistic Palaces*, Aarhus, 1999², pp. 133–4.

[34] For the meaning of *'eden'*, see *Al kanfei Jonah: collected studies of Jonas C. Greenfield on Semitic philology*, eds S. Paul et al, 2 vols, Leiden, 2001, II, pp. 750–55 ('A Touch of Eden' 1984¹).

[35] G. Husson, 'Le Paradis de délices (Genèse 3, 23–24)', *Revue des Études Grecques*, 101, 1988, pp. 64–73.

[36] Cf. P. Briant, 'Histoire et idéologie. Les Grecs et la "décadence perse"', in *Mélanges Pierre Lévêque*, eds M.-M. Mactoux and E. Geny, II, Paris, 1989, pp. 33–47; S. Stelluto, 'Il motivo della *tryphê* in Filarco', in *Seconda Miscellanea Filologica*, ed. I. Gallo, Naples, 1995, pp. 47–84.

description. Yet there are some interesting similarities and differences with our Greek and Persian material. First, as in Persia, though not in ancient Greece, paradise seems to be enclosed, since an angel can guard the way to the Tree of Life.[37] Second, the most important vegetation seems to be trees (2:8), on which fruit grows which the first couple can eat (2:16, 3:2), with the exception, of course, of the Tree of the Knowledge of Good and Evil. Third, contrasting with Greece, there seem to be animals present, not only the serpent (2:19–20). Fourth, as in Persia, there is a river flowing through it to water the garden (2:10), which later branches out into four rivers, of which two, the Euphrates and the Tigris, are in the real world, whereas the other two, the Pishon and the Gihon, cannot be located.[38] Finally, whereas Greek Elysium is situated in the west and the Persian *paradeisoi* could be anywhere, Israelite paradise could hardly have been in the west in the sea, but seems to be thought of in the east beyond Mesopotamia, if we take the references to the Euphrates and Tigris seriously. Given the similarities between the contemporary *paradeisoi* and the *gan Eden*, the choice of the Jewish translators was not a bad one, we must admit.

Yet how new was the term *paradeisos* for the Jews of the earlier third century BC? It certainly was not unknown. We find the Hebrew equivalent *pardes*, another example of Median influence, in the book of Nehemiah (2:8). Here Nehemiah requests wood from the *pardes* of the Persian king's overseer, Asaph, presumably a high Jewish official. The wood for carpentry clearly points to the trees of the *paradeisos*, which was perhaps located in Lebanon, but we get a much pleasanter idea from the Songs of Songs, which mentions a '*pardes* of pomegranates with all choicest fruits, henna with nard, nard and saffron, calamus and cinnamon, with all trees of frankincense, myrrh and aloes, with all chief spices' (4.13–15, trans. *New Revised Standard Version*). Although the dating of both books is much debated, it seems reasonable to accept their existence before the translation of the Septuagint.

We find a more interesting and important reference to paradise, however, in *I Enoch*. This composite book had a special place for the elect, but they had to reside in a corner of a great and high mountain in the west (22:1), where they were separated from the other dead by a spring of water with light upon it (22:9). The water looks traditional, but the light is clearly a new element that we have not yet met, and to which we shall return. We come much closer to our subject when we follow Enoch on his travels. With the angel Raphael he travelled, by flying it seems, to the east beyond the Red Sea, that is, beyond the known world, till he came to

> the *Paradise* of Righteousness and I saw from afar, superior to those (former) trees, trees, more numerous and large, growing there, sweet smelling, tall, exceedingly beautiful and glorious; and

[37] The implicit enclosure of Genesis is made explicit in the *Apocalypse of Moses* 17.1; *Vita Adae et Evae* 31.2, 40.2; *Testament of Levi* 18.10; Origen, *Contra Celsum*, 6.33 as well as in tractates of the Babylonian Talmud; *bKetubbot* 77b and *bShabbath* 119b.

[38] For the four rivers, see most recently E. Noort, 'Gan-Eden in the Context of the Mythology of the Hebrew Bible', in *Paradise Interpreted*, ed. G. Luttikhuizen, Leiden, 1999, pp. 21–36.

> (I saw) the tree of knowledge, of the fruit of which those partake understand great wisdom. That tree is in height like the cypress and its leaves are (like the leaves) of the carob tree, and its fruit is like the clusters of the vine, shining brightly; and its fragrance penetrates far away from the tree.[39]

Once again we have here a *paradeisos* with many trees, but apparently selected for their pleasure-giving qualities, not for hunting. In other words, this *paradeisos* makes the impression of a man-made park, just like the one in the Song of Songs. The reference to the paradise myth of Genesis is clear, as is also apparent from the following verses that mention Adam and Eve's expulsion from paradise after eating from the Tree of Knowledge. The 'Paradise of Righteousness' is also mentioned in the oldest part of *I Enoch* (77:3), the *Astronomical Book* (72–82), but in none of these cases do we find an eschatological function of paradise.[40] We may even wonder if the expression 'Paradise of Righteousness/Truth' does not draw on Persian traditions, but there seems to be no linguistic evidence for connections between *pardes* and *asha/arta* which is, of course, not surprising given the dearth of ancient Persian texts.

It certainly seems safe to say that neither the location nor the future function of the Garden of Eden seems to have exerted a great fascination on the various Jewish groups of the last two centuries BC. This can also be seen from Book Three of the *Sibylline Oracles*, which was written by a Greek-speaking Jew, as the many contacts with the Septuagint show, and is now dated to the middle of the first century BC by its latest students.[41] Although the books of the later Old Testament constitute its main frame of reference,[42] the author was also well acquainted with Greek culture. Thus, twice, when referring to paradisiacal situations, he made use of Greek motifs of the Golden Age,[43] if the second time in combination with the famous prophesy of animal peace from Isaiah (11:6–8).[44] Yet he did not refer to the paradise story. In the end, it is hard to disagree

[39] *1 Enoch* 32:3–5, ed. M. Black, *Apocalypsis Henochi Graece,* Leiden, 1970, transl. M. Black, *The Book of Enoch or I Enoch: a new English edition*, Leiden, 1985, p. 41: καὶ ἦλθον πρὸς τὸν παράδεισον τῆς δικαιοσύνης, καὶ ἴδον μακρόθεν τῶν δένδρων τούτων δένδρα πλείονα καὶ μεγάλα δύο μὲν ἐκεῖ, μεγάλα σφόδρα καλὰ καὶ ἔνδοξα καὶ μεγαλοπρεπῆ, καὶ τὸ δένδρον τῆς φρονήσεως, οὗ ἐσθίουσιν <ἅγιοι τοῦ καρποῦ αὐτοῦ> καὶ ἐπίστανται φρόνησιν μεγάλην. <ὅμοιον τὸ δένδρον ἐκεῖνο στροβιλέᾳ τὸ ὕψος, τὰ δὲ φύλλα> αὐτοῦ κερατί(ᾳ) ὅμοια, ὁ δὲ καρπὸς αὐτοῦ ὡσεὶ βότρυες ἀμπέλου ἱλαροὶ λίαν, ἡ δὲ ὀσμὴ αὐτοῦ διέτρεχεν πόρρω ἀπὸ τοῦ δένδρου.

[40] *Contra* Hultgård, 'Das Paradies' (n. 1 above), p. 33, cf. E. J. C. Tigchelaar, 'Eden and Paradise: the Garden Motif in Some Early Jewish Texts (*1 Enoch* and Other texts Found at Qumran)', in *Paradise Interpreted*, ed. Luttikhuizen (n. 38 above), pp. 37–62 (42); K. Coblentz Bautch, 'Situating the Afterlife', in *Paradise Now. Essays on Early Jewish and Christian Mysticism*, ed. A. Deconick, Leiden, 2006, pp. 249–64.

[41] J.-D. Gauger, *Sibyllinische Weissagungen,* Darmstadt, 1998, pp. 440–51; R. Buitenwerf, *Book III of the* Sibylline Oracles *and its Social Setting*, Leiden, 2003, pp. 126–30.

[42] This is now demonstrated by J. Lightfoot, *The Sibylline Oracles*, Oxford, 2007, pp. 220–42.

[43] *Oracula Sibyllina*, 3.619–23, 744–61, cf. J. N. Bremmer, *The Rise and Fall of the Afterlife*, London and New York, 2002, pp. 120–27 (originally published in 2001 as 'Paradise in the Oracula Sibyllina') (120–24).

[44] H. Lutterbach, '"Tiere: In allem gehorsam wie Mönche..." Die Vorstellung vom kosmischen Frieden im Christentum', *Saeculum*, 51, 2000, pp. 294–331.

with a recent study that 'the later identification of Eden with the future paradise, the transposition of paradise to heaven, or the distinction between a heavenly paradise or Eden and its earthly counterpart are not yet made in the Early Jewish texts'.[45]

4. Jewish-Christian transition

Given this lack of early Jewish interest in paradise, it is the more striking that from the turn of the Christian era onwards we can see a sudden increase of passages that connect paradise with heaven. This development can be noted in the first century AD both in Jewish and Christian traditions, which clearly were not much differentiated at this point. The oldest testimonies seem to be the *Parables of Enoch* (*I Enoch* 37–71) and the apostle Paul. The *Parables* are clearly later than the *Book of Watchers* (*I Enoch* 1–36) and the *Astronomical Book*, and generally dated now to the first century AD.[46] Here we find 'all the elect ones in the Garden of Life' (61:12) and 'the Garden of Eden, wherein the elect and the righteous ones dwell' (60:8), but Enoch is also lifted up and placed 'between two winds, between the north-east and the west, where the angels took a cord to measure for me the place for the elect and righteous ones' (70:3–4). In *I Enoch* 60 paradise is located on earth, but this chapter derives from a lost *Apocalypse of Noah*, which seems to explain its location: consistency is not a highly valued quality in antiquity. However this may be, it seems clear that in this Enochic writing there is already a paradise for the elect and righteous which is no longer located on earth.

Another early testimony is the enigmatic verse in Paul's Second Letter to the Corinthians (12:4), in which he informs us that he was 'carried off to the third heaven ... into paradise', the first unequivocal Jewish reference to a multiple-heaven cosmology.[47] This location of paradise already seems to be traditional to some extent, as we find the same connection between paradise and the third heaven in *II Enoch* (8.1; see also *I Enoch* 71.1), the *Urtext* of which is dated to the first century AD,[48] where we read: 'And the men took me from there. They brought me up to the third heaven. And they placed me in the midst of paradise'. Similarly, according to the Jewish *Apocalypse of Moses* (37.5, ed. Dochhorn = *Vita Adae et Evae*, ed. Tromp;[49] see also 40.1), which has recently been dated

[45] Tigchelaar, 'Eden and Paradise' (n. 40 above), p. 57.

[46] M. Knibb, *Essays on the Book of Enoch and Other Early Jewish Texts and Traditions*, Leiden, 2009, pp. 5–6, 19, 143–60 ('The Date of the Parables of Enoch: A Critical Review', 1978/79[1]).

[47] For the Near Eastern background to this scheme, see P. Schäfer, 'In Heaven as it is in Hell: The Cosmology of *Seder Rabbah di-Bereshit*', in *Heavenly Realms and Earthly Realities in Late Antique Religions*, eds R. S. Boustan and A. Y. Reed, Cambridge, 2004, pp. 233–74 (259–61).

[48] Thus F. I. Andersen, in *The Old Testament Pseudepigrapha*, ed. J. Charlesworth, I, London, 1983, pp. 94–7. Although *II Enoch* has survived only in a Slavonic translation, there are strong indications of an early medieval Latin translation or epitome, cf. P. Dronke, *Imagination in the Late Pagan and Early Christian World*, Florence, 2003, pp. 142–5.

[49] See the editions of J. Dochhorn, *Die Apokalypse des Mose: Text, Übersetzung, Kommentar*, Tübingen, 2005; J. Tromp, *The Life of Adam and Eve in Greek*, Leiden, 2005.

to late first- or early second-century AD Palestine,[50] God tells the Archangel Michael to lift Adam 'into paradise until the third heaven'. Around AD 100, paradise in heaven is also mentioned in *IV Ezra* (4:7–8, 7:123) and the Gospel of Luke (23:43), where Jesus says to the robber on the cross: 'Today you will be with me in paradise', which in all probability was meant to be in heaven. Our final example comes from *II Baruch* (51:11), which is written around the same time, but probably after *IV Ezra*.[51] Here the righteous will be like angels and 'the extents of paradise will be spread out for them, and to them will be shown the beauty of the majesty of the living beings under the throne'. In other words, the righteous will be in a vast paradise (see also 59:8) where God resides too, and the vastness of paradise is also commented upon by Christian authors and martyrs, as we shall see shortly. There is hardly a reference to the Persian *paradeisoi* here, as Hultgård suggests:[52] these must have mostly disappeared from the collective memory four centuries after the fall of the Persian Empire.

What several of these places have in common is that God is mentioned in paradise too, as He is in the book of Revelation (2:7). Yet we are also struck by the fact that all these references are without any informative detail. The notion of paradise in heaven is clearly advancing, but it does not yet exert its later fascination. How do we explain this gradual rise? One factor must be the growing interest in an afterlife, which we can note from the second century BC onwards.[53] As this development entailed punishments for the wicked and rewards for the righteous, a corresponding division of the afterlife could develop. Given that the Jewish tradition was not helpful in this respect, it is not surprising that some ancient Jews looked elsewhere and borrowed material from the Greeks, who already had such a division, in particular in the case of the Orphics. The latter's 'heaven' consisted, as we have seen, in lush meadows and roses. We may perhaps see the heavenly paradise as a Jewish adaptation of this idea to its own tradition. It is clear, though, that it is easier to pose the question of the origin than to answer it.

5. *Early Christians*

We come much closer to the popular idea of a paradise in early Christian literature. This becomes already visible in the *Apocalypse of Peter*, which was probably composed around AD 130.[54] It is not always easy to cut a path through its textual jungle, but it seems certain

[50] O. Merk and M. Meiser, *Das Leben Adams und Evas*, Gütersloh, 1998, pp. 764–9; Dochhorn, *Die Apokalypse des Mose* (n. 49 above), pp. 149–72.

[51] Thus A. F. J. Klijn, in *Old Testament Pseudepigrapha*, ed. Charlesworth, I (n. 47 above), pp. 616–17.

[52] *Contra* Hultgård, 'Das Paradies' (n. 1 above), p. 37.

[53] The literature on the subject is enormous. See recently J. Day, 'The Development of the Belief in Life after Death in Ancient Israel', in *After the Exile*, eds J. Barton and D. J. Reimer, Macon GA, 1996, pp. 231–57; J. Blenkinsopp, 'Post-Mortem Existence in the Old Testament', in *Lebendiger Hoffnung – ewiger Tod?!*, eds M. Labahn and M. Lang, Leipzig, 2007, pp. 33–51.

[54] For the Apocalypse, see *The Apocalypse of Peter*, eds J. N. Bremmer and I. Czachesz, Leuven, 2003; J. N. Bremmer, 'Orphic, Roman, Jewish and Christian Tours of Hell: Observations on the *Apocalypse of Peter*', in *Other Worlds and Their Relation to This World*, eds E. Eynikel et al, Leiden, 2010, pp. 305–21.

that at this point the original Jewish-Christian text was considerably changed for theological reasons.[55] In the Rainer fragment, a Greek papyrus with a fragment of the oldest available version of the text, we read:

> I will give to my called and my chosen whomsoever they will ask me for, out of punishment, and I will give them a fine baptism in the salvation of what is called the Acherousian Lake in the Elysian Plain, a part of the justice with my holy ones.[56]

This is a very surprising appearance of Greek notions in this early Christian Apocalypse heavily influenced by Jewish ideas. The Acherousian Lake derives ultimately from Plato's *Phaedo* (113a–114a), where the Lake is the place to which come the souls of the dead; it is situated near Okeanos where we also found the Elysian Plain as in Homer (see section 1 above). We may doubt, though, whether the author took the Lake straight from Plato, as in the *Apocalypse of Moses* (37.3), Adam is also washed in the Acherousian Lake before being lifted up to paradise in the third heaven.[57] As Plato does not mention any washing in our Lake, the author of the *Apocalypse of Peter* is probably dependent on the *Apocalypse of Moses*, unless both depend on the same, unknown, source. As the Acheron was traditionally connected with the underworld, our author 'upgraded' the Lake by situating it in the Elysian Plain, the traditional Greek place for the elect, as we have seen. The combination of Acherousian Lake and Elysian Plain is used by several treatises that used the *Apocalypse of Peter*. In the Greek *III Baruch*, which probably also dates to the second century AD, the angel takes Baruch to the third heaven where he sees the righteous in 'an unbroken plain and in the middle of it was a lake of water' (10.2), another connection, then, of paradise with the third heaven. Whereas here the author has omitted the Greek names, which he probably considered to be offensive, Book Two of the *Sibylline Oracles* (335–8) closely follows our text with both names, whereas the *Apocalypse of Paul* (22–3) just mentions the Acherousian Lake (see section 6 below).

From a traditional Greek point of view, the geographical location is rather curious, since the Acherousian Lake was located in northern Thesprotia (Pausanias 1.17.5), whereas the Elysian Plain was at the ends of the earth. Apparently, the close combination derives from the belief that after baptism in the Acheron a straight transition to paradise was possible, such as we already saw in the *Apocalypse of Moses* (37.3), which was perhaps imitated by the late Coptic *Book of the Resurrection of Jesus Christ*, by Bartholomew the Apostle (46.3, ed. Westerhoff). As is the case in the *Apocalypse of Moses*, the Acherousian Lake of the *Apocalypse of Peter* is not yet identified with paradise. In the Greek version,

[55] See most recently Lightfoot, *The Sibylline Oracles* (n. 42 above), pp. 521–7.

[56] *Papyrus Vindobonensis. Graecus* 39756, cf. *Das Petrusevangelium und die Petrusapokalypse*, ed. and transl. in German and English by T. J. Kraus and T. Nicklas, Berlin and New York, 2004, pp. 124–6.

[57] For parallels, see E. Peterson, *Frühkirche, Judentum und Gnosis*, Freiburg, 1959, pp. 310–32; Dochhorn, *Die Apokalypse des Mose* (n. 49 above), pp. 490–93; Lightfoot, *The Sibylline Oracles* (n. 42 above), pp. 521–8.

which, at this point, is more detailed than the Ethiopic version, the place for important figures, like Moses, Elijah, the patriarchs and 'the other righteous fathers', is in:

> ... a very large country outside this world, exceedingly bright with light, and the air there was lighted with the rays of the sun, and the earth itself blooming with unfading flowers and full of spices and plants, fair-flowering and incorruptible, and bearing blessed fruit. And so great was the perfume that it carried there even to us.[58]

Interestingly, paradise is not located very precisely here, except that it is outside this world, but it is a very vast country, as in the already quoted passages from *II Baruch*. The stress on its size seems to be a Jewish innovation, as I have not found this characterization in the Greek material. The light, on the other hand, is already found in the Orphic underworld, as is the wondrous smell.[59] Later in the second century AD, the apologist Theophilus of Antioch also characterises paradise as 'marked out by light, illumined by shining air, with very beautiful plants'.[60] Yet Christianity had trumped the Orphic tradition by putting more stress on both light and fragrance, of which especially light, *lux perpetua*, will become *the* striking characteristic of heaven for the Christian faithful.[61]

6. From persecuted minority to established Church

We arrive at a new stage in the history of paradise when the persecutions intensify and the Church begins to keep records of the bravery of the martyrs to inspire the faithful. It is from these memories that we can see to what extent Christian believers had interiorized the literary depictions.[62]

Our oldest source is a vision of Perpetua, a young Christian woman with a reasonable education, who was executed on 7 March 203, but kept a diary of the period preceding

[58] *Apocalypse of Peter* (Greek), 15–16, *Das Petrusevangelium und die Petrusapokalypse*, eds and tr. Kraus and Nicklas (note 56 above), pp. 106–7 (Greek text), 118 (translation): καὶ ὁ κύριος ἔδειξέ μοι μέγιστον χῶρον ἐκτὸς τού[τ]ου τοῦ κόσμου ὑπέρλαμπρον τῷ φωτί, καὶ τὸν ἀέρα τὸν ἐκεῖ ἀκτῖσιν ἡλίου καταλαμπόμενον | καὶ τὴν γῆν αὐτὴν ἀνθοῦσαν ἀμαράντοις ἄνθεσι καὶ ἀμωμάτων πλήρη καὶ φυτῶν εὐανθῶν καὶ ἀφθάρτων καὶ καρπὸν εὐλογημένον φερόντων. τοσοῦτον δὲ ἦν τὸ ἄνθος ὡς καὶ ἐφ᾽ ἡμᾶς ἐκεῖθεν φέρεσθαι.

[59] Pindar fr. 129, ed. Maehler; Aristophanes, *Frogs*, 455; Ps.-Plato, *Axioch*, 371d; Virgil, *Aeneid* 6.640–1; Valerius Flaccus 1.842; Plutarch fr. 178, 211, ed. Sandbach; Lucian, *True Histories* 21.12; *Visio Pauli* 21; Claudian, *De raptu Proserpinae* 2.283–4; Lightfoot, *The Sibylline Oracles* (n. 42 above), p. 528.

[60] Theophilus 2.19, cf. Theophilus of Antioch, *Ad Autolycum*, ed. and transl. R. M. Grant, Oxford, 1970, p. 58: Μετὰ δὲ τὸ πλάσαι τὸν ἄνθρωπον ὁ θεὸς ἐξελέξατο αὐτῷ χωρίον ἐν τοῖς τόποις τοῖς ἀνατολικοῖς, διάφορον φωτί, διαυγὲς ἀέρι λαμπροτέρῳ, φυτοῖς παγκάλοις, ἐν ᾧ ἔθετο τὸν ἄνθρωπον.

[61] G. Sanders, *Licht en duisternis in de Christelijke grafschriften*, 2 vols, Brussels, 1965, II, pp. 826–59 (light).

[62] In this part I freely borrow from my earlier studies: 'Perpetua and Her Diary: Authenticity, Family and Visions', in *Märtyrer und Märtyrerakten*, ed. W. Ameling, Stuttgart, 2002, pp. 77–120; 'The Vision of Saturus in the *Passio Perpetuae*', in *Jerusalem, Alexandria, Rome. Studies in Ancient Intercultural Exchanges in Honour of A. Hilhorst*, eds F. García Martínez and G. P. Luttikhuizen, Leiden, 2003, pp. 55–73 and 'Contextualizing Heaven in Third-Century North Africa', in *Heavenly Realms and Earthly Realities*, eds Boustan and Reed (n. 47 above), pp. 159–73.

her death. This was published less than a decade after her martyrdom,[63] in combination with a report of a vision of her teacher Saturus, by an unknown editor, who also supplied an introduction and an epilogue. In it she tells that, via a ladder, she climbed to heaven where God awaited her: 'And I saw an immensely large garden, and in the middle a white-haired man sitting in shepherd's garb, tall, milking sheep. And many thousands of people dressed in white garments stood around him' (4.8). Once again we note the stress on the vastness of paradise, but otherwise Perpetua completely focuses on God as the Good Shepherd, who is more important to her than the afterlife. This is rather different in the vision of her teacher Saturus:

> 11. ... 'We had been martyred', he said, 'and left our flesh behind, and we began to be taken eastward by four angels whose hands did not touch us. And we did not go lying on our backs and looking upwards, but as if walking up a gentle hill. And when we were free of the first world we saw an enormous light, and I said to Perpetua (for she was right beside me), "This is what the Lord promised us: Now we've got it!" And while we were being carried by the same four angels, a great space opened up before us that was just like a park with rose trees and all kinds of flowers. The trees were as tall as cypresses, and their leaves kept falling and falling ... And the four angels who were carrying us got afraid and put us down; and on our own feet we went across along a broad street to an arena ... The angels said to us: "First come in and greet the Lord." 12. "And we came near a place whose walls were just as if they were made of light, and before the gate of that place stood four angels; and they went in and put on white robes. And we went in and heard a voice chanting in unison, "Holy, holy, holy," over and over again ... 13. All of us together inhaled a scent that was so satisfying I can't describe it. Then I woke up, rejoicing.'[64]

This is a highly interesting description, which starts with the traditional direction of paradise, the east (*in orientem*), which remained a popular direction in intertestamental literature and would continue to do so in early Christianity.[65] At the same time, heaven is situated beyond 'the first world', just as paradise lies *extra mundum* in the *Passion of*

[63] For the date, see Bremmer, 'The Motivation of Martyrs', in *Religion im kulturellen Diskurs. Festschrift für Hans G. Kippenberg zu seinem 65. Geburtstag*, eds B. Luchesi and K. von Stuckrad, Berlin and New York, 2004, pp. 535–54 (540).

[64] I quote from the new translation and working text by J. Farrell and C. Williams, in *Perpetua's Passions*, eds J. N. Bremmer and M. Formisano, Oxford, 2012, pp. 14–32: 11. '"Passi" inqui "eramus, et exivimus de carne, et coepimus ferri a quattuor angelis in orientem, quorum manus nos non tangebant. Ibamus autem non supini sursum versi, sed quasi mollem clivum ascendentes. Et liberato primo mundo vidimus lucem inmensam", et dixi Perpetuae (erat enim haec in latere meo): "Hoc est quod nobis Dominus promittebat: percepimus promissionem". Et dum gestamur ab ipsis quattuor angelis, factum est nobis spatium grande, quod tale fuit quasi viridiarium arbores habens rosae et omne genus flores. Altitudo arborum erat in modum cypressi, quarum folia cadebant sine cessatione ... Et expavescentes quattuor illi angeli, qui gestabant nos, deposuerunt nos. Et pedibus nostris transivimus ad stadium via lata ... Angeli dixerunt nobis: "Venite prius, introite, et salutate Dominum". 12. Et venimus prope locum, cuius loci parietes tales erant quasi de luce aedificati; et ante ostium loci illius angeli quattuor stabant, qui introeuntes vestierunt stolas candidas ... 13. Universi odore inenarrabili alebamur, qui nos satiabat. Tunc gaudens experrectus sum'.

[65] Genesis 2:8; *Testament of Job* 52:10; *1 Enoch* 32:2; *2 Enoch* 31:1, 42:3; *4 Ezra* 5:21, cf. F. Dölger, *Sol salutis*, 2nd edn, Münster, 1925, pp. 173–81; S. de Blaauw, *Met het oog op het licht. Een vergeten principe in de oriëntatie van het vroegchristelijk kerkgebouw*, Nijmegen, 2000.

Lucius and Montanus (7.5), African martyrs who were executed in the spring of 259.[66] No further details are provided regarding its geography, except that Saturus already saw *lucem immensam* before his actual arrival in heaven. As we noted, light was perhaps the most striking feature of Christian paradise; in this way its brightness is particularly stressed.

Like Perpetua in her first vision, Saturus saw a great open space, which looked like a *viridiarium*, 'park', which probably refers to the wonderful parks of the Roman grandees in North Africa, which we can still admire in mosaics and which ultimately derived from the Persian *paradeisoi*.[67] In this case the park was full of flowers, with very tall roses in particular. Roses are the spring flowers *par excellence* in the Mediterranean, and their mention suggests a mild climate in heaven: spring also became *de rigueur* in Roman descriptions of the Golden Age and the *locus amoenus*.[68] Roads were probably typical of these parks, as *paradeisoi* also provided possibilities for walking.[69]

From the park the martyrs arrived at God's palace on which Saturus is rather reticent, as he mentions only that its walls were built of light, another example of the prominence of light. In *II Enoch* (8 and 20), paradise is situated in the third heaven and God's palace in the seventh, but in the African imagination the two are clearly in the same space and their close connection might be another Persian legacy (see section 2 above); similarly, in the *Passion of Marian and Jacob*,[70] two martyrs executed shortly before Montanus and his group, Marian heads for the *praetorium* of the heavenly judge through a wonderful landscape:

> Our road lay through a place with lovely meadows, clad with the joyous foliage of flourishing woods, shaded by tall cypresses and pine trees that beat against the heavens, so that you would think that that spot all round was crowned with flourishing groves. In the centre was a hollow of a crystal spring that abounded with fertilizing watercourses and pure water.[71]

[66] For the most recent and best edition, see F. Dolbeau, *Sanctorum societas*, 2 vols, Brussels, 2005, I, pp. 83–129 ('La Passion des saints Lucius et Montanus', 1983[1]).

[67] For these Roman wild parks, see F. Olck, 'Gartenbau', in *Paulys Realencyclopädie*, VII, Stuttgart, 1912, pp. 768–841 (838); M. Blanchard-Lemée et al, *Mosaics of Roman Africa*, transl. K. D. Whitehead, London, 1996, pp. 167–77; M. Guggisberg, 'Vom Paradeisos zum "Paradies": Jagdmosaiken und Gartenperistyle in der römischen Herrschaftsarchitektur Nordafrikas und Siziliens', *Hefte des Archäologische Seminar Universität Berns*, 17, 2000, pp. 21–39.

[68] Spring flowers: Nisbet and Hubbard on Horace, *Odes*, 1.38.4. Golden Age: Virgil, *Georgics*, 2.149–50; Ovid, *Metamorphoses*, 1.107 with the commentary of F. Bömer *ad loc.*, *Fasti*, 5.207–8; Lucian, *True Histories*, 2.12; Claudian, *Epithalamium*, 55.

[69] Bremmer, *Greek Religion and Culture* (n. 8 above), pp. 48–9.

[70] For the text, see P. Franchi de' Cavalieri, *La Passio SS. Mariani et Iacobi*, Rome, 1900, pp. 42–63. For a recent French translation, introduction and helpful notes, see P. Maraval, *Actes et passions des martyrs chrétiens des premiers siècles*, Paris, 2010, 211–27. For the date, see the excellent discussion by A. R. Birley, 'A Persecuting Praeses of Numidia under Valerian', *Journal of Theological Studies*, NS, 42, 1991, pp. 598–610 (603).

[71] *Passion of Marian and Jacob* 6.12–13: 'iter autem nobis erat per locum pratis amoenum et virentium nemorum laeta fronde vestitum, opacum cupressis consurgentibus in excelsum et pinis pulsantibus caelum, ut putares eum locum per omneni circuitus ambitum lucis virentibus coronatum. sinus autem in medio pellucidi fontis uberantibus venis et puris licoribus redundabat.' For an even more detailed description in, probably, Carthage in the early 250s,

The landscape recurs in an imitation of Plato's *Symposium* by a Christian author of the later third century, perhaps Bishop Methodius of Lycian Olympus, but certainly an author geographically far removed from North Africa. Its description of the Garden of Virtue, the place where a number of virgins discuss themes of Christian theology, contains several themes that are now familiar to us, such as light, fragrance, trees and shade, but it also says:

> in the very centre was a spring from which there bubbled up, as gently as though it were oil, the most delicious water; and the crystal-clear water formed into little rivulets. These, overflowing their banks, as rivers do, watered the ground all round with their abundant streams.[72]

As this place is explicitly situated towards the east and one of the virgins calls it a 'second paradise', its description is clearly modelled on that of biblical paradise. This must mean that the spring in Marian's vision is the spring in paradise from which the four great rivers (Pishon, Gihon, Tigris and Euphrates: Genesis 2:10–14) originated.[73] The spring eventually goes back to the Ancient Near East, but the fountain in the Temple will also have played a role.[74] Although the spring is not yet mentioned in Genesis, the early Jewish *Apocalypse of Abraham* already mentions 'the source and the river flowing from it' (21.6). This convergence of Christian and Jewish texts suggests that these descriptions with the fountain go back to a common source, which may well be a description in a first-century AD lost Jewish Apocalypse.

The descriptions by the martyrs are much more detailed than those by the leading Carthaginian theologians of the time, Tertullian and Cyprian. Although the former alludes to paradise dozens of times in his writings, his descriptions are rather reticent, and he clearly was more interested in hell. It seems significant that the only detail he gives from his lost treatise *De paradiso* concerns the fate of the dead *apud inferos* until the Last Judgement.[75] Yet when he has to explain paradise to the pagans, he mentions that it is separated from this world by a wall of fire and resembles the Elysian Plain.[76] Cyprian,

which also contains the spring, see Pseudo-Cyprian, *De laude martyrii*, 21, transl. R. E. Wallis, *The Writings of Cyprian* II = *Ante-Nicene Christian Library*, XIII, Edinburgh, 1894, pp. 245–6, cf. J. Amat, *Songes et visions*, Paris, 1985, pp. 155–6. For date and place, see the concise discussion by J. Doignon, in *Handbuch der lateinischen Literatur der Antike IV*, eds R. Herzog and P. L. Schmidt, Munich, 1997, p. 578.

[72] Methodius, *Symposium*, Prelude, 5–8, ed. Débidour-Musurillo, transl. H. Musurillo, *St. Methodius, The Symposium. A Treatise on Chastity*, Westminster and London, 1958, pp. 40–41: καὶ πηγὴ κατὰ τὸ μεσαίτατον ἡσύχως καὶ καθαρὸν ῥέον ἐποίει κρήνας. Αἱ δὲ ποταμηδὸν ὑπερχεόμεναι ἐπότιζον ἅπαντα τὸν χῶρον πλούσια παρέχουσαι νάματα.

[73] For a representation of the source on a Tunisian vessel, see A. Grabar, *Christian Iconography*, Princeton NJ, 1968, fig. 39.

[74] See most recently B. Janowski, 'Die heilige Wohnung des Höchsten. Kosmologische Implikationen der Jerusalemer Tempeltheologie', in *Gottesstadt und Gottesgarten*, eds O. Keel and E. Zenger, Freiburg, 2002, pp. 24–68 (53–57) with rich bibliography; J. van Ruiten, 'The Four Rivers of Eden in the *Apocalypse of Paul* (*Visio Pauli*): The Intertextual Relationship of Genesis 2.10–14 and the *Apocalypse of Paul* 23', in *The Visio Pauli and the Gnostic Apocalypse of Paul*, eds J. N. Bremmer and I. Czachesz, Leuven, 2007, pp. 50–76 (59–63).

[75] Tertullian, *De anima*, 55.

[76] Tertullian, *Apologeticum*, 47.13. For the Elysian Plain, see section 1 above.

too, is very reticent and when he refers to paradise, it is only to compare it to the Church and the four Gospels to the four rivers of paradise.[77] Yet the representation of heaven as the paradisiacal garden will now be unstoppable, and we find it in Latin Church Fathers like Lactantius and Ambrose, in Christian poets like Prudentius, Sedulius and Dracontius as well as in Syrian and Persian theologians like Ephrem and Aphraates.[78]

Let me conclude my survey with the *Apocalypse of Paul*, also called *Visio Pauli*, perhaps the last late antique detailed description of paradise.[79] This, originally, Egyptian, late fourth-century Greek text was translated into Latin in Rome at the end of the sixth century,[80] where at the same time Gregory the Great was writing his *Dialogues*, whose depictions of the afterlife were to have a great impact on the Western world.[81] The *Apocalypse* is clearly meant to satisfy the curiosity of believers as to their future fate. It performs this task not very economically, and we hear twice of Paul's visit to heaven and hell. After a first description of the place for the damned, the *angelus interpres* takes Paul to the third heaven, a clear reference to his own writing (see section 4 above). Having entered the gate he first meets Enoch, the 'scribe of justice' (20), a nice indication of the apocalyptic interests of the author.[82] But then the angel brings Paul down to the second heaven, from where he goes to the firmament and then proceeds to the doors that are right above the river Okeanos, perhaps an allusion to Homer's description of the Elysian Plain (see section 1 above).[83] This is the Land of Promise where the souls of the righteous await the second coming of Christ, as they do in the *Testament of Abraham* (11.10, 14.8, 20.14). When Paul looks at the *Terra repromissionis* he sees:

> a river flowing with milk and honey; and at the bank of the river were planted trees full of fruit. Each tree was bearing twelve times fruits in the year, various and different. And I saw the nature of that place and all the work of God. And I saw there palm trees, some of twenty cubits and others of ten cubits. Now that land was seven times brighter than silver. And there were trees full of fruit from the root to the top ... And after that he took me up away from that place where

[77] Amat, *Songes et visions* (n. 71 above), pp. 148–53 (Tertullian), and pp. 153–4 (Cyprian).

[78] For a useful collection of passages, see S. Bregni, '"Paradisus, locus amoenus": immagini del paradiso nei primi secoli dell'era cristiana', *Rivista di storia e letteratura religiosa*, 41, 2005, pp. 297–328 (318–19).

[79] Unless otherwise indicated I quote the best-preserved version (L^1) from T. Silverstein and A. Hilhorst, *Apocalypse of Paul. A New Critical Edition of Three Long Latin Versions*, Geneva, 1997; add now M.-F. Damongeot-Bourdat, 'Un Nouveau Manuscrit de l'Apocalypse de Paul (Paris, BnF, nouv. acq. lat. 2676)', *Bulletin du Cange* 67, 2009, 29–64. The translations are adapted from H. Duensing and A. de Santos Otero, 'Apocalypse of Paul', in *New Testament Apocrypha*, ed. W. Schneemelcher, 2 vols, 2nd edn, Cambridge, 1992, II, pp. 712–48.

[80] For date and place, see J. N. Bremmer, 'Christian Hell: From the *Apocalypse of Peter* to the *Apocalypse of Paul*', *Numen*, 56, 2009, pp. 298–325. For studies and full bibliography, see *The Visio Pauli*, eds Bremmer and Czachesz (n. 74 above).

[81] G. Gracco, 'Gregorio e l'oltretomba', in *Grégoire le Grand*, eds J. Fontaine et al, Paris, 1986, pp. 255–66.

[82] This is a quote from *I Enoch* (12:4) and/or the *Testament of Abraham* (B 11.3), books that were still read in Egypt at the time that the *Apocalypse of Paul* was composed; cf. G. Nickelsburg, 'Two Enochic Manuscripts: Unstudied Evidence for Egyptian Christianity', in *Of Scribes and Scrolls*, eds H. W. Attridge et al, Lanham MD and London, 1990, pp. 251–60; D. Allison, Jr., *Testament of Abraham*, Berlin and New York, 2003, 32–3.

[83] K.B. Copeland, 'Thinking with Oceans: Muthos, Revelation and the *Apocalypse of Paul*', in *The Visio Pauli*, eds Bremmer and Czachesz (n. 74 above), pp. 77–104.

> I had seen these things and, behold, I saw a river whose waters were very white, whiter than milk. And I said to the angel: 'What is this?' He said to me: 'This is the Acherousian Lake where the City if Christ is, but not every man is allowed to enter into that city'.[84]

Having said that Michael baptizes repentant sinners in the Lake, the angel brings Paul in a boat to the City of Christ. It had not only twelve gates, like the New Jerusalem in Revelation (21:12), but also

> four rivers which encircled it. Now there was a river of honey and a river of milk and a river of wine and a river of oil. And I said to the angel: 'What are these rivers which encircle this city?'And he said to me: 'These are the four rivers which flow abundantly for those who are in the Land of Promise; as for their names: the river of honey is called Pishon, and the river of milk Euphrates, and the river of oil Gihon and the river of wine Tigris'.[85]

Although this heaven is clearly modelled on paradise, we hear that name only once (20). This is different when Paul visits paradise for the second time, which, at the beginning, is already described as *paradisus in quo erravit Adam et eius mulier* (45). This time the righteous he meets, such as Jacob, Moses and Elijah, are described in a detailed manner. Yet the passage has something superfluous, and it will hardly be chance that it is lacking in the other manuscripts with a long version of the Latin text. Still, for our purpose, it is useful to quote the passage, as it offers some new aspects. I will start after the angel has told Paul that the just are eagerly awaiting him and shown him again the four rivers:

> and going in further I saw a tree planted out of whose roots water flowed, and from this was the beginning of the four rivers. And the Spirit of God rested over that tree and when the Spirit blew the waters flowed. And I said: 'Lord, is it this tree itself which makes the waters flow?' And he said to me: 'Because from the beginning, before heaven and earth appeared, everything was invisible, the Spirit of God hovered over the waters (Genesis 1:2); but since the commandment of God brought to light heaven and earth, the Spirit rests over this tree. Therefore, whenever the Spirit has breathed, the waters flow from the tree. And he held me by the hand and led me near the Tree of Knowledge of Good and Evil and said: 'This is the tree through which death entered into the world, and Adam receiving from his wife ate of it and death came into the world.

[84] *Visio Pauli* 22: 'Et circumspexi terram illam et vidi flumen currentem lac et mel. Et erant ad litus fluminis ipsius arbores plantate plene fructibus. Una quoque autem arbor erat adferens fructos xii in anno varios et diversos fructos abentes. Et vidi creaturam loci illius et omnem facturam dei. Et vidi ibi palmas cubitorum xx, alias autem cubitorum x. Terra autem illa clari[or] argento septies. Et era [nt] arbores plene fructi[bus] arradice usque ad sum [mum], ordines x milium fruc[tum] palmarum super x milia fructose ... Et post haec adsumpsit me ex apeliote loci illius. Et ecce flumina aque eius erant aque candide valde super lac. Et dixi angelo: Quis est hoc? Dixit mihi: Haec est Aceriosus lacus ubi est civitas Christi, sed non omnis homo permitetur ingredi in civitatem illam.'

[85] *Visio Pauli* 23: 'Et xii porte erant in circuitu [ci]vitatis pulcritudine magna et quatuor fl[u]mina qui circuibant. Erat autem flumen mellis et [flu]men lactis et flumen vini [et] flumen olei. Et dixi angel[o]: Que sunt flumina haee ci[r]cuiencia civitatem histam? Et ait mihi: Haec sunt iiii^or flumina que decurrunt sufficienter. his qui sunt in ac terra repro[mis]sionis, quorum nomina sunt h[ec]: flumen mellis dicitur Fison. et flumen lactis Eufrat[es] et flumen oley Gion et flumen vini Tigris.'

> And he showed me another tree in the middle of paradise, and he said to me: 'This is the Tree of Life'.[86]

It is clear that in some respects the second description is much more traditional than the first. Its four rivers are enumerated in the biblical order without specific qualities,[87] and we even hear of the two special trees in paradise. In other respects, though, it is less traditional. We have already seen the fountain in African paradises, but the just in paradise are usually not waiting for others and the Spirit of God does not rest on the tree from which the waters flow, nor do the waters flow when the Spirit chooses to blow at them.[88] Yet we can see these details as elaborations of the first chapters of Genesis and as answers to questions which we perhaps no longer ask, but which the early Christians certainly did, such as what the Spirit was doing after the creation.

The first description is in a way more interesting. Once again we hear of the Acherousian Lake, which is a reminder of the fact that our author used the *Apocalypse of Peter*. But another early Christian treatise was more important for him. The bearing of fruit twelve times derives from Revelation (22:2), and is a sign of the enormous influence of that book on our author,[89] which is also manifested by the presence of the City of Christ in paradise (Revelation 21–2);[90] the description in Revelation was so influential in antiquity that we can see its echoes in *True Stories* of the pagan Lucian of Samosata.[91] In addition to the *Apocalypse of Peter* our author clearly read other Apocalypses as well,[92] as the presence of honey, milk, wine and oil in paradise first occurs in *II Enoch* where we read: 'And two streams come forth, one a source of honey and milk, and a source which

[86] *Visio Pauli* 45: 'Ingressus autem interius et vidi arborem plantatum de cuius radicibus aque emanabant et erat ex ac inicium iiiior fluminum. Spiritus autem dei requiescebat super arborem illam et cum flasset spiritus, efflabant aque. Et dixi: Domine, arbor aec ipsa est que fluet aquas? Et dixit mihi quia Ab inicio, priusquam celum et terram manifestarentur, erant autem omnia invisibilia, spiritus autem dei ferebatur super aquas. Ex quo autem preceptum dei apparuit celum et terram, spiritus requievit super arborem hunc. Propterea cum flaverit spiritus, emanant aque ex arbore. Et tenuit mihi manum et duxit me iuxta arborem cognoscende bone et male et dixit: Haec est arbor per quem mors ingre[s]sa est in seculo et ex ea accipiens a muliere sua Adam et manducavit et ingressa est mors in mundo. Et ostendit mihi aliam arborem in medio paradiso et ait ad me: Haec est arbor vite.'

[87] Van Ruiten, 'The Four Rivers of Eden' (n. 74 above), pp. 57–8.

[88] The 'innovations' are noted by A. Hilhorst, 'A Visit to Paradise: Apocalypse of Paul 45 and Its Background', in Luttikhuizen, *Paradise Interpreted* (n. 38 above), pp. 128–39 (129–30).

[89] Hilhorst, 'The *Apocalypse of Paul*', in *The Visio Pauli*, eds Bremmer and Czachesz (n. 80 above), pp. 1–22 (16).

[90] C. Markschies, 'Himmlisches und irdisches Jerusalem im antiken Christentum', in Hengel et al, *La Cité de dieu* (n. 1 above), pp. 303–50, well notes the growing distance between depictions of the Heavenly Jerusalem and those of the historical Jerusalem, but mistakenly comments regarding our passage: 'der Name "Jerusalem" fällt nicht' (p. 318, note 53): c. 29: *haec est Hierusalem civitas* (L^1, Escorial, Arnhem).

[91] For Lucian's knowledge of Revelation and other early Christian Apocalypses, see P. von Möllendorff, *Auf der Suche nach der verlogenen Wahrheit. Lukians Wahre Geschichten*, Tübingen, 2000, pp. 427–32 and 'Christliche Apokalypsen und ihr mimetisches Potential in der paganen Bildungskultur. Ein Beitrag zu Lukians Wahren Geschichten', in *Die Bibel im Dialog der Schriften*, eds S. Alkier and R. B. Hays, Tübingen and Basel, 2005, pp. 179–94.

[92] For the four rivers in early Christian iconography, see N. Stone, 'The Four Rivers that Flowed from Eden', in Schmid and Riedweg, *Beyond Eden* (n. 1 above), pp. 227–50.

produces oil and wine' (8.4–6), and the reversed order honey and milk (above) instead of the proverbial order milk and honey occurs here also for the first time. As we are in paradise (see section 3 above), the reference to the four rivers of Genesis (2:10–14) seems obvious, but we find four streams of pure water also on Calypso's island, a kind of paradise, in the *Odyssey* (4.70) and four rivers in the Homeric underworld (*Odyssey*, 10.513–14),[93] a common origin for Israel and Greece in Mesopotamia, which liked to divide the world in quadrants, seems most likely.[94]

From the liquids mentioned, milk and honey are a traditional couple in the Old Testament (Exodus 3:8, 17; 13:5; 33:3; Leviticus 20:24, etc.),[95] and together with wine they occur in Book Three of the *Sibylline Oracles* (622–3) in a kind of earthly paradise. Yet they also occur early in ancient Greece, where they, together with wine, are mentioned first in Euripides's *Hypsipyle* in a Dionysiac context.[96] In Greece, then, their presence is closely associated with the god Dionysos, whose typical drink was wine. Oil, together with milk, honey and wine, is mentioned first by Onesicritus (*FGrH* 134 F 17), the head steersman of Alexander the Great, who lets a gymnosophist say that in 'olden times' there were fountains with milk, wine, honey, water and oil, fluids which, without water, also occur in a paradisiacal state in the same Book Three of the *Sibylline Oracles* (744–61) between the coming of the Last Judgement and the eternal Kingdom. As the author of Book Three was a hellenized Jew, it seems reasonable to conclude that the combination of these four substances eventually derives from Greek tradition.

Conclusion

Let us try to formulate some conclusions. Both Greece and Israel knew of a special area, a *locus amoenus*,[97] which was reserved for important people at the end of their lives in the one case and for the first human couple in the other. The Israelite Garden of Eden probably continued a Near Eastern tradition, which the Persians had appropriated as a hunting park and called 'paradise'. The term was taken over by the translators of the Septuagint, who apparently saw a connection between the Persian *paradeisos* and the

[93] Thesleff, 'Notes on the Paradise Myth in Ancient Greece' (n. 26 above), p. 134.

[94] Thus M. L. West, *The East Face of Helicon*, Oxford, 1997, p. 422.

[95] See more recently C. Grottanelli, *Sette storie bibliche*, Brescia, 1998, pp. 138–45; H. Ausloos, '"A Land Flowing with Milk and Honey": Indicative of a Deuteronomistic Redaction?', *Ephemerides Theologicae Lovanienses*, 75, 1999, pp. 297–314; M.-J. Pierre, 'Lait et miel, ou la douceur du Verbe', *Apocrypha*, 10, 1999, pp. 139–76; E. Levine, 'The Land of Milk and Honey', *Journal for the Study of the Old Testament*, 87, 2000, pp. 43–57 (hardly persuasive).

[96] Euripides, *Hypsipyle*, fr. 758a, ed. Kannicht (wine, milk, honey); see also Euripides, *Bacchae*, 142–3 (milk, wine, honey), 704–11 (water, wine, milk, honey); H. Usener, *Kleine Schriften*, 4, Leipzig, 1913, pp. 398–417 (with mention of our *Apocalypse* on p. 401, note 16); F. Graf, 'Milch, Honig und Wein', in *Perennitas. Studi in onore di Angelo Brelich*, ed. G. Piccaluga, Rome, 1980, pp. 209–21.

[97] In addition to the literature mentioned in note 3, see G. Schönbeck, *Der Locus Amoenus von Homer bis Horaz*, diss., Heidelberg, 1962; H. Thesleff, 'Man and *locus amoenus* in Early Greek Poetry', in *Gnomosyne*, eds G. Kurz et al, Munich, 1981, pp. 31–45; Amat, *Songes et visions* (n. 71 above), pp. 117–20; P. Hass, *Der* locus amoenus *in der antiken Literatur*, Bamberg, 1998.

Israelite Garden of Eden. When the Jews started to develop their afterlife in the second century BC, they had to fill a void and did it by promoting the earthly paradise as the place for the righteous and moved it to heaven, probably under Greek (Orphic) influence. The early Christians took over this Jewish legacy, which gained a special importance in time of persecution. They developed the vegetation side of paradise, but in the end the connection with the earthly paradise limited somewhat the possibilities for imagination. This is perhaps also the reason why we hear so little of women in paradise. Mohammed, who was much more interested in paradise than in hell, was also indebted to Christian Apocalypses for his eschatology and, surely, somewhat overreacted in this respect.[98]

The enormous popularity of the *Apocalypse of Paul* in the earlier Middle Ages widely promoted the lush picture of paradise, which culminated in Dante's *Paradiso*.[99] All the time, though, earthly paradise kept attracting the interest of Christians, and expeditions and attempts to locate it have lasted into the twenty-first century.[100] These pastimes are relatively innocent. Much more worrying are the attempts to establish a paradise on earth as we see them in modern utopian movements. This is of course another story, but it is good to remember always that the notion of paradise remains an ambivalent legacy of the Judaeo-Christian tradition.

[98] T. Andrae, *Der Ursprung des Islams und das Christentum*, Uppsala and Stockholm, 1926, pp. 234–7.

[99] For the Middle Ages, see the bibliography in Silverstein and Hilhorst, *Apocalypse of Paul* (n. 79 above), note 4 pp. 19–20; add now L. Jiroušková, *Die Visio Pauli. Wege und Wandlungen einer orientalischen Apokryphe im lateinischen Mittelalter unter Einschluß der alttschechischen und deutschsprachigen Textzeugen*, Leiden, 2006; N. H. Trunte, *Reiseführer durch das Jenseits: die Apokalypse des Paulus in der Slavia Orthodoxa*, Munich, Berlin and Washington DC, 2013.

[100] For the most recent attempts, see Scafi, *Mapping Paradise* (n. 1 above), pp. 342–73.

Sumerian Paradise Lost

Markham J. Geller

All conventions and beliefs surrounding the location of paradise and the hereafter are clouded by a confusing array of contradictions, as one might expect about a place from which no one, except in literature, has ever returned. The same is true for the very earliest of accounts regarding these mysterious places, from third millennium BC Sumer and from slightly later accounts from Semitic (Akkadian) sources from the subsequent millennium. The problem is that many of these contradictions have escaped modern notice.

The first and perhaps most obvious contradiction is between the locations of paradise and the hereafter, the latter term referring to the place where the departed go after death. There are two basic problems which immediately catch our attention: the first is that paradise and the hereafter may or may not be the same place, and the second is that the hereafter may not always be the same as the netherworld or underworld, although these terms are traditionally used as clichés for the world of the dead. In fact, the entire question of the location of the hereafter relates to ancient notions of cosmology and perceptions of the geography of the universe.

Let us first consider the question of paradise, best known from the Bible as the Garden of Eden. In fact, later Jewish (post-biblical) sources equated the Garden of Eden (*gan Eden*) as the residence of the righteous departed, whereas the wicked would descend to *Gehenom*, which actually was the *wadi* extending below the walls of Jerusalem. Neither term is satisfactory as descriptions for the hereafter, and the Bible itself is equally vague; the dead descend to a dark abyss or Sheol, and underworld, whence the prophet Samuel's ghost was brought up by the Witch of Endor in a necromantic ritual (I Samuel 28:8, 11), while the righteous patriarchs are simply 'gathered unto their forefathers', ultimate destination unstated (Genesis 25:17; 35:29). Nevertheless, Genesis 2:10–14 appears to locate Eden in Mesopotamia, as a garden watered by four rivers, two of which being the Tigris and Euphrates.

To resolve some of these terminological dilemmas, it is useful to reach back into literary records for earliest references to the world of the dead, namely from Sumerian sources, where we find a clear distinction between 'paradise' and the hereafter. The Sumerian view, however, is far from what we might expect: paradise was to be found as an island in the Persian Gulf called Dilmun, firmly identified as Bahrain,[1] which in ancient times was thought to be an emporium specially fed by underground channels of sweet waters. A

Abbreviation used in this essay:
ETCSL = Electronic Text Corpus of Sumerian Literature

[1] B. Alster, 'Dilmun, Bahrain, and the Alleged Paradise in Sumerian Myths and Literature', in *Dilmun, New Studies in the Archaeology and Early History of Bahrain*, ed. D. Potts, Berlin, 1983, p. 44.

popular Sumerian myth, Enki and Ninhursag, describes Dilmun as the place suitable for a union between the god Enki and his daughter Ninsikila; Dilmun is transformed in this story from virgin territory into a great emporium, watered by a special underground channel provided by Enki.

> Dilmun drank water, a plenty from them
> her pools of salt water became pools of sweet water only,
> her fields and glebe, the plowlands, [produced] grain for her,
> her city became (an emporium), a (store)house on the quay for the country's produce.[2]

Dilmun was also the place where the immortal Ziusudra, hero of the Sumerian flood story, was to reside, although described as 'toward the east over the mountains'.[3] The Gilgamesh Epic, in its later Akkadian form, reiterates this identification of the 'remote, mythical east',[4] where the hero of the Flood spends his immortality at the 'mouth of the rivers' (*Gilgamesh* XI 205), reminiscent of the four rivers emanating from a single source forming biblical Eden. Katz finds it decisive that Dilmun is not mentioned in this passage of Gilgamesh, and she is right to question whether Dilmun was indeed a 'paradise' or not, comparable to biblical Eden. Katz argues that the biblical paradise represented a place where man need not work, and this paradigm has no parallel within Mesopotamian mythology.[5] Nevertheless, there is more than one definition of paradise, for which the biblical example serves as only one model; in Mesopotamian terms, in Alster's sober view,

> the unique position of Dilmun as a prosperous trade centre gave the island a unique place in Sumerian literature, one comparable to Delos, the sacred birthplace of Apollo and Artemis, and the 'islands of the blessed along the shore of deep swirling Ocean', where, according to Hesiod, the selected sons of the fourth generation of mankind live untouched by sorrow.[6]

Furthermore, there is also little doubt in Sumerian literature that Dilmun was distinguished from the world of the dead,[7] hence hardly *gan Eden* [cf. p. 31, para 3]; but what we can infer is that Dilmun, like Eden, was a distant ideal place located on the earth (rather than in heaven), and was originally the habitation of immortals, not dead spirits.

We move now to the actual location of the hereafter, which turns out to be elusive because of our own cultural preconceptions and prejudices, since we are immediately inclined to speak of the 'netherworld' or 'underworld', which definitely rules out other possible locations within the cosmos. The Sumerian hereafter is particularly difficult to

[2] T. Jacobsen, ed. and transl., *The Harps that Once: Sumerian Poetry in Translation*, New Haven CT and London, 1987, p. 190.

[3] Ibid., p. 150.

[4] C. Woods, 'At the Edge of the World: Cosmological Conceptions of the Eastern Horizon in Mesopotamia', *Journal of Ancient Near Eastern Religions*, 9, 2009, p. 200.

[5] D. Katz, 'Enki and Ninhursaga, Part One: the Story of Dilmun', *Bibliotheca Orientalis*, 64, 2007, pp. 580–81.

[6] Alster, 'Dilmun, Bahrain, and the Alleged Paradise' (n. 1 above), pp. 59–60.

[7] Despite the presence of some 150,000 burial mounds, which are not easily explained, see ibid., p. 52.

assess within this framework.[8] On one hand, it may be logical to consider the concept of a netherworld because of common burial practices, suggesting that the dead descend into a subterranean afterlife; but, on the other, the idea of the hereafter is often more closely related to cosmology in general and how the universe was conceived. The location of the land of the dead needed to be placed within a more general scheme of the cosmos, much as the way medieval maps often found a place for paradise within a *mappa mundi*.[9]

From the ancient world there are two basic cosmological schemes, which turn out to be either a 'flat-earth' universe or a layer cake, depending upon various seemingly contradictory propositions. Essentially, a 'flat-earth' scheme differentiated only between heaven and earth: some gods resided in heaven, others (usually associated with nature) resided on earth; men inhabited the earth, as did demons, and the dead resided in a distant place, also on earth.[10] In the more multi-dimensional layer-cake cosmology, described by W. G. Lambert, the heavens above were differentiated from both the earth and a subterranean netherworld, which also somehow shared its space with the subterranean ocean of sweet water.[11] As it happens, Lambert's description of the cosmos works well for the Semitic world but is less convincing for Sumer.

The core problem is how to interpret cosmological terminology which relates to the hereafter, and this problem occupied Assyrian and Babylonian scholars as well. The most significant terms are Sumerian terms, *kur* and *ki*, both of which can refer to the place of the dead, but both of which have primary meanings which do not relate to the dead. The word *kur* meant 'mountain', and it came to be used generally for 'land' with the sense of foreign territory, since regions surrounding Mesopotamia were often mountainous (e.g. the Iranian plateau); Sumer itself was known as the *kalam*, the alluvial plain. The term *kur* as applied to the dead was built into a common turn of phrase, *kur-nu-gi*, 'the mountain of no return', since the journey to the hereafter was a single direction only. The word *ki*, on the other hand, was a generic term for 'place', which was translated by Akkadian scribes as *erṣetu*, 'earth',[12] and then by extension, 'underworld'; the term could apply to the habitation of both living and dead as there was not a specific Akkadian term meaning 'netherworld' or 'underworld', with any clear geographical subterranean designation. In effect, Akkadian scholars also translate *kur* as *erṣetu*, by simply ignoring

[8] The title of the useful book by Dina Katz, *The Image of the Nether World in the Sumerian Sources*, Bethesda MD, 2003, deals just with this problem, although her discussion of the image of the netherworld within Sumerian sources offers a rather confusing picture of the location of the dead.

[9] See A. Scafi, *Mapping Paradise: A History of Heaven on Earth*, London and Chicago IL, 2006, and W. Horowitz, *Mesopotamian Cosmic Geography,* Winona Lake IN, 1998, 20–22, showing *nagû* or regions on the Babylonian *mappa mundi* where immortals reside; this is the earliest of world maps, dating from the 5th century BC. See also F. Badalanova Geller, 'Recasting the Bible, Recapturing Eden: Parascriptural Cosmologies in the Byzantine Commonwealth' in *Old Testament Pseudepigrapha and the Scriptures*, ed. E. Tigchelaar, Leuven, 2014, pp. 463–89.

[10] See, for example Thorkild Jacobsen's comments on the god Dumuzi and his earthly activities in his 'Towards the Image of Tammuz', *History of Religions*, 1, 1962, p. 190.

[11] W. G. Lambert, 'The Cosmology of Sumer and Babylon' in *Ancient Cosmologies*, eds C. Blacker and M. Loewe, London, 1975, pp. 42–65.

[12] Cognate to the Hebrew term used in the biblical creation account for 'earth'.

the basic meaning of 'mountain'. The lack of clarity in these terms was obviously unsatisfactory, so that a more focused vocabulary referring to the land of the dead developed: Sumerian used the terms *ki-gal* 'large place' and *urugal / irigal* 'large city', as designations for the hereafter. Both terms were then loaned into Akkadian (*kigallu, irkallu*), with the latter term *urugal* being adopted by Akkadian scribes as a logogram for the 'grave' (*qabru*); the underlying meaning was that the habitation of the dead was a great city, reflecting the highly urbanized culture of Mesopotamia. Other terms were associated with this place, such as the *arali*-steppe which (being distant) one had to transverse to arrive there, and the fact that the place of the dead had a palace (called *ganzir*) at its entrance, and a series of gates. All of these terms have been previously studied in detail.[13]

In the end, however, where is it? There is nothing here to suggest a subterranean place, since its three main geographical designations are that it is a mountain, a 'great city', and that it is distant (across the steppe); arriving might also involve crossing water, since according to one idiom, one 'sails' (u_5)[14] to this place.

The crux in deciding this matter is in the well-known opening lines of the myth relating to the goddess Inanna's journey to the world of the dead, to rival her sister Ereshkigal, the queen of the hereafter (*ki-gal*).[15] In the opening lines of the myth, we are told that Inanna 'went up/down to the *Kur*' (*kur-ra ba-e-a-*e_{11}); the verbal root $\boldsymbol{e}_{11}$ indicates motion either upwards or downwards. A parallel text, also referring to Inanna, gives the identical phrase with a slight variation (*kur-ra ba-e-a-íl*), in which the verb *íl* implies 'raising, lifting up';[16] in this second instance Inanna 'raises herself' up to the *Kur*(-mountain). Katz's treatment of these lines is equivocating;[17] she acknowledges that the second text clearly shows Inanna going up to the *Kur*, but she does not carry the argument to its logical conclusion, that the land of the dead was located in the mountains, to which Inanna ascended. This idea conforms nicely to the flat-earth cosmology outlined above, in which gods (like Dumuzi) and mankind live on earth, while demons and the dead inhabit the *Kur*, a dangerous place.

There is a logical problem with this flat-earth cosmology, namely how one explains the path of the sun through the heavens: it rises in the east, sets in the west, and then

[13] Cf. Katz, *The Image of the Nether World* (n. 8 above), *passim*; M. J. Geller, 'The Landscape of the "Netherworld"', in *Landscapes, Territories, Frontiers and Horizons in the Ancient Near East*, eds L. Milano et al, Padua, 1999, pp. 41–9.

[14] See for instance, the example cited by Katz, *The Image of the Nether World* (n. 8 above), p. 40, from Inanna's lament over Dumuzi, '[*mu-l*]*u-zu alim-e kur-šè ba-*u_5', 'the bison has sailed your man to the "mountain"'.

[15] *Inanna's Descent to the Netherworld* line 5: '^{d}inanna an mu-un-šub ki mu-un-šub kur-ra ba-e-a-e_{11}', 'Inanna left heaven, she left earth, she went up/down to the Kur'. For convenience cf. the ETCSL (http://etcsl.orinst.ox.ac.uk/section1/b141.htm).

[16] Inanna and Shukaletuda, line 16: '^{d}inanna [an m]u-un-šub ki mu-un-šub kur-ra ba-e-a-íl', 'Inanna left [heaven], she left earth, she took herself up to the Kur'. For convenience cf. ETCSL (http://etcsl.orinst.ox.ac.uk/section1/c133.htm).

[17] Katz, *The Image of the Nether World* (n. 8 above), pp. 251–4.

returns to the east; how is this accomplished? A flat-earth scenario would suggest a two-tiered world in which the heaven (and its divine bureaucracy) more or less mirrors the earth, but no dimension is specified for a subterranean space; the sweet waters of the *Abzu* (abyss) belong to the earth.

Despite any possible contradictions, the mountains – probably to the east of Mesopotamia – served as the most likely location for the Sumerian land of the dead, but this idea was not shared by Akkadian scribes, who intentionally translated Inanna's journey to Ereshkigal with the term *urid*, 'descended', clearly indicating that in their view Inanna was descending to a netherworld. This changes the cosmology considerably, since it reflects a universe that has to accommodate heaven, earth, and netherworld on three distinct levels, which this leads to another possible contradiction.

Central to the netherworld or layer-cake [cf. *supra*] scheme is the fact that the sun makes a full orbit, emerging from the horizon in the east, entering the horizon in the west, and traversing the netherworld during the night in order to return to the east by morning. There is, however, a problem here, namely the contradiction of the sun's orbit including a passage through the netherworld, which was by its very nature defined as a dark place.[18] It is difficult to imagine how the sun can travel anywhere without lighting it up. This contradiction, in fact, can already be found within the terminology associated with the solar passage through day and night.[19] Sumerian and Akkadian expressions referring to the sun are as follows:

Sum. *utu-è* = Akk. *ṣīt šamši* 'sunrise' (lit. 'coming out of sun')
Sum. *utu-šú* = Akk. *erēb šamši* 'sunset' (lit. 'entering of the sun')

The terms describe the sun's relations to the horizon, exiting from below the horizon at dawn and entering the horizon again at sunset, only to travel through the netherworld to return to the eastern horizon. Although oft repeated in bilingual texts, these correspondences do not accurately reflect the basic meanings of the Sumerian words. Sumerian *šú*, for example, never actually means 'enter' (in contrast to 'exit') since, in this particular context referring to the passage of the sun, the word *šuš* (written *šú*) means 'to darken' (Akkadian *erēpu*); in other contexts it means to 'cover' or 'overwhelm' a place. In fact, a better Akkadian translation for the expression *utu-šú* would no doubt be *erēp šamši*, but the concept of *erebu* 'entering, sunset'[20] became fixed within the language, corresponding to *erev* 'evening' in Hebrew and other Semitic languages as well. The image of sunset as the time when the sun 'enters' the horizon dominated Semitic thinking, but is not represented in the Sumerian expression; it is a re-interpretation of Sumerian terminology to conform to a fixed idea stemming from a preconceived cosmology.

[18] See Woods, 'At the Edge of the World' (n. 4 above), p. 224.
[19] See Geller, 'The Landscape of the "Netherworld"' (n. 13 above), p. 47.
[20] Corresponding to the Sumerian *ku*$_4$, 'enter'.

The same type of re-interpretation of the Sumerian by Akkadian speakers can be seen in the complementary expression for sunrise, *utu-è*, which was certainly understood by Semitic speakers as the sun breaking out of the horizon; in iconography the sun god is accompanied by his saw, probably instrumental in cutting his way to the early morning sky.[21] Nevertheless, Sumerian offers consistent lexical and bilingual evidence for the verb *è* corresponding to Akkadian *napāhu*, 'to light a fire', such as the hymnic phrase, *ᵈutu an-šà-ta è*, translated by the Akkadian as *šamaš ina išid šamê tappuha*, 'O Sun(-god), you are kindled on the horizon'.[22] The sun is treated in these expressions as a lamp to be lit every morning in the east, and 'darkened' or extinguished in the west, without either exiting or entering the horizon at all. This shows the fundamental difference between the 'flat-earth' cosmology and that of the orbiting sun in a layer-cake cosmos.[23]

Is there any way of resolving any of these contradictions? How, for instance, does the sun as a lamp return from west to east every day? We are never told. Perhaps this unresolved problem was one reason for Akkadian scribes to reject this cosmology. What about the alternative proposition, in which the orbiting sun must traverse the netherworld in order to return to the east? It may be that Egyptian cosmology offers a solution which was never realized in Mesopotamia.

From a Mesopotamian perspective, things in Egypt appear to be somewhat upside down, especially the concept of Mother Sky (*Nut*) and Father Earth (*Geb*).[24] The *Amduat*, the Egyptian 'Book of the Hidden Chamber' of the netherworld, describes the nocturnal travels of Re, the sun-god, through the netherworld on his journey from west to east, describing every hour of Re's nightly voyage. The descriptions evoke a desert landscape full of serpents, but also of darkness, a place of the damned, beyond the fields of the more blessed dead. Egyptian funerary literature, however, has a unique resolution to the conundrum of the sun visiting the darkness of the netherworld: the sun is reborn daily on his journey, which is described as a daily resurrection of the sun-god; in the text of *Amduat*, the dead king was identified with the sun-god in his resurrection.[25] This offers a brilliant solution to the cosmic contradiction which was never resolved in Mesopotamia: the sun actually dies every day in the west, and hence his light is extinguished, only to be reborn when

[21] See the illustrations of Tessa Rickards of the sun-god with his saw in J. Black and A. Green, *Gods, Demons, and Symbols of Ancient Mesopotamia*, London, 1992, p. 183.

[22] Cf. *Chicago Assyrian Dictionary*, II, '*N*', part 1, ed. E. Reiner, Chicago IL, 1980, p. 263.

[23] My argument that the Sumerian view of the netherworld (beyond the mountains in a horizontal universe) was different from the Akkadian (beneath the Land of the Living in a vertical universe) has been criticized by P. Steinkeller, in 'Of Stars and Men: The Conceptual and Mythological Setup of Babylonian Extispicy', in *Biblical and Oriental Essays in Memory of William L. Moran*, ed. A. Gianto, Rome, 2005, p. 19 and n. 18; and 'On Sand Dunes, Mountain Ranges, and Mountain Peaks', in *Studies presented to Robert D. Biggs*, Chicago IL, 2007, pp. 231–2. Steinkeller's opinion, however, is not accompanied by any new convincing evidence nor does he refute any of the specific arguments above.

[24] See W. Forman and S. Quirke, *Hieroglyphs and the Afterlife in Ancient Egypt*, London, 1996, p. 9. See also E. Hornung, *Valley of the Kings: Horizon of Eternity*, New York, 1990, plates 44 and 45.

[25] Ibid., p. 117.

arriving at the east. The transfiguration of Re in the *Amduat*, however, still resembles the orbiting sun of Mesopotamia, while at the same time reflecting the lamp-like nature of the sun in Sumerian cosmology.

In retrospect we see that the orbiting sun and netherworld eventually carried the day, whether in biblical Sheol or Classical Hades. Still, it is worth bearing in mind that alternative world views existed but had to be literally interpreted out of the literary record.

Around, Inside and Beyond the Walls: Names, Ideas and Images of *Paradise* in Pre-Islamic Iran. With an Appendix on Old Persian <p^a-r^a-d^a-y^a-d^a-a-m^a>

Antonio C. D. Panaino

Any discussion about 'paradise' in the Iranian framework must focus on two aspects: the first one concerns the well-known fact that our modern designations of this wonderful place descend (mostly via Latin *paradīsus*)[1] from a Greek word, *παράδεισος* (*parádeisos*),[2] which, in its turn, was clearly a loanword from an Iranian formation, a fact well recognized also by some Greek grammarians.[3] In fact, if the word were originally Greek, it should

Author's Note: In the preparation and in final redaction of this work I had the opportunity to consult some friendly colleagues, who gave me various suggestions and remarks. In particular I would like to thank G. P. Basello, J. N. Bremmer, A. Hultgård, A. Iannucci, V. Sadovski, R. Schmitt and N. Sims-Williams. The present research has been developed in the framework of the Project of National Interest (PRIN), entitled *Organizzazione territoriale e ideologia nello stato achemenide: gli insediamenti di Persepoli*, directed by A. V. Rossi (Università di Napoli, l'Orientale), and sponsored by the Italian Ministry of the University. This study is one of many other final results produced by the local Unity of the University of Bologna (Branch of Ravenna).

Abbreviations used in this essay:

Av. = Avestan; O.Av. = Old Avestan; L.Av. = Late Avestan; O.P. = Old Persian; N.P. = New Persian; Pahl. = Pahlavi; Ved. = Vedic; Sum. = Sumerian; IE. = Indoeuropean; *Y.* = *Yasna*; *Vd.* = *Widēwdād*; Gr. = Greek; Lt. = Latin; PF = Persepolis Fortifications (Tablets); Sept. = Septuaginta; m. = masculine; f. = feminine; n. = neuter; nom. = nominative; acc. = accusative; sg. = singular; par = paragraph; pl. = plural; subj. = subjunctive; lit. = literally.

KEWA = M. Mayrhofer, *Kurzgefaßtes etymologisches Wörterbuch des Altindischen* (*A Concise Etymological Sanskrit Dictionary*).

EWA = M. Mayrhofer, *Etymologisches Wörterbuch des Altindoarischen.*

[1] The Latin reception (*paradīsus*, m.) first appears in Aulus Gellius, but its diffusion with the meaning of 'paradise' dates back to the first Latin translation of the Old Testament. When Tertullian (see M. Ernout and A. Meillet, *Dictionnaire étymologique de la langue latine. Histoire des mots*, 4th edn, Paris, 1959, repr. 2001, p. 481; A. Walde and J. B. Hoffmann, *Lateinisches Etymologisches Wörterbuch*, 3, II, '*M–Z*', Heidelberg, 1954, p. 251; W. Meyer-Lübke, *Romanisches etymologisches Wörterbuch*, 3rd edn, Heidelberg, 1935, p. 512) made use of it, the word was already common (A. Hultgård, 'Das Paradies: vom Park des Perserkönigs zum Ort der Seligen', in *La Cité de Dieu/Die Stadt Gottes*, eds M. Hengel, S. Mittmann and A. M. Schwemer, Tübingen, 2000, pp. 1–43 (1, n. 3)). For a brief discussion concerning the Roman *paradisi*, see J. N. Bremmer, 'The Birth of Paradise', in Bremmer, *Greek Religion and Culture, the Bible and the Ancient East*, Leiden and Boston MA, 2008, pp. 35–55 (44, 49–50). A larger treatment of the image of the garden in medieval Iran, but also with references to pre-Islamic Iran, has been offered by M. Subtelny, *Le Monde est un jardin: Aspects de l'histoire culturelle de l'Iran medieval*, Paris, 2002.

[2] P. Chantraine, *Dictionnaire étymologique de la langue grecque. Histoire des mots*, with a *Supplément*, eds A. Blanc, C. de Lamberterie and J.-L. Perpillou, Paris, 1999, p. 857; C. Tuplin, *Achaemenid Studies*, Stuttgart, 1996, pp. 120–28.

[3] See Pollux Grammaticus, *Onomasticon*, ed. E. Bethe, 9, 13, Leipzig, 1931, p. 150: οἱ δὲ παράδεισοι, βαρβαρικὸν εἶναι δοκοῦν τοὔνομα ἥκει κατὰ συνήθειαν εἰς χρῆσιν Ελληνικήν, ὡς καὶ ἄλλα πολλὰ τῶν Περσικῶν; '*parádeisoi*: thought to be barbarian, the word has been adapted by custom to Greek usage, like, indeed, many other words of the Persians'. See also K. Latte and P. A. Hansen, *Hesychii Alexandrini Lexicon*, III, '*Π–Σ*', Berlin, 2005, p. 25: παράδεισος· τίθεται ἐπὶ τῶν ἀναισθήτων. οἷς ἐστιν ἐμπεριπατεῖν. ἢ τόπος [ἔνυδρος ἤτοι] εὔυδρος, ἐν ᾧ περίπατοι. καὶ ἡ βασιλέως κατάλυσις; '*parádeisos*: it denominates one of the stupid ones; to them to whom it is proper to walk about; or a place [truly well-watered] abounding in waters, in which there are covered walks. Also the resting-place of the

have been *περίτοιχος[4] or περίτειχος (n.).[5] This Iranian antecedent, **paridaiza-*, i.e. 'a (garden) surrounded (**pari°*) by walls (**daiza*)', 'having walls around' (a *bahuvrīhi*-compound of the ἔνθεος type),[6] should be, for phonetic reasons, of so-called 'Median'[7]

king'. Cf. P. de Lagarde, *Gesammelte Abhandlungen*, Leipzig, 1866, pp. 210–11. Hesychius confirms the fact that it was the name of a royal place, although the presence of *Perípatoi* does not appear as a standard character of any *parádeisos*. The first meaning, apparently peculiar, i.e. that of 'stupid fellow' (H. G. Liddell, R. Scott and H. S. Jones, *A Greek-English Lexicon*, rev. H. Stuart Jones with R. McKenzie, *With a Supplement*, Oxford, 1968, p. 1127) seems to be attested in the Comic writers. Hesychius makes reference to the fact that it refers to a person who 'walks about' (probably doing nothing), and A. Meineke and H. Jacobi, eds, *Fragmenta Comicorum Graecorum*, V, *Comicae Dictionis Indicem et Supplementa continentis*, Pars Prior, '*A–Γ*', Berlin, 1857a, p. 124; Pars Altera, '*K-Ω*', 1857b, p. 799) explained it with reference to Photius (*Lexicon*), s.v., as follows: '*nomine* παράδεισος *significari hominem cui propter stupore impune insultatur*'. Actually Photius wrote (ed. R. Porson, Cambridge, 1822, 2nd part, p. 382): Παράδεισος: τὸ ἐνεστώς· σημαίνει δὲ οἷον τὸ ἐμπεριπατεῖσθαι τεθειμένως διὰ τὴν ἀναισθησίαν· ὁ γὰρ παράδεισος ἐπὶ τοῦ περιπάτου δένδρα καὶ ὕδατα ἔχοντος· καὶ χρῶνται συνεχῶς οἱ κωμικοί, τιθέντες τὸν παράδεισον ἐπὶ τῶν ἀναισθητῶν· ἔστι δὲ τοὔνομα Περσικόν, καὶ λέγεται φαρδαιθί; '*parádeisos*: that which gives entertainment; it means something like to walk about in a confused fashion; in fact, *parádeisos* (refers) to a covered walk with trees and waters; and the Comics continuously use (it), putting the (person named) *parádeisos* among the stupid ones; but the name is Persian, and it is pronounced *phardaithi*'. We can wonder if there was, at least originally, a kind of etymological understanding of the fact that the word meant 'place encircled by walls'. This way, 'stupid' would apparently be also one who has his mind encircled or closed, exactly like a *parádeisos*, and then blocked in his astonishment or pleasure. On the other hand, it is necessary to note that the interpretation of Photius's gloss as a *vox comica* is still debated by Classicists, and Meinecke's explanation does not appear definitive. This gloss is in any case inserted among the *fragmenta adespota* by T. Kock, *Comicorum Atticorum Fragmenta*, III, *Novae Comoediae Fragmenta*, part II, *Comicorum Incerta Aetatis Fragmenta. Fragmenta Incertorum Poetarum. Indices. Supplementa*, Leipzig, 1888, p. 590 [fr. 1102], followed by R. Kassel and C. Austin, eds, *Poetae comici graeci*, VIII, Berlin and New York, 1995. On this subject cf. also Bremmer, 'The Birth of Paradise' (n. 1 above), p. 49. I must thank for the help received in the discussion of this subject A. Iannucci (Bologna – Ravenna). With regard to the possibility that some Greeks had a clear idea that a 'paradise' should have been an 'enclosed' place, we can mention Xenophon's *Helleniká*, 4, 1, 15, where this fact is explicitly stated (Bremmer, 'The Birth of Paradise', n. 1 above, pp. 44–5). It would be also worthwhile to recall that, as late as in the 1st century AD, Quintus Curtius Rufus (*Historia Alexandri*, VIII, 1, 12; *Historiae Alexandri Magni*, II, ed. J. E. Atkinson, transl. into Italian by T. Gargiulo, Milan, 2000, pp. 176–7) mentioned the presence in Sogdiana of woods surrounded by walls *(muris nemora cinguntur*); cf. Hultgård, 'Das Paradies' (n. 1 above), pp. 4–5. About Gr. παράδεισος see also the important discussion offered by M. Brust, *Die indischen und iranischen Lehnwörter im Griechischen*, Innsbruck, 2005, pp. 506–14, which I discovered only when this article was ready.

[4] Cf. also H. Frisk, *Griechisches Etymologisches Wörterbuch*, 2 vols, Heidelberg, 1960, II, p. 473.

[5] 'A surrounding wall' (Septuaginta, IV Kings 25.1) in Liddell, Scott and Jones, *A Greek-English Lexicon* (n. 3 above), p. 1200; cf. also, for instance, παρατείχιον, corresponding to Lat. *pomoerium*, 'the open space along a city wall' (Constantine VII Porphirogenitus, *De cerimoniis,* 109. 9; Liddell, Scott and Jones, *A Greek-English Lexicon* (n. 3 above), p. 1144).

[6] See also C. Werba, '*mavāred-rā na-bāyad ziyād kard be joz-e ehtiyāj*. (Indo-)Iranische Rekonstrukte als textkritisches Korrektiv in der Altiranistik', in *Iranistik in Europa: Gestern, Heute, Morgen*, eds H. Eichner, B. G. Fragner, V. Sadovski and R. Schmitt with H. Hofmann and V. Giesen, Vienna, 2006, pp. 262–306 (279–80, n. 90; 281, n. 100), who suggests that the Avestan form is the result of a *figura etymologica*, in principle different from the Median one.

[7] The reference to a Northern dialect variety with respect to the so-called standard Old Persian can be conventionally defined as 'Median', but this denomination should be taken in a generic sense, and we cannot deny the possibility that Old Persian knew some sociolinguistic and diastratic varieties; see R. Schmitt, 'Die Sprache der Meder: eine große Unbekannte', in *Continuity of Empire (?): Assyria, Media, Persia*, eds G. B. Lanfranchi, M. Roaf and R. Rollinger, Padua, 2003, pp. 23–36; cf. also A. V. Rossi, 'La varietà linguistica nell'Iran achemenide', in *Atti del convegno «La varietà linguistica nel mondo antico»*, Naples, 1981, pp. 141–96 (155–60) and I. Gershevitch, 'Dialect Variation in Early Persian', *Transactions of the Philological Society* [1965], 1964, pp. 1–29. See also M. Mayrhofer, *Die Rekonstruktion des Medischen*, Vienna, 1968, pp. 15–18.

origin or, in any case, not of strictly Old Persian derivation, because the Greek form παράδεισος continues a sibilant instead of a dental stop. At present we know two cognate forms of **paradaiza-*, both belonging to the early Iranian linguistic period: the first, *pairi.daēza-*,[8] is attested only twice in Avestan (*Vd.* 3, 18; 5, 49),[9] while the second one, until now isolated, seems to have been preserved in the framework of a later Old Persian inscription belonging to the time of Artaxerxes II (Susa D), where it can be read as *paradayadām*, but other (in my opinion, better) readings, like *paradaydā̆m* or already *pardaydăm* ([p^{a}-r^{a}-d^{a}-y^{a}-d^{a}-a-m^{a}]),[10] are possible. Thus, with a reasonable degree of probability, notwithstanding a long and intriguing discussion, we can postulate an Old Persian entry like **paridaida-* [11] (Later O.P. **pardaida-/*pardēda-/pardēdə* or **pardēd*), m.[12] The existence of an Iranian stem like **pardaiza-* is later continued also in Christian and Manichaean Sogdian, respectively as *prdyz*[13] / *prδyz*[14] 'garden, orchard' (to be probably read *parδēz*), and in Modern Persian as *pālēz*[15] (contemporary pronunciation

[8] C. Bartholomae, *Altiranisches Wörterbuch*, Strasbourg, 1904, p. 865. L.Av. *pairi.daēza-*, like *uzdaēza-*, m., 'wall' (ibid., pp. 411–12), is a cognate of O.P. *didā-*, f., 'wall, stronghold, fortress', Gr. τεῖχος 'wall', Vedic *dehī́-*, f., 'wall', Pahl. *diz* and N.P. *diz*, 'town', Thracian –διζα/-διζος, Oscan *feíhuss* (acc. pl. = *muros*), 'wall', from Indo-European **dʰeiǵʰ-* (J. Pokorny, *Indogermanisches Etymologiches Wörterbuch*, I/4, Tübingen and Basel, 2002, pp. 244–5; T. Zehnder in H. Rix, ed., with M. Kümmel, R. Lipp, Br. Schirmer, *Lexicon der indogermanischen Verben. Die Wurzeln und ihre Primärstammbildungen*, 2nd edn, Wiesbaden, 2001, 140–41; cf. KEWA, II, '*D-M*', Heidelberg, 1963, 65–6; EWA, I/10, Heidelberg, 1992, 746–7. See also Latin *fingere* (Ernout and Meillet, *Dictionnaire étymologique* [n. 1 above], pp. 235–6). Contrariwise, no evidence attests the existence of an O.P. **daida-* "Wall" (see Werba, '(Indo-) Iranische Rekonstructe' [n. 6 above], p. 281, n. 100), although we cannot exclude its theoretical presence. It is also possible that Old Persian had a verb from the same root (like Av. *daēz-/diz-* [Bartholomae (*supra*), pp. 673–4; J. Kellens, *Liste du verbe avestique*, Wiesbaden, 1995, pp. 31–2]), but in this case we cannot any longer register an entry such as [*d*]*išta-* 'built', p.p.p. of **did-* 'to build' (W. Brandenstein and M. Mayrhofer, *Handbuch des Altpersischen*, Wiesbaden, 1964, pp. 116–17), because such an earlier reading has been recently rejected by R. Schmitt, *Die altpersischen Inschriften der Achaimeniden*, Wiesbaden, 2009, p. 132 for [*p*]*išta-*. Cf. in any case also Avestan °*dišta-*, and Sogdian *δšt'y* 'built' (B. Gharib, *Sogdian Dictionary. Sogdian – Persian – English*, Tehran, 2004, p. 150); Khwaresmian *parδys-* 'to build' (D. N. MacKenzie,'The Khwarezmian Glossary', *Bulletin of the School of Oriental and African Studies*, 33/3, 1970, pp. 540–59 (549)); Ossetic (Iron) *fældīsyn : fældyst* / (Digor) *fældesun : fældist*: (1) 'to create, make', (2) 'to consecrate (something) to a dead' (V. I. Abaev, *Istoriko-Ėtimologičeskij Slovar' Osetinskogo Jazyka*, I, '*A-K*', Moscow and Leningrad, 1958, pp. 435–6). For a number of many other cognates in the Iranian framework, see also V. S. Rastorgueva and D. I. Edel'man, *Ėtimologičeskij Slovar' Iranskich Jazykov*, II, '*B-D*', Moscow, 2003, p. 312, and J. Cheung, *Etymological Dictionary of the Iranian Verb*, Leiden and Boston MA, 2007, pp. 51–2. See again Werba '(Indo-) Iranische Rekonstructe' (n. 6 above), 279–83 and notes.

[9] See Hultgård, 'Das Paradies' (n. 1 above), pp. 2–3. Bremmer's statement, 'The Birth of Paradise' (n. 1 above), p. 36, that 'The Median form does recur in later Avestan [...]' is misleading. In the *Widēwdād* we find a genuine Avestan stem, which allows us to postulate also a North-Western stem like **pari-daiza-*, not the contrary.

[10] See the 'Appendix'.

[11] For this form, see R. Schmitt, *Beiträge zu altpersischen Inschriften*, Wiesbaden, 1999, p. 82.

[12] Schmitt, *Die altpersischen Inschriften* (n. 8 above), p. 195. Cf. W. Brandenstein and M. Mayrhofer, *Antiguo Persa. Gramatica, Inscripciones*, Madrid, 1958, p. 137.

[13] N. Sims-Williams, *The Christian Sogdian Manuscript C2*, Berlin, 1985, p. 220.

[14] Gharib, *Sogdian Dictionary* (n. 8 above), p. 285, with bibliography. See also the title *prδyzp'n* [*parδēzpān*], 'chief of parks', in secular Sogdian.

[15] F. Wolff, *Glossar zu Firdosis Schahname*, Berlin, 1935, p. 181; Cf. Lagarde, *Gesammelte Abhandlungen* (n. 3 above), pp. 75–6.

pāliz;[16] cf. also Kurdish *parēs*).[17] We will soon enter again other problems related to this form.

The second subject regards the image and the conception of what ideologically and religiously can actually be compared with such a special place, located somewhere, where the souls of the pious Mazdeans were, and are still, supposed to go in the afterlife. In fact, the Western reception of such an iconographic representation of a beautiful garden, which was used in order to indicate the terrestrial paradise, is undoubtedly different from the Mazdean one. The latter, in its turn, used, as we will see very soon, different expressions that fittingly designated the final place of resting and of beatification for the followers of the so-called 'good religion' (Mazdeism or Zoroastrianism).

The architectural solution of building large areas for the pleasure and amusement of the kings, their family and guests, is surely pre-Iranian.[18] As Gignoux has noted, already Aššurnasirpal II (883–859 BC) ordered the edification of a 'garden of delights',[19] but some ancient reports regarding the marvellous hanging gardens of Babylon could, at least in part,[20] represent a Mesopotamian background for these places. In particular, Diodorus Siculus[21] wrote that Semiramis would have built up one of these special gardens in Media, on the river Xoaspes, then in a clear Iranian framework. Although the legends concerning Semiramis should be, for various reasons,[22] considered as referring to Nebuchadnezzar, the cultural implications of this tradition appear in any case relevant. Gignoux,[23] following

[16] See A. A. Dehxodā, *Loγatnāme*, 14 vols, Tehran, 1993–94, s.v. *pāliz*: 1. 'garden' (*fāliz*), 'vegetable garden' (*ǰāliz*), 'garden' (*bāγ, bustān*), 'rosegarden' (*golestān*); 2. 'cropped field', in general, and nowadays, a field where watermelons, melons and cucumbers are cultivated; cf. also M. Mo'in, *Farhang-e fārsi*, 6 vols, Tehran, 1371/1992–93⁸, s.v. *pāliz*: 1. 'garden' (*bāγ, bustān*); 2. 'farming field' (*keštzār*); 3. 'a field where watermelons, cucumbers *et similia* are grown'. Cf. Also G. Lazard, *Dictionnaire persan-français*, Leiden, 1990, p. 72, sub *pāliz* 'melonnière, champ de melon, de pastèque ou de cocombre'.

[17] See A. Jaba, *Dictionnaire Kurde-Français*, St Petersbourg, 1879, p. 71; F. F. Omar, *Kurdisch-Deutsches Wörterbuch (Zentralkurdisch/Soranî)*, Berlin, 2005, p. 174, s.v. [1]*perêz* (Soranī dialect); cf. W. Fauth, 'Der königliche Gärtner und Jäger im Paradeisos. Beobachtungen zur Rolle des Herrschers in der vorderasiatischen Hortikultur', *Persica*, 8, 1979, pp. 1–53 (5).

[18] Cf. A. L. Oppenheim, 'On Royal Gardens in Mesopotamia', *Journal of Near Eastern Studies*, 24/4, 1965, pp. 328–33.

[19] P. Gignoux, 'Quelques réflexions sur la représentation du Paradis', in *From Byzantium to Iran. Armenian Studies in Honour of Nina G. Garsoïan*, eds J.-P. Mahé and R. W. Thomson, Atlanta GA, 1997, pp. 3–15 (5); J. J. Glassner, 'À propos des jardins mésopotamiens', in *Jardins d'Orient*, Paris, 1991, pp. 1–17 (12–13). See also Hultgård, 'Das Paradies' (n. 1 above), pp. 12–13. Cf. Fauth, 'Der königliche Gärtner' (n. 17 above), pp. 15–16, 25, passim.

[20] Cf. Tuplin, *Achaemenid Studies* (n. 2 above), pp. 106–8, passim. See also D. Stronach, 'The Royal Garden at Pasargadae: Evolution and Legacy', in *Archaeologia Iranica et Orientalis. Miscellanea in honorem Louis Vanden Berghe*, II, Ghent, 1989, pp. 476–502 (476–80); and 'The Garden as a Political Statement: Some Case Studies from the Near East in the First Millenium B. C.', *Bulletin of the Asia Institute*, 4, 1, 1992, pp. 171–80 (171–4).

[21] Diodorus 2, 13; Fauth 'Der königliche Gärtner' (n. 17 above), pp. 14–15; Glassner, 'Á propos des jardins mésopotamiens' (n. 19 above), p. 9. See also Tuplin *Achaemenid Studies* (n. 2 above), p. 106.

[22] See W. Nagel, *Ninus und Semiramis in Sage und Geschichte*, Berlin, 1982; cf. Gignoux, 'Quelques réflexions' (n. 19 above), p. 5.

[23] Gignoux, 'Quelques réflexions' (n. 19 above), pp. 5–6.

Glassner,[24] rightly insisted also on the existence of an early Sumerian myth describing the creation of the first garden in Mesopotamia, with a reference in particular to the origin of the date-palm (^{GIŠ}*gišimmar*), a kind of plant which will turn out to be very important in the framework of the present discussion. Although Tuplin,[25] in the framework of a remarkable and comprehensive study dedicated to the Achaemenid paradise, denies that its introduction and its functions strictly followed a Mesopotamian antecedent (as differently suggested by Fauth,[26] who explicitly considered it an Assyrian invention), we may prudently advance a little general remark about the whole subject. It is quite reasonable to suppose, as already Gignoux and Hultgård did, that, stimulated by Babylonian and Assyrian precedents, the Medes[27] (and then the Persians, too) firstly adopted an earlier tradition and then adapted it (with the respective changes) to their own convenience, thus creating, for certain aspects *ex novo*, special parks, which they denominated 'paradises', thus making use of an original compound[28] taken from their Iranian *Sprachgut*, but with a special technical meaning. I do not want to venture into a more precise historical reconstruction of this side of the problem, simply stating that the Iranian paradise was already Mesopotamian, a subject about which Tuplin has dedicated a detailed research with a negative answer[29] – but I would simply note that a radical negation of any influence from the Akkadian and Babylonian background seems to me an extreme conclusion.

The Iranian name of the 'encircled garden' and the linguistic history of its fortune

If the presence in Avestan of the already mentioned *pairi.daēza-*, functionally '*circumvallatio*',[30] doubtless shows that this kind of compositional stem was genuine and existed independently also in eastern Iran, its functions result, contrariwise, very differently. The Avestan *pairi.daēza-*, in fact, did not refer to a beautiful park dedicated to various kinds of amusements, but, against any romantic expectation, the result was to have been, at least according to *Vd.* 3, 18, a place of definitive seclusion, where a particular

[24] Ibid., pp. 9–10.

[25] Tuplin, *Achaemenid Studies* (n. 2 above), pp. 80–88.

[26] Fauth 'Der königliche Gärtner' (n. 17 above), pp. 23–9, passim. Fauth recalled that F. Delitzsch, *Wo lag das Paradies? Eine biblisch-assyriologische Studie*, Leipzig, 1881, p. 97, wrongly wondered about the real Iranian origin of the word. About the Achaemenian gardens, see Stronach, 'The Royal Garden' (n. 20 above), 1989, pp. 483–6; 1992, pp. 174–8.

[27] About the possible relation between Median traditions and the Urartean background, see Hultgård, 'Das Paradies' (n. 1 above), p. 15; see also 17–20.

[28] See J. Duchesne-Guillemin, *Études de morphologie iranienne*, I, *Les Composés de l'Avesta*, Liège and Paris, 1936, p. 67, § 101.

[29] Tuplin, *Achaemenid Studies* (n. 2 above), p. 87. For the Persian 'paradise' as a space of re-creation, also with eschatological implications, see B. Lincoln, 'À la recherche du paradis perdu', *History of Religions*, 43/2, 2003, pp. 139–54 (153–4).

[30] See already M. Müller, *Chips from a German Workshop*, IV, *Essays chiefly on the Science of Language*, New York, 1895, 21–3.

kind of sinner (he who looks at human corpses)[31] had to be enclosed and purified. In the case of *Vd.* 5, 49, *pairi.daēza-* is a construction to be used for the provisional[32] isolation and purification of a woman who has given birth to a dead baby. Both passages are identical and run as follows: *aētaδa hē aēte mazdaiiasna aŋhå zəmō pairidaēząn pairi.daēzaiiąn*[33] 'There, on that place (lit. 'earth'), these Mazdeans shall build around (*pairi.daēzaiiąn* [3rd pl. pres. subj.]) an enclosure made of surrounding walls (lit. 'enclosures [etc.]',[34] *pairidaēząn* being acc. pl. m.)'. While this passage shows that such an enclosure was not particularly pleasant, we must remark the presence of an interesting *figura rhetorica* (viz. *fig. etymologica*), with an accusative of the internal object, *pairidaēza- pairi.daēz-* 'to build around a perimetrical building (or an enclosure made from surrounding walls)', which possibly could also imply a technical meaning in the language of masonry. In any case, the *Widēwdād* passages (here already quoted) insist on the fact that these places should be far away from fire and water, and Hultgård[35] is right when he notes that these *pairidaēza*-s are practically the opposite of what we could imagine in the case of a joyful 'paradise'. The concept of this kind of construction will have a different development in western Iran, and from a place of purification it will have the good chance to be used in order to represent the Jewish-Christian paradise.

Probably only in a well-advanced political milieu, where strong monarchies had emphasized the image of their power and developed social rituals at the courts, as happened in Mesopotamia and western Iran,[36] was it possible to exalt the role of the *parádeisoi* as joyful places, where kings and nobles used also to spend their time, and in some circumstances also to hunt wild animals.[37] Thus, paradise became a mark of distinction for royalty and nobility,[38] although the impression they made on Greek authors seems to represent a very 'mundane' phenomenon.[39] Later, Parthians and

[31] See *Vd.* 3, 15.

[32] See Hultgård, 'Das Paradies' (n. 1 above), pp. 7–8.

[33] K. F. Geldner, *Avesta. The Sacred Books of the Parsis*, III, *Vendîdâd*, Stuttgart, 1896, p. 18; F. Wolff, *Avesta. Die heiligen Bücher der Parsen übersetzt auf der Grundlage von Chr. Bartholomae's Altiranischem Wörterbuch*, Strasbourg, 1910, pp. 328, 348. Cf. H. Reichelt, *Avesta Reader. Texts, Notes, Glossary and Index*, Strasbourg, 1911, pp. 142–3.

[34] In the Pahlavi version *pairidaēza-* was translated as *pērāmōn-dahišnīh* 'creation/construction around': see Hultgård, 'Das Paradies' (n. 1 above), p. 28.

[35] Ibid., p. 8 with reference to *Vd.* 3, 15. On the contrary, the most beautiful places are, according to *Vd.* 3, 4, those where there are plants and waters (ibid., p. 20).

[36] The archaeological material regarding the Achaemenid garden has been presented by Tuplin, *Achaemenid Studies* (n. 2 above), pp. 88–92.

[37] The whole of this complex subject has been discussed in detail by Tuplin, ibid., pp. 108–9, who has analysed both positive and negative data about the *realia* existing behind the sources at his disposal. For later periods cf. P. Gignoux, 'La Chasse dans l'Iran sassanide', in *Orientalia Romana*, ed. G. Gnoli, Rome, 1983, pp. 101–18.

[38] Hultgård, 'Das Paradies' (n. 1 above), p. 6, passim. Hultgård in particular notes that the Iranian paradise should have been a kind of natural microcosm; see also P. Briant, *Histoire de l'empire perse de Cyrus à Alexandre*, Paris, 1996, p. 215. Cf. Fauth 'Der königliche Gärtner' (n. 17 above), pp. 2–3.

[39] As Tuplin, *Achaemenid Studies* (n. 2 above), p. 131, concluded.

Sasanians[40] also continued this Oriental custom, and their example was significantly important also in Armenia, as Garsoïan[41] has underlined. In Armenia, the 'garden' was designated as *partēz*, although it was not specifically used in the case of the royal parks.[42]

If we do not postulate its earlier attestation in the works of Xanthus,[43] or take into consideration its (genuine) presence already in a fragment attributed to Ctesias (*FGrH* 688 F 34a),[44] the first certain Greek use[45] of the word παράδεισος apparently occurs only with Xenophon, who, for instance, describes it as follows:

> At this point Critobulus said: 'Well, Socrates, if the Great King does this, it seems to me that he pays as much attention to husbandry as to warfare.' 'Yet further,' continued Socrates, 'in all the districts he resides in and visits he takes care that there are gardens, "paradises", as they call them, full of all the good and beautiful things that the soil will produce, and in this he himself spends most of his time, except when the season precludes it.' 'Then it is of course necessary, Socrates, to take care that these paradises in which the king spends his time shall contain a fine stock of trees and all other beautiful things that the soil produces.'[46]

Thus, it is very probable that a 'Median' or 'Northern' stem such as *paridaiza-* (perhaps adopted [together with, or instead of, an *echtpersisch* stem like **paridaida-*] also by

[40] According to V. Minorsky, 'Geographical Factors in Persian Art', *Bulletin of the School of Oriental and African Studies*, 9/3, 1938, pp. 621–52 (622), the pavilions known at least since Sasanian times were a derivation of the Persian 'paradise'. See also Bremmer, 'The Birth of Paradise' (n. 1 above), p. 50.

[41] N. Garsoïan, 'Prolegomena to a Study of the Iranian Aspects in Arsacid Armenia', *Handes Amorya*, 90, 1976, cols. 177–234, repr. in Garsoïan, *Armenia between Byzantium and the Sasanians*, London, 1985, pp. 27–9; the importance of the royal residence (named in Armenian *banak*) and of its *parádeisoi*, has been underlined by Garsoïan in '*Tagaworanist kayeank' kam Banak ark'uni*: Les Résidences royales des Arsacides arméniens', *Revue d'Etudes Arméniennes*, NS 21, 1988–89, pp. 251–69.

[42] As noted by Hultgård, 'Das Paradies' (n. 1 above), pp. 21–2.

[43] On this possibility, see Bremmer, 'The Birth of Paradise' (n. 1 above), pp. 40–41, who suggests that a quotation made by Clearchus, a pupil of Aristoteles, was derived from a lost work of the Lydian historian Xanthus.

[44] F. Jacoby, *Die Fragmente der griechischen Historiker*, III, C, Leiden, 1958, 688, F. 34°; see Bremmer, 'The Birth of Paradise' (n. 1 above), p. 45. Cf. Ctésias de Cnide, *La Perse*; *L'Inde*; *Autres fragments*, ed. and transl. D. Lenfant, Paris, 2004, p. 166 (quoted by Aelianus, *De natura animalium*, 7, 1).

[45] We must nevertheless recall that Herodotus did not use the word in his whole work. If this loanword was already well embedded into Greek, we should expect to find it, although we cannot exclude apriori the possibility that Herodotus knew it. The problem, in any case, is open.

[46] Xenophon, *Oeconomicus*, IV, 12–15, in *Opera Omnia*, ed. E. C. Marchant, II, Oxford, 1921, pp. 164–5 [my emphasis]: ἐκ τούτων ὁ Κριτόβουλος εἶπεν: οὐκοῦν εἰ μὲν δὴ ταῦτα ποιεῖ βασιλεύς, ὦ Σώκρατες, οὐδὲν ἔμοιγε δοκεῖ ἧττον τῶν γεωργικῶν ἔργων ἐπιμελεῖσθαι ἢ τῶν πολεμικῶν. ἔτι δὲ πρὸς τούτοις, ἔφη ὁ Σωκράτης, ἐν ὁπόσαις τε χώραις ἐνοικεῖ καὶ εἰς ὁπόσας ἐπιστρέφεται, ἐπιμελεῖται τούτων ὅπως κῆποί τε ἔσονται, οἱ **παράδεισοι** καλούμενοι, πάντων καλῶν τε κἀγαθῶν μεστοὶ ὅσα ἡ γῆ φύειν θέλει, καὶ ἐν τούτοις αὐτὸς τὰ πλεῖστα διατρίβει, ὅταν μὴ ἡ ὥρα τοῦ ἔτους ἐξείργῃ. νὴ Δί', ἔφη ὁ Κριτόβουλος, ἀνάγκη τοίνυν, ὦ Σώκρατες, ἔνθα γε διατρίβει αὐτός, καὶ ὅπως ὡς κάλλιστα κατεσκευασμένοι ἔσονται οἱ **παράδεισοι** ἐπιμελεῖσθαι δένδρεσι καὶ τοῖς ἄλλοις ἅπασι καλοῖς ὅσα ἡ γῆ φύει. For a detailed presentation of the Greek sources regarding the Persian 'paradise', see again Tuplin, *Achaemenid Studies* (n. 2 above), pp. 100–131, and passim. Bremmer, 'The Birth of Paradise' (n. 1 above), pp. 42–4, tries to give a significant relative chronology of Xenophon's use of this word. Cf. also Liddell, Scott and Jones, *A Greek-English Lexicon* (n. 3 above), p. 1308. Xenophon mentioned paradises in some other places, like *Anabasis*, 1, 2, 7; 2, 4, 1, 4; *Cyropaedia*, 1, 3, 14; *Hellenikà*, 4, 1, 15, etc.; see Tuplin, *Achaemenid Studies* (n. 3 above), pp. 118, 120; Hultgård, 'Das Paradies' (n. 1 above), pp. 3–7, 19. Cf. also M. A. Dandamayev, 'Royal *Paradeisoi* in Babylonia', in *Orientalia J. Duchesne-Guillemin Emerito Oblata*, Leiden, 1984, pp. 113–117 (114); Fauth 'Der königliche Gärtner' (n. 17 above), pp. 1–4.

Persians, who in many cases accepted in their lexicon many dialectal varieties of Northern origin), entered the Greek language at the latest about the end of the fifth century or at the beginning of the fourth century.[47] The existence in Old Persian of a genuine form like *_pardayda-_/*_pardēda-_ occurs, as we have already noted, only once and unfortunately in a debated place, where it is apparently engraved as <p^{a}-r^{a}-d^{a}-y^{a}-d^{a}-a-m^{a}> (_paradaydām_ [?]). Anyway, its formal status cannot be used in order to explain the Greek adaptation with παρά°, instead of παρί° (or, better, περί°), because it is probable that its spelling was already *_pardēdăm_ (in the accusative,[48] the nominative being simply *_pardēdə_, with a very short vowel in _Auslaut_). Then, the second _-a-_ in παρά° seems to have been the fruit of a popular adaptation for a foreign word.[49] With regard to the actual spelling of the Greek loanword παράδεισος we must note that, very probably, it was not <_parádeisos_>, as it could seem _prima facie_, but <_parádēsos_> (-ει- being already the written way to represent an _-ẹ̄-_ in Xenophon's dialect and time),[50] as already Jacobsohn[51] and later Schwyzer[52] suggested, although it is unnecessary to derive this strictly from a Middle Iranian phase. More prudently, we can recall that spoken Median and Old Persian had a number of pronunciation forms in the living language that anticipated[53] many of the phenomena later fully documented only in Middle Persian and Parthian. Thus, behind the attested orthography of the later Old Persian form engraved as <p^{a}-r^{a}-d^{a}-y^{a}-d^{a}-a-m^{a}>, we can easily postulate a form spelt (in nominative, of course) as *_pardēdə_, and later as *_pardēd_ (cf. again Christian Sogdian _prdyz_, Khwarezmian _parδyzk_ 'kitchen garden',[54] Armenian _partēz_). Of course, it remains theoretically open, but in my opinion less probable, the possibility that the Greek loanword was earlier, and that it still reflected an older pronunciation like *_par^{i}dai̯za-_, then, with Greek -ει- actually corresponding to North-

[47] See Bremmer, 'The Birth of Paradise' (n. 1 above), pp. 40–45.

[48] The long _-ā-_ in _–ām_ is not etymological, but it is one of the various peculiar spellings attested in this later inscription; it is impossible to imagine that the accusative ending of a masculine stem in _-a-_ had a real lengthening of its vowel.

[49] See H. Jacobsohn, 'Σκυθικά', _Zeitschrift für vergleichende Sprachforschung_, 54, 1927, pp. 254–86 (258). Cf. already O. Weise, 'Volksetymologische Studien, I', _Bezzenbergers Beiträge_, 5, 1880, pp. 68–94 (91).

[50] M. Lejeune, _Traité de phonétique grecque_, 2nd edn, Paris, 1955, p. 199.

[51] Jacobsohn, 'Σκυθικά' (n. 49 above), 257–9. Rightly, on pp. 257–8 and n. 1, he insists on the fact that, although C. Bartholomae, 'Arica XVII', _Indogermanische Forschungen_, 38, 1917/1920, pp. 1–26 (25, n. 2) was ready to admit that Xenophon had still heard a word with an open diphthong, so that 'sein παράδεισος [würde] ebenfalls auf _i_-Farbung des _a_-Vokals im _i_-Diphthongen hinweisen', Xenophon already pronounced -ει- (_-ei̯-_) as a closed _-ē-_; so, also a secondary closed long _-ẹ̄-_ would be transcribed by means of an ει. The later form attested in the 3rd-century BC papyri is παράδισος (R. Schmitt, _Die iranischen und Iranier-Namen in den Schriften Xenophons_, Vienna, 2002, p. 49).

[52] E. Schwyzer, _Griechische Grammatik, auf der Grundlage von Karl Brugmanns Griechischer Grammatik bearbeitet_, I, _Allgemeiner Teil – Lautlehre – Wortbildung – Flexion_, Munich, 1939, p. 193.

[53] About this subject, see also M. Allegri and A. C. D. Panaino, 'On the _š_-ending in O.P. _akunauš_ and similar forms, with a contribution by I. Gershevitch', in _Proceedings of the Second European Conference of Iranian Studies_, Rome and Bamberg, 1995, pp. 1–33.

[54] D. N. MacKenzie, _A Concise Pahlavi Dictionary_, London, 1990, p. 120 (from Avestan _paridaēza-_ + _-ka-_).

West Iranian *-ai-*, as suggested by Schmitt[55] and Bremmer.[56] With regard to Armenian *partēz* Meillet[57] already noted that its introduction should be reasonably ancient (i.e. belonging to the Achaemenid period), because this loanword doubtless presents the transformation of *-rd-* to *-rt-* (Iranian **par*^{i}*dēza-* > Old Armenian **pardēz* > Armenian **partēz*), an evidence meaning that, diachronically, it was earlier than the Armenian consonantal *Lautverschiebung*. It would be theoretically possible that the Iranian word were still pronounced as **paridēza-*, but, as Jacobsohn[58] suggested, this *-i-* (in the second syllable) would have regularly disappeared in Armenian. Here *partēz* was used with the basic meaning of 'garden' (so corresponding to Gr. κῆπος) already in the earliest translations of the Bible, but the word designating the eschatological paradise was *draxt* 'date-palm'.[59]

Furthermore, we must note that also the Hebrew form *pardēs*[60] (see, too, Syriac *pardēsā*)[61] was clearly derived from Iranian, but in this case too we may postulate that it did not derive from an Early Iranian stem like **paridai̯za-*, but from a Proto- or Early-Middle Iranian form like **pardēz*. Jacobsohn[62] suggested that the early presence in Nehemia 2:8 of this word should prove that the Jewish people heard an Iranian stem with a syncope of the *-i-* in the second syllable (*par°* instead of *pari°*) and with a monophthongization of *-a -* (> *-ē-*).[63] According to Hübschmann,[64] Modern Persian *pālēz* (see also *pālēzbān* 'gardener'),[65] in its turn, cannot derive from a Western Iranian stem (interestingly in Middle Persian **pardēd*[66] or **pardēz* are not attested),[67] but from

[55] Schmitt, *Die iranischen und Iranier-Namen* (n. 51 above), pp. 34, 49, 75, n. 95, 139; R. Schmitt, *Iranische Anthroponyme in den erhaltenen Resten von Ktesias' Werk*, Vienna, 2006, pp. 60, 132, 284.

[56] Bremmer, 'The Birth of Paradise' (n. 1 above), p. 36, with reference to R. Schmitt.

[57] A. Meillet, 'Sur les mots iraniens empruntés par l'arménien', *Mémoires de la Société de Linguistique de Paris*, 17, 1911/12, pp. 242–50 (250), repr. in Meillet, *Kleinen Schriften zum Armenischen*, Hildeseim, 1976, pp. 300–307; see also G. Bolognesi, *Le fonti dialettali degli imprestiti iranici in Armeno*, Milan, 1960, p. 50, n. 1. Cf. Jacobsohn, 'Σκυθικά' (n. 49 above), p. 259.

[58] Jacobsohn, 'Σκυθικά' (n. 49 above), p. 259.

[59] See Hultgård, 'Das Paradies' (n. 1 above), p. 21.

[60] Cf. Tuplin, *Achaemenid Studies* (n. 2 above), p. 119.

[61] See R. Payne Smith, *Thesaurus Syriacus*, eds S. M. E. Quatremère, G. H. Bernstein, II, Oxford, 1901, cols 3239–3240; cf. O. Klein, *Syrisch-Griechisches Wörterbuch zu den Vier Kanonischen Evangelien nebst einleitenden Untersuchungen*, Giessen, 1916, p. 85: *pardēsa* (Luke 23:43, *Peshitta* and *Sinaiticus*); p. 37: *ganath 'eden* 'garden of Eden' (Luke 23:43, *Vetus Syrus Curetonianus*); cf. C. Brockelmann, *Lexicon Syriacum*, Edinburgh and Berlin, 1895, pp. 285, 460. About Syriac *prdys'*, see also C. Ciancaglini, *Iranian Loanwords in Syriac*, Wiesbaden, 2008, p. 237.

[62] Jacobsohn, 'Σκυθικά' (n. 49 above), p. 259.

[63] Then, the idea suggested by Bartholomae, 'Arica XVII' (n. 51 above), p. 25, n. 2, that Armenian *partēz* was adopted when the Iranian stem still had a diphthong *(-ai̯-)* results in being very doubtful.

[64] H. Hübschmann, *Armenische Grammatik*, I, *Armenische Etymologie*, Leipzig, 1897, p. 229.

[65] We may recall the existence of Greek παραδεισάριος 'gardener', which was translated in the Armenian *Gospel of John* as *partizpan* 'guardian of the garden' (or as *partizapan* by Ephrem in his *Lives of the Fathers*); this word has been adapted in Syriac as *pardayspana*. See Hübschmann, *Armenische Grammatik* (n. 64 above), p. 229; Fauth, 'Der königliche Gärtner' (n. 17 above), p. 5; Bremmer, 'The Birth of Paradise' (n. 1 above), p. 38, and n. 17.

[66] Schmitt, *Die iranischen und Iranier-Namen* (n. 51 above), p. 49, n. 33.

[67] The reference made by Fauth, 'Der königliche Gärtner' (n. 17 above), p. 5, to a Middle Persian form is misleading.

an unattested Parthian **pardēz*;[68] but, as Nicholas Sims-Williams kindly remarks to me, an *-l-* from *-rd-* is strange. Maybe it would be better to assume an earlier 'Median' loanword.

Although Herodotus (IX, 116) seems to have been informed about the function and importance of Persian gardens and parks, he did not use the word παράδεισος, and Tuplin[69] has expressed some sceptical remarks about his real acquaintance with true *parádeisoi*. On the other hand, as again Tuplin underlined,[70] *pardēsu*[71] was adopted also in Akkadian from an Iranian stem within the first five years of Cyrus in Babylon. This fact is *per se* interesting because it shows that the Iranian terminology gained ground also in Akkadian, probably because it designated a special kind of garden or park. In addition, Akkadian *pardēsu* [<par-de-su>; <pa-ar-de-e-su>] seems to confirm that Iranian **parida za-* was pronounced already in (relatively) earlier times **par*^{i}*dēsa-*, with a monophthongization of *-a -*. Thus, we find another piece of supporting evidence that with *-e -* Greek παράδεισος could represent a spoken *-ẹ̄-*, as we have already remarked. An open question is that concerning the connection of the Elamite word *partetaš* (or *bartetaš* [<h.par-te-taš>; <h.par-te-da>]) with O.P. **pari-dai̯da-*. It was originally proposed by Benveniste,[72] while Hinz and Koch,[73] accepting this comparison, suggested that it should be translated more fittingly as 'Domäne' or 'Herrengut'. In a further work H. Koch[74] has advanced the translation of *partetaš* as royal 'Vorratslager' or 'Sammelstelle'. Cameron,[75] and more

[68] O. Szemérenyi, 'Contributions to Iranian Lexicography', *Journal of the American Oriental Society*, 70/4 (Oct.-Dec.), 1950, pp. 226–36 (234) assumes that 'NPers. *-l–*, in general, continues the OPers. cluster *-rd-*' is right, but he gave a misleading example when he wrote: 'One of the best known and clearest examples is NPers. *palēz* 'park' deriving from Old Iranian *pari-daiza-* as attested by Avestan *pairidaēza-*'. In fact, in this very case, the derivation cannot be from a Western Iranian dialect.

[69] Tuplin, *Achaemenid Studies* (n. 2 above), pp. 120–21. It is in fact possible, as noted, that this word was used in order to represent an Oriental garden, which looked larger or more sumptuous than a garden.

[70] Ibid., p. 120.

[71] *Chicago Assyrian Dictionary*, XII, '*P*', ed. M. T. Roth, Chicago IL, 2005, p. 182. See also J. N. Strassmaier, *Inschriften von Cyrus, König von Babylon*, Leipzig, 1890, Nr. 212: 3; W. von Soden *Akkadisches Handwörterbuch*, II, '*M-S*', Wiesbaden, 1972, 2, p. 833, s.v. *pardēsu*; W. Hinz, *Neue Wege im Altpersischen*, Wiesbaden, 1973, p. 73. Cf. also Dandamayev, 'Royal *Paradeisoi*' (n. 46 above), pp. 113–17; Glassner, À propos des jardins mésopotamiens' (n. 19 above), p. 10; J. Black, A. George and N. Postgate, *A Concise Dictionary of Akkadian*, 2nd edn, Wiesbaden, 2000, p. 266. Cf. also Bremmer, 'The Birth of Paradise' (n. 1 above), p. 37.

[72] E. Benveniste, 'Eléments perses en araméen d'Egypte', *Journal Asiatique*, 242, 1954, pp. 297–310 (308); 'Notes sur les tablettes élamites de Persepolis', *Journal Asiatique*, 246, 1958, pp. 49–65 (57–8).

[73] W. Hinz, and H. Koch, *Elamisches Wörterbuch*, 2 vols, Wiesbaden, 1987, I, p. 160, s.v. <h.bar-te-taš>; cf. <h.bar-te-taš-nu-iš-ki-ip> 'Domänenhüter' (ibid.). See also Hinz, *Neue Wege* (n. 71 above), pp. 72–3; Hinz, with P.-M. Berger, G. Korbel and A. Nippa, *Altiranisches Sprachgut der Nebenüberlieferungen,* Wiesbaden, 1975, p. 179 sub **pardaiδa-*, m. We can recall that also Dandamayev, 'Royal *Paradeisoi*' (n. 46 above), p. 113, accepted the identification of Elamite *partetaš* as 'paradise'.

[74] H. Koch, 'Steuern in der achämenidischen Persis?', *Zeitschrift für Assiriologie und Vorderasiatische Archäologie*, 70, 1980, pp. 105–37 (119, n. 73); Koch, *Verwaltung und Wirtschaft im persischen Kernland zur Zeit der Achämeniden*, Wiesbaden, 1990, p. 21; Koch, *Es kündet Dareios der König* [...] *Vom Leben im persischen Großreich*, Mainz, 1992, pp. 264–75.

[75] G. G. Cameron, 'New Texts from the Persepolis Treasury', *Journal of Near Eastern Studies*, 24/3, 1965, *Erich F. Schmidt Memorial Issue*, I, July, pp. 167–92 (in particular pp. 169 and 177); see also Cameron, *Persepolis Treasury Tablets*, Chicago IL, 1948, p. 282, where Bar-te-taš was translated 'a gateway (?)'.

recently Lecoq,[76] contrariwise, have expressed a negative opinion about this solution, while Hallock[77] simply translated it as 'a kind of storage place for commodities'. Although Tuplin's investigation has not offered a final solution,[78] I can endorse his remark[79] about the fact that 'from a functional-analytic point of view' the interpretation of *partetaš* proposed by Hinz and Koch appears 'understandable'. From a linguistic perspective such a comparison seems to me fitting and very clear; it is also very important, because an orthography like Elamite <h.par-te-da> (PF 158: 2f)[80] seems to confirm the existence of a South-Western Iranian form like **pardēda-* (or, possibly, **pardēdə-*).[81] I must also draw attention to a recent article by Uchitel,[82] who has shown that, in the Fortification Archive from Persepolis, the word *partetaš* clearly refers to lands managed directly by a land-owner, although, in a 'totally different socio-economic situation [...] its usage became restricted to parks, gardens and orchards'.[83]

It is clear, as Bremmer remarked,[84] that of the two dialectologically different forms attested, the Northern one (**paridaiza-*) was adopted by Babylonians, Greek and Jews, while the genuine Old Persian one (**paridaida-*) was Elamiticized in *partetaš* (*bartetaš*), although all these loanwords were probably received through a later spelling like **pardēz(a)-/pardēd(a)-*.

The beauty of these remarkable gardens did not attract only the interest of the Greeks, but Jewish biblical literature also knew this Iranian word for 'garden, park'. Hebrew *pardēs*,[85] in fact, is attested in three different Old Testament books: in

[76] P. Lecoq, 'Paradis en vieux-perse?', in *Contributions à l'histoire de l'Iran. Mélanges offerts à Jean Perrot*, ed. F. Vallat, I, Paris, 1990, pp. 209–11 (211, n. 23).

[77] R. T. Hallock, *Persepolis Fortifications Tablets*, Chicago IL, 1969, p. 742.

[78] Tuplin, *Achaemenid Studies* (n. 2 above), pp. 93–7. He has also investigated (pp. 116–18) the possibility that these parks were used for ritual religious ceremonies, but the evidence seems to be limited. On the other hand, Hultgård, 'Das Paradies' (n. 1 above), pp. 9–10, has collected some data supporting the existence of ritual ceremonies in the framework of the paradises. He also underlined the fact that, already in the Mesopotamian tradition, part of the *akītu*-festival took place in a garden (pp. 12–15); cf. Fauth, 'Der königliche Gärtner' (n. 17 above), p. 18; the ritual importance of the *parádeisoi* is in any case attested in the Greek Papyri of Egypt (Hultgård, 'Das Paradies', n. 1 above, pp. 24–5; Bremmer, 'The Birth of Paradise', n. 1 above, pp. 46–7). About the symbolic role of the garden we must mention G. Widengren, *The King and the Tree of Life in Ancient Near Eastern Religion*, Uppsala, 1951.

[79] Tuplin, *Achaemenid Studies* (n. 2 above), p. 96.

[80] Hallock, *Persepolis Fortifications Tablets* (n. 77 above), p. 116.

[81] See J. Tavernier, *Iranica in the Achaemenid Period (ca. 550–330 B.C.). Lexicon of Old Iranian Proper Names and Loanwords, Attested in Non-Iranian Texts*, Leuven, Paris and Dudley MA, 2007, pp. 445–7; cf. also Schmitt, *Beiträge* (n. 12 above), p. 84.

[82] A. Uchitel, 'Persian Paradise: Agricultural Texts in the Fortification Archive', *Iranica Antiqua, 32*, 1997, pp. 137–44. See also Bremmer, 'The Birth of Paradise' (n. 1 above), pp. 37–8.

[83] Uchitel, 'Persian Paradise' (n. 82 above), p. 144.

[84] Bremmer, 'The Birth of Paradise' (n. 1 above), p. 39.

[85] W. Gesenius, with F. Buhl, *Hebräisches und aramäisches Handwörterbuch über das Alte Testament*, Berlin, Göttingen and Heidelberg, 1915, p. 657 (cf. also *A Hebrew and English Lexicon of the Old Testament with an Appendix containing the Biblical Aramaic*, based on the Lexicon of W. Gesenius as transl. by E. Robinson, ed F. Brown, with S. R. Driwer and C. A. Briggs, 7th repr. of the 1907 ed., Oxford, 1972, p. 825). Cf. Fauth, 'Der königliche Gärtner' (n. 17 above), p. 5.

Nehemiah 2:8,[86] it refers to the royal garden of Artaxerxes, placed under the responsibility of Asaph;[87] in Canticles 4:13–14,[88] to a garden (Septuagint παράδεισος)[89] full of precious plants, while in Ecclesiastes 2:5,[90] Qohelet declares that he has built gardens (*gannôt*) and parks (*pardēsim*; Septuagint παράδεισους).[91] The Jewish translators of the Septuagint, in their turn, called the garden in Eden (*gan*[92] *be ʿēden*), appearing in Genesis 2:8[93] (καὶ ἐφύτευσεν κύριος θεὸς παράδεισον ἐν ᾿Εδεμ),[94] παράδεισος,[95] while in later Jewish texts belonging to the Talmudic tradition, like *Ḥagigah*, 14b, the word *pardēs* assumed also the meaning of 'enclosure (in heaven)',[96] but as a metaphoric expression corresponding to 'esoteric philosophy'. It is, however, reasonable to admit with Bremmer[97] that, when, in the third century BC, the translators of the Septuagint decided to translate *gan be ʿēden* with παράδεισος and not with κῆπος 'garden', they were not referring to a properly Persian royal park, but to the contemporary royal παράδεισοι of the Hellenistic kings (particularly the Ptolemaic ones) 'with many trees and less wild than their Persian predecessors'.[98]

Finally, in the Christian framework, the Gospel of Luke (23:43: 'ἀμήν σοι λέγω, σήμερον μετ᾿ ἐμοῦ ἔσῃ ἐν τῷ παραδείσῳ'; 'Amen dico tibi: hodie mecum eris in paradiso')[99] patently attests that the word παράδεισος was used in order to make reference to the heavenly abode of the blessed ones in the afterlife. In II Corinthians 12:3–4, it is referred to as a celestial place (connected with the third heaven) where a living man was able to go ([3]

[86] *Biblia Hebraica Stuttgartensia*, eds K. Elliger, W. Rudolph, H. P. Rüger, and G. E. Weil, Stuttgart, 1966–7, p. 1432. See Hultgård, 'Das Paradies' (n. 1 above), p. 4.

[87] On the importance of this passage, see Bremmer 'The Birth of Paradise' (n. 1 above), pp. 39–40.

[88] *Biblia Hebraica Stuttgartensia,* eds Elliger, Rudolph, Rüger, and Weil (n. 86 above), p. 1330. See also Tuplin, *Achaemenid Studies* (n. 2 above), p. 115, who notes that in this place sexual activity is also meant. See also Bremmer, 'The Birth of Paradise' (n. 1 above), pp. 40, 52.

[89] *Septuaginta, Id est Vetus Testamentum grace iuxta LXX interpretes,* Duo volumina in uno, ed. A. Rahlfs, Nördlingen, 2004, II, p. 265.

[90] *Biblia Hebraica Stuttgartensia,* eds Elliger, Rudolph, Rüger, and Weil (n. 86 above), p. 1338; Hultgård, 'Das Paradies' (n. 31 above), pp. 22–3.

[91] *Septuaginta*, ed. Rahlfs (n. 89 above), p. 240.

[92] For the comparison between γάνος (i.e. *gan*) and παράδεισος in Greek glossography, see Tuplin, *Aechemenid Studies* (n. 2 above), p. 93. Cf. also Hultgård, 'Das Paradies' (n. 1 above), pp. 2, 15–16, 22–3, and in particular p. 26; very important is the final *Excursus* offered by Bremmer, 'The Birth of Paradise' (n. 1 above), pp. 54–5, about 'Paradise' in Cyprus.

[93] *Biblia Hebraica Stuttgartensia,* eds Elliger, Rudolph, Rüger, and Weil (n. 86 above), p. 3. See J. Hamburger, *Real-Encyclopädie für Bibel und Talmud: Wörterbuch zum Handgebrauch für Bibelfreunde, Theologen, Juristen, Gemeinde- und Schulvorsteher, Lehrer etc.*, II, Leipzig, 1878, pp. 892–6.

[94] *Septuaginta*, ed. Rahlfs (n. 89 above), I, p. 3.

[95] See G. Caracausi, *Lessico Greco della Sicilia e dell'Italia Meridionale (secoli X–XIV),* Palermo, 1990, p. 437.

[96] See M. Jastrow, *Dictionary of the Targumin, the Talmud Babli and Yerushalmi, and the Midrashic Literature*, I, New York, 1950, p. 1216. Very important is the discussion of this passage by M. Subtelny, 'The Tale of the Four Sages who Entered the *Pardes*. A Talmudic Enigma from a Persian Perspective', *Jewish Studies Quarterly*, 11, 2004. Cf. again Subtelny, *Le Monde* 2002 (n. 1 above).

[97] Bremmer, 'The Birth of Paradise' (n. 1 above), pp. 46–54.

[98] Ibid., p. 53.

[99] According to *Novum Testamentum Graece et Latine*, ed. S.J. Merk, 9th edn, Rome, 1964, p. 298.

καὶ οἶδα τὸν τοιοῦτον ἄνθρωπον – εἴτε ἐν σώματι εἴτε χωρὶς τοῦ σώματος οὐκ οἶδα, ὁ θεὸς οἶδεν – [4] ὅτι ἡρπάγη εἰς τὸν παράδεισον καὶ ἤκουσεν ἄρρητα ῥήματα ἃ οὐκ ἐξὸν ἀνθρώπῳ λαλῆσαι; 'Et scio huiusmodi hominem – sive in corpore sive extra corpus nescio, Deus scit – [4] quoniam raptus est in paradisum et audivit arcana verba, quae non licet homini loqui').[100] The eschatological meaning is clearly present also in Revelation 2:7 (ὁ ἔχων οὖς ἀκουσάτω τὶ τὸ πνεῦμα λέγει ταῖς ἐκκλησίαις. τῷ νικῶντι δώσω αὐτῷ φαγεῖν ἐκ τοῦ ξύλου τῆς ζωῆς, ὅ ἐστιν ἐν τῷ παραδείσῳ τοῦ θεοῦ; 'Qui habet aurem, audiat quid Spiritus dicat ecclesiis. Vincenti dabo ei edere de ligno vitae, quod est in paradiso Dei').[101] In any case, the word *paradīsus* became the standard designation of this marvellous abode already before Jerome's redaction of the Vulgate,[102] at least in the framework of Christian and Jewish cultures;[103] it was a divine and fantastic place but, now, without walls.

Doubtless we can recognize a certain influence also on the Islamic tradition, where paradise is a garden, called *djanna* 'garden', *ǎdn* 'Eden', *dar al-salām* 'Abode of peace', but also *firdaws*. This word, attested in two Koranic Suras (18, 107; 23, 11 [*jannat al-firdaws* 'the gardens of paradise']), was borrowed, probably from the Christian Syriac tradition (*pardaisā*),[104] in the form *faradis*, which, in its turn, was considered to be a plural.[105] Then, a new singular, *firdaws*, was derived, and in this way the name passed on to Modern Persian as *firdaus*[106] (or contemporary *ferdows*), which is at the origin of the family name of the great Persian poet Firdowsī.[107]

From the present overview of the sources we can confirm the conclusion of Bremmer, who noted that the Iranian 'paradise' was a kind of *vox media*, because it referred to different *realia*, spanning from that of royal park for hunting and pleasures to 'storage-place, vineyard, orchard, stable, forest or nursery of trees'.[108] We must again underline that its secular meaning differed strongly from that attested in Avestan sources, where it was charged with negative implications. No Persian king would have appreciated his own

[100] Ibid., p. 615.

[101] Ibid., p. 796.

[102] The Latin term *paradisus* first occurred in the *Vetus Latina*, the collection of fragmentary texts that preceded Jerome's early 5th-century translation, and became common in Christian literature since the writings of Tertullian (c. AD 160–c. 220): see Hultgård, 'Das Paradies' (n. 1 above), p. 1, n. 3.

[103] With regard to the Qumranic tradition and the book of Henoch, see Hultgård, 'Das Paradies' (n. 1 above), pp. 26, 32–6; Bremmer, 'The Birth of Paradise' (n. 1 above), p. 48.

[104] See Hultgård, 'Das Paradies' (n. 1 above), pp. 39–41 with large bibliography on the subject. It is impossible to discuss in this article the problem of the patent relations existing between the Iranian image of paradise and the Islamic one, for instance in the representation of the *houris*, etc.

[105] See already W. Muss-Arnolt, 'On Semitic Words in Greek and Latin', *Transactions of the American Philological Association*, 23, 1892, pp. 35–156 (104, n. 1).

[106] F. Wolff, *Glossar zur Firdosis Schahname*, Berlin, 1935, p. 606. See also Dehxodā, *Loγatnāme* (n. 16 above), s.v. *ferdows*: 1. 'The best place in paradise, paradise' (*behešt*); 2. 'vineyard' (*bāγ-e angur*), 'garden' (*bostān*), 'garden where any type of grapes, of flowers and fruits are cropped'; Mo'in, *Farhang-e fārsi*, (n. 18 above), s.v. *ferdows*: 1. 'garden' (*bāγ*, *bustān*); 2. 'paradise' (*behešt*).

[107] See P. Lecoq, *Les Inscriptions de la Perse achéménide*, Paris, 1997, p. 116, n. 4; Wolff, *Glossar* (n. 106 above), p. 606.

[108] See Bremmer, 'The Birth of Paradise' (n. 1 above), p. 39.

seclusion in an Avestan *pairi.daēza-*. This remarkable difference shows an important lexical distinction between eastern and western Iran. In any case, after the fall of the Persian Empire, the functions of the 'paradises' also partly changed: in particular, that of hunting parks belonging to a special elite was abandoned (with some exceptions for the Iranian(ate) regions). Contrariwise, their utilitarian functions increased and 'in the course of time the Graeco-Roman *paradeisos* became more and more artificial', as Bremmer underlined.[109]

The Iranian paradise

In the second part of this contribution I do not intend to propose a new investigation regarding the final abode of the souls in the Mazdean afterlife, but have limited myself to summarizing some of the data we know of, adding some useful remarks for specialists of other fields who, maybe, are interested in this subject.

The Zoroastrian tradition has developed an important doctrine regarding the destiny of the individual concerning the postmortem condition, while the general eschatology of humanity as a whole, with the final resurrection of the dead and the defeat of Ahreman are the fruits of a longer (and partly later) elaboration.[110] Any human being, male or female, possesses a complex spiritual structure, with a masculine (*uruuan-*) and a feminine (*daēnā-*) soul, which meet in the afterlife.[111] In particular, the *daēnā-* appears as the visible manifestation of human behaviour on earth: beautiful if the dead were a righteous person, monstrous if he (or she) were a sinner. The Iranian world knows also the symbolic image of the bridge, named *cinuuatō pərətu- or cinuuat.pərətu-* (Pahl. *činwad puhl*),[112] which, according to the Zoroastrian tradition, becomes wider under the feet of the righteous Mazdean, or, contrariwise, very thin, if he (or she) were a follower of Ahreman and the demons. Bartholomae explained this denomination as 'the bridge of the separator', with a reference to the fact that it divided the right ones from the sinners.[113] But *cinuuaṇt-* could have a more prosaic explanation, if Kellens[114] is right, when he tries to connect it

[109] Ibid., pp. 46–52.

[110] See J. Kellens, 'Characters of Ancient Mazdaism', *History of Anthropology*, 3, 1987, pp. 239–62; Kellens, 'Yima et la mort', in *Languages and Cultures. Studies in Honor of Edgar C. Polomé*, eds M. A. Jazayery and W. Winter, Berlin and New York, 1988, pp. 329–34; Kellens, 'L'Eschatologie mazdéenne ancienne', in *Irano-Judaica III. Studies relating to Jewish Contacts with Persian Culture throughout the Ages*, eds S. Shaked and A. Netzer, Jerusalem, 1994, pp. 49–53; Gignoux, 'Quelques réflexions' (n. 19 above), p. 8. Cf. also P. Gignoux, 'L'Enfer et le paradis d'après les sources pehlevies', *Journal Asiatique*, 256, 1968, pp. 219–45.

[111] See *Hādōxt Nask II. Il racconto zoroastriano della sorte dell'anima*, ed. and transl. A. Piras, Rome, 2000. Cf. also C. Pavry, *The Zoroastrian Doctrine of the Future Life from Death to the Individual Judgment*, New York, 1928; A. C. D. Panaino, '*nāirikā-* e *jahikā-* nell'aldilà zoroastriano', in *Bandhu. Scritti in onore di Carlo della Casa in occasione del suo settantesimo compleanno*, eds R. Arena, M. P. Bologna, M. L. Mayer Modena, A. Passi, II, Turin, 1997, pp. 831–43.

[112] See, e.g., H. S. Nyberg, *A Manual of Pahlavi*, II, *Ideograms, Glossary, Abbreviations, Index, Grammatical Survey, Corrigenda to Part I*, Wiesbaden, 1974, II, p. 55. Cf. Hultgård, 'Das Paradies' (n. 1 above), p. 30.

[113] Bartholomae, *Altiranisches Wörterbuch* (n. 8 above), pp. 596–7.

[114] Kellens, 'Characters of Ancient Mazdaism' (n. 110 above), p. 332.

with the verbal root *cin* 'to build, to pile'. Thus, it could refer to the builder of the bridge ('the bridge of the compiler'),[115] who piled up the necessary stones. Only after this crucial crossing can the soul reach his destination, paradise, hell or an intermediate place, which could be compared with a sort of purgatory (Pahl. *hamiyastagān*,[116] *hammistagān*;[117] Pāzand *hamēstagą* ; lit. 'raised together [at the same height]').[118] A very intricate matter concerns the possible impact of the Mazdean tradition regarding the future life and the judgment of the soul on the Hebrew world and its vision of the afterlife.[119] This is the subject of a number of studies that cannot be rehearsed here; but I would like to remark that, in my opinion, the Iranian model could have attracted the interest of many foreign peoples conquered by the Persians, among whom we must also consider the Jewish community, liberated from the Babylonian captivity. In any case, the importance of the afterlife is well preserved in the Pahlavi Zoroastrian literature, in particular in the famous *Ardā Wīrāz Nāmag*,[120] a text where the voyage of the Righteous Wīrāz to paradise, purgatory and hell is described with some baroque features. Its potential comparison with Dante's *Divina Commedia* has attracted the attention of many scholars.[121] Although this and other Pahlavi sources[122] concerning individual and universal eschatology are a later production of the Zoroastrian community, we must recall that already in the third century AD the High Priest Kirdīr described paradise (*wḥšty*) and hell (*dwšḥwy*) in his inscriptions,[123] which seem to contain the description of a kind of esoteric journey.

[115] See M. Moazami, 'The Dog in Zoroastrian religion: *Vidēvdād* Chapter XIII', *Indo-Iranian Journal*, 49, 2007, pp. 127–49 (128, n. 10). Moazami (ibid.) notes that, according to P. O. Skjærvø, such a denomination 'refers to the picking up and gathering of the "records" of good deeds, etc., hence the "Bridge of the Accountant".' Other earlier solutions have been summarized by J. Kellens, *Les Noms-racines de l'Avesta*, Wiesbaden, 1974, II, 28–9, n. 3.

[116] P. Gignoux, ed. and transl., *Le Livre d'Ardā Vīrāz*, Paris, 1984, p. 160, n. 1.

[117] D. N. MacKenzie, *A Concise Pahlavi Dictionary*, London, New York and Toronto, 1971, p. 41.

[118] See G. Klingenschmitt, 'Avestisch *hāməmiiāsaitē* und Pahlavī *hamystk'n*'', *Münchener Studien zur Sprachwissenschaft*, 30, 1972, pp. 79–92. Incorrect etymology in Nyberg, *A Manual of Pahlavi* (n. 112 above), p. 91.

[119] A presentation of this problem has been offered by A. C. D. Panaino, 'Trends and Problems concerning the Mutual Relations between Iranian Pre-Islamic and Jewish Cultures', in *Schools of Oriental Studies and the Development of Modern Historiography. Proceedings of the Fourth Annual Symposium of the Assyrian and Babylonian Intellectual Heritage Project*, eds A. C. D. Panaino and A. Piras, Milan, 2004, pp. 209–36.

[120] Gignoux, ed. and transl., *Le Livre d'Ardā Vīrāz*, (n. 116 above); F. Vahman, *Ardā-Virāz-Nāmag, The Iranian «Divina Commedia»*, London and Malmö, 1986.

[121] See A.V. Rossi, 'Il libro di Ardâ Wîrâz', in *Lectura Dantis 2001*, ed. V. Placella, Naples, 2005, pp. 255–72; and A. C. D. Panaino, 'L'aldilà zoroastriano e quello dantesco. Appunti per una riflessione comparativa e tipologica su forme e motivi ricorrenti nei viaggi ultraterreni', in *Dante e la fabbrica della Commedia*, ed. A. Cottignoli, Ravenna, 2008, pp. 171–87, with a detailed bibliography.

[122] See Gignoux, 'Quelques réflexions' (n. 19 above), pp. 8–12; P. Gignoux and A. Tafazzoli, *L'Anthologie de Zādspram*, Paris, 1993, p. 431, sub *wahišt*.

[123] See P. Gignoux, 'La Signification du voyage extra-terrestre dans l'eschatologie mazdéenne', in *Mélanges d'histoire des religions offerts à Henri-Charles Puech: sous le patronage et avec le concours du Collège de France et de la Section des Sciences Religieuses de l'École Pratique des Hautes Études*, Paris, 1974, pp. 63–9; idem, *Les quatre inscriptions du mage Kirdīr. Textes et concordances*, Paris, 1991, pp. 73–99, 102 [dwšḥwy], 105 [(wḥ)št(y)]; idem, *Man and Cosmos in Ancient Iran*, Rome, 2001; cf. also P. O. Skjærvø, 'Kirdir's Vision: Translation and Analysis', *Archaelogische Mitteilungen aus Iran*, N.S. 16, 1983, pp. 269–306; and F. Grenet, 'Pour une nouvelle visite à la «vision de Kerdīr»', *Studia Asiatica*, 3/1–2, 2002, pp. 2–27; D. N. MacKenzie in G. Hermann, D. N. MacKenzie, and R. Howell Cadelcott, *The Sasanian Rock Reliefs at Naqsh-i Rustam*, Berlin, 1989, pp. 59–61, 66–72.

The Zoroastrian paradise was called in Avestan *vahišta- ahu-*, 'the best existence', a denomination well attested in Pahlavi as *wahišt* (and from Modern Persian onwards as *bihišt*);[124] see also Sogdian *wəštmāx* [<wštm'x>], etc.,[125] while the derived adj. *wəštmāxčīk* [<wštm'xcyk>] is apparently attested only in Manichaean Sogdian.[126] It is the abode of Ahura Mazdā. Frequently, this place, which was located above the fourth heaven, that of the infinite lights (*raocā̊ anaγra*),[127] is described with the expression *garō nmāna-*, m., in Later Avestan (but in the *Gāθās* we find both sequences *garō dəmānē* (*Y.* 50, 4; 51, 15) and *dəmānē garō* (*Y.* 45, 6),[128] meaning, as Kellens[129] has explained, 'the house of the song of *welcome*', and not simply 'the house of the song'.[130] The syntagm attested in Later Avestan was probably treated as a compound and, in fact, it appears as a univerbation in Pahl. *garōdman*.[131] Some Pahlavi sources, as Hultgård noted,[132] distinguish between *wahišt* (or *pāšom axwān* 'the excellent existence'), as the lowest part of paradise (the atmosphere, sphere of the stars and of the moon),[133]and *garōdman*, as the highest one, including the sphere of the sun, and the infinite lights.

In Manichaean sources of Central Asia we find many denominations for paradise. For instance, we have in Middle Persian *bārist* [<b'ryst, b'rysṯ, b'rst>], with a superlative meaning 'the highest (place)', 'heights',[134] in Parthian and Middle Persian *padišt* 'place, home'[135] and *wahišt* and *wahištāw* 'paradise',[136] the latter of Mazdean origin. Buddhist Sogdian has *βaγestān* [<βγ'yst''n,[137] βγ'yst'n>][138] 'place of the Gods', but also *r(u)xušna-*

[124] Nyberg, *A Manual of Pahlavi* (n. 112 above), p. 201.

[125] Gharib, *Sogdian Dictionary* (n. 8 above), p. 415, n° 10236 [wštm'γw]. Thus Gharib, but, as Sims-Williams kindly informed me, the modern transliteration is wštm'xw.

[126] Ibid., n° 10239. In Manichaean Sogdian see also *wištmax* [wyštm'x]; ibid., p. 427, n° 10520.

[127] A. C. D. Panaino, 'Uranographia Iranica I. The Three Heavens in the Zoroastrian Tradition and the Mesopotamian Background', in *Au carrefour des religions. Mélanges offerts à Philippe Gignoux*, Bures-sur-Yvette, 1995, pp. 205–25.

[128] *dəmānē* is the locative sg. of O.Av. *dəmāna-*, n., 'house'; L.Av. *nmāna-*; cf. Ved. *mā́na-*; J. Kellens and E. Pirart, *Les Textes vieil-avestiques*, II, *Répertoires grammaticaux et lexique*, Wiesbaden, 1990, p. 262; cf. Bartholomae, *Altiranisches Wörterbuch* (n. 9 above), pp. 1090–92.

[129] Kellens, *Les Noms-racines* (n. 115 above), p. 27–9. See also Kellens and Pirart, *Les Textes vieil-avestiques* (n. 128 above), p. 238, sub *gar-*, f., 'chant de bienvenue', with reference to Ved. *gír-*. Cf. Hultgård, 'Das Paradies' (n. 1 above), pp. 28–9.

[130] See Bartholomae *Altiranisches Wörterbuch* (n. 8 above), pp. 512–13.

[131] Nyberg *A Manual of Pahlavi* (n. 112 above), p. 81.

[132] Hultgård, 'Das Paradies' (n. 1 above), p. 29.

[133] M. West, 'Darius' Ascent to Paradise', *Indo-Iranian Journal*, 45, 2002, pp. 51–7 (56) has recently explained the reliefs of the rock tombs of Darius I and his successors at Naqš-e Rustam as referring to the ascent of the king to the Moon, the Sun, and, of course, the paradise of A^{h}uramazdā.

[134] D. Durkin-Meisterernst, *Dictionary of Manichaean Texts*, III, *Texts from Central Asia and China*, Part 1, *Dictionary of Manichaean Middle Persian and Parthian*, Turnhout, 2004, pp. 104–5.

[135] Ibid., p. 275.

[136] Ibid., pp. 341–2.

[137] Gharib, *Sogdian Dictionary* (n. 8 above), p. 101, n° 2567.

[138] Ibid., p. 103, n° 2609.

γarδmān [<rxwšn’γrδmn>].[139] A similar compound occurs also in Manichaean Sogdian.[140] In this latter case we can also see the Zoroastrian background, because we have a derivation from the Avestan adjective *raoxšna-* ‘bright, luminous’ plus the sequence *garō nmāna-*.

In the framework of the present discussion we must also mention the existence of a special refuge, where Yima escaped with part of humanity when, for the first time, the winters and the winds brought by Aŋra Mainiiu and the demons arrived on the earth (*Vd.* 2, 229); this place is the *vara-* (cf. Vedic *vála-*), which was probably located underground, although there the stars, the sun and the moon can be seen (*Vd.* 2, 40) rising and setting. The *vara-* seems to be like Noah’s Ark, where the best part of the living world was saved, but it is also a dimension without suffering and demons. It is not properly a paradise,[141] because its origin is due to the arrivals of the demonic winters; thus, it resembles a place of resistance and defence against these new dangers, although it can be an earthly *dimora beata*, spared from the fight with the forces of Aŋra Mainiiu. Furthermore, it is necessary to recall that, according to a later Avestan source (*Aogəmadaēcā*, 28), Aŋra Mainiiu’s dwelling is eventually located into (another?) *vara-*.[142] Such an isolated reference puts the representation of any place called *vara-* in an ambiguous position, in particular if we consider that also in the *R̥gveda* the function that the *vála-* played is doubtless negative.[143]

It would be also relevant to recall that Christian sources, too, belonging to the Church of Persia, both in Middle Persian and Sogdian[144] did not use the Iranian name *pardēz* as the normal designation for their ‘paradise’, but have adopted the same terminology which the Zoroastrians used.[145] In Sogdian, for instance, we find *uštmāx* [<’wštm’x>],[146] which corresponds to Av. *vahištəm ahūm* (acc.) ‘the best existence’, but not *prdyz*, which, in any case, was known to Christian Sogdians as the usual designation of ‘garden’. This choice

[139] Ibid., p. 341, n° 8480. Note that *ruxšna-γarδmānčīk* [<rwxšn’γrδmncyk>], (mis)quoted by Gharib (ibid., n° 8481, with reference to secular sources) is actually from a text in Manichaean script. See also *roxšna-γarδmān* [rwxšn’γrδmn(h)], ibid., p. 345, n° 8572.

[140] Ibid., p. 346, n° 8605: *roxšna-γarδmān* [rwxšn’γrδmn].

[141] Cf. Hultgård, ‘Das Paradies’ (n. 1 above), 25–6, who, on the contrary, accepts the equation with paradise. Cf. Kellens, *Liste du verbe avestique* (n. 8 above), p. 48.

[142] B. Lincoln, ‘The House of clay’, *Indo-Iranian Journal*, 24, 1982, pp. 1–12 (3), mentions also ‘the cave of Lie’ (*gərəδa-*) on Mount Arəzūra (*Vd.* 3, 7), a subterranean enclosure contaminated with death and putrefaction, ‘whence all demons issue’.

[143] See W. W. Malandra, *An Introduction to Ancient Iranian Religion*, Minneapolis MN, 1983, p. 178; cf. P. Jackson, *The Transformations of Helen. Indo-European Myth and the Roots of the Trojan Cycle*, Dettelbach, 2006, pp. 71, 75.

[144] E. Benveniste, ‘Le Vocabulaire chrétien dans les langues d’Asie centrale’, in *Atti del Convegno Internazionale sul Tema: L’Oriente Cristiano nella Storia della Civiltà*, Rome, 1964, pp. 85–91; G. Widengren, ‘The Nestorian Church in Sasanian and Early-Post-Sasanian Times’, in *Incontro di Religioni in Asia tra il III e il X secolo d.C.*, ed. L. Lanciotti, Florence, 1984, pp. 1–30 (14–15).

[145] See Gignoux, ‘Quelques réflexions’ (n. 19 above), pp. 3–15; Benveniste, ‘Le Vocabulaire chrétien’ (n. 144 above), p. 87.

[146] Gharib, *Sogdian Dictionary* (n. 8 above), p. 79, n° 2008.

could be due to the force of a long tradition, and shows the cultural links of the Iranian Christians to their religious antecedents. Furthermore, I wonder if they were unwilling to use a word that still had a strong secular meaning, while its negative[147] implications, at least in the Avestan lexicon, were potentially present in the basic knowledge of the newly converted. We cannot forget that, in Modern Persian, *pālēz* refers only to a very modest kind of garden.[148] Thus, it is worthy to note that, while the Iranian name of this special garden had enormous success outside its borders, its meaning did not gain any speculative or eschatological implication in its homeland. It is very remarkable that also in Armenia, where the massive conversion to Christianity had an enormous impact, the word *partēz* was not used for the Christian 'paradise', which was called *draxt*, i.e. with another Iranian name designating a tree, more precisely the date-palm. This association is very important because, in the Iranian framework, the date-palm is the protagonist of a beautiful dialogue of Parthian origin, surviving in Pahlavi, the *Draxt ī asūrīg* 'The Assyrian date-palm'.[149] 'Paradise' and 'hell' were also designated in Armenian sources as *anoyšk*' (cf. Pahl. *anōšag* 'immortal', *anōš* 'idem')[150] and *džoxk*' (< cf. Pahl. *dušox* 'hell'),[151] as underlined by Gignoux.[152] Nevertheless, the image of the garden was kept in Christian Sogdian, but with a different denomination, *βōstăn* [<βwstn>][153] 'garden, paradise' (cf. Pahl. *bōyestān* 'garden [of flowers]' with reference to the 'perfume' (²*bōy*).[154] Also in this case, the attested terminology is not completely extraneous to the Mazdean world; in fact, the image of the garden of paradise appears in some Pahlavi sources, like, for instance,[155] *Wizīdagīhā ī Zādspram* 30, 52, 55, 59 (*bōyestān*),[156] or *Bundahišn* 30, 15[157] (*bōyestān kerb* 'the form of the garden'), although these passages contain later speculation about the afterlife. But it is certain that the Mazdean paradisiacal dimension was announced by the good smell of a perfumed wind, and that a stench[158] was, contrariwise, considered a symbol of hell.

[147] Hultgård, 'Das Paradies' (n. 1 above), p. 8.

[148] J. Duchesne-Guillemin, 'Coup d'œil sur la linguistique iranienne', in *La posizione attuale della linguistica storica nell'ambito delle discipline linguistiche*, Rome, 1992, pp. 131–45 (138).

[149] See C. G. Cereti, *La letteratura pahlavi. Introduzione ai testi con riferimenti alla storia degli studi e alla tradizione manoscritta*, Milan, 2001, pp. 185–6 with detailed bibliography.

[150] According to chapter 10 of the *Ardā Wīrāz Nāmag*, the Right Wīrāz, when arriving to paradise, receives a drink called *anōš*.

[151] Pahlavi *dušox* derives from an Avestan secondary stem like *daožaŋᵛha-* or *daožaŋhuua-*, n., literally '(place of the) bad existence', and, then, 'hell', in its turn, from *dužaŋhauu-*, m., (< **duš-ahu-*, 'having a bad existence, life'); cf. Bartholomae, *Altiranisches Wörterbuch* (n. 8 above), pp. 675, 756.

[152] Gignoux and Tafazzoli, *L'Anthologie de Zādspram* (n. 122 above), p. 7.

[153] Gharib, *Sogdian Dictionary* (n. 8 above), p. 115, n° 2928.

[154] This stem must be distinguished from ¹*bōy* 'perception, sense'.

[155] See Hultgård, 'Das Paradies' (n. 1 above), pp. 30–32.

[156] Gignoux and Tafazzoli, *L'Anthologie de Zādspram* (n. 122 above), pp. 108–11. With reference to the Manichaean tradition, see also Fauth, 'Der königliche Gärtner' (n. 17 above), p. 8.

[157] F. Pakzad, *Bundahišn. Zoroastrische Kosmogonie und Kosmologie*, I, *Kritische Edition*, Tehran, 2005, p. 347.

[158] See A. C. D. Panaino, '"Smell", "God Smell", "Stench" and "Stinking" in Avestan', in *Parfumes d'Orient*, eds R. Gyselen with F. Aubaile-Sallenave, S. Aufrère, Al. Avanzini, Bures-sur-Yvette, 1998, pp. 167–76.

Appendix

About Old Persian <p^a-r^a-d^a-y^a-d^a-a-m^a>

The *Problemwort* we try to discuss in this final note is a so-called *hapax gegrammenon* (p^a-r^a-d^a-y^a-d^a-a-m^a). It is, in fact, attested only in a later trilingual inscription probably belonging to the time of Artaxerxes II[159] from Susa (A^{2}Sd),[160] engraved on a severely damaged column.[161] Unfortunately, such a document is very fragmentary and – worst of all – the Elamite and Akkadian versions do not offer any enlightening explanation with specific regard to the dubious parallel passage containing O.P. p^a-r^a-d^a-y^a-d^a-a-m^a.[162] The Old Persian text of the inscription occupies four lines, containing two paragraphs, edited as follows (although there are minor variants in the attested copies[163] [A^{2}Sd^a; A^{2}Sd^b; A^{2}Sd^c; A^{2}Sd^d; A^{2}Sd^e (only three recognizable lines in Akkadian)]; three other fragments published by Vallat[164] (A^{2}Sd^{Vx}): A^{2}Sd^{V1}; A^{2}Sd^{V2}; A^{2}Sd^{V3}; plus one in Elamite: A^{2}Sd^{Vel}; and two again in Akkadian: A^{2}Sd$^{Vak.1}$; A^{2}Sd$^{Vak.2}$):

1) a-d^a-m^a : a-r^a-t^a-x^a-š^a-ç^a-a : XŠ: v^a-z^a-r^a-k^a : XŠ : XŠ-y^a-a-n^a-m^a : XŠ : DH$_{1-}$ y^a-u-n^a-a-m^a: XŠ : a-h^a-y^a-a-y^a-a : BU-
2) y^a-a : d^a-a-r^a-y^a-v^a-u-š : XŠ : p^a-u-ç^a : h^a-x^a-a-m^a-n^a-i-š^a-i-y^a : θ^a-a-t^a-i-y^a : a-r^a-t^a-x^a-š^a-ç^a-a : XŠ : v^a-š^a-n^a-a : AM^{1-}

[159] F. Vallat, 'Fragments de Suse', *Revue d'Assyriologie*, 64, 1970, p. 171, for instance, wonders if the basements of columns, firstly published by V. Scheil (*Inscriptions des Achéménides à Suse*, Paris, 1929), actually belong to a building of Artaxerxes II or Artaxerxes III.

[160] The Old Persian inscriptions are referred to by abbreviations referring to the name of the king (A^2 = Artaxerxes), and the location (S = Susa), sometimes followed by a small letter indicating a particular inscription or an ancient copy of it. See R. G. Kent, *Old Persian Grammar Texts Lexicon*, 2nd edn, New Haven CT, 1953, pp. 4–5, and Schmitt, *Die altpersischen Inschriften* (n. 8 above), pp. 6–32.

[161] Schmitt, *Beiträge* (n. 11 above), p. 80; Schmitt, *Die altpersischen Inschriften* (n. 8 above), pp. 194–5; G. Schweiger, *Kritische Neuedition der Achaimenidischen Keilinschriften*, I, *Textband*, II, *Katalog*, Taimering, 1998 (II, pp. 505–6). Cf. Scheil, *Inscriptions des Achéménides* (n. 159 above), pp. 91–3; V. Scheil, *Inscriptions des Achéménides. Supplément et Suite*, Paris, 1933, pp. 126–9; R. G. Kent, 'The Recently Published Old Persian Inscriptions', *Journal of the American Oriental Society*, 51/3 (Sept.), 1931, pp. 189–240 (228–9); Kent, *Old Persian Grammar* (n. 160 above), pp. 154–5; M.-J. Steve, *Nouvelles Mélanges Epigraphiques: Inscriptions royales de Suse et de la Susiane*, Nice, 1987, pp. 96–7.

[162] See already F. H. Weissbach, 'Achämenidisches. 3. Nochmals zu den "Inschriften unbestimmter Herkunft". 4. Die dreisprachige Inschrift Artaxerxes II. Susa d.', *Zeitschrift der Deutschen Morgenländischen Gesellschaft*, 91, NS 16, 1937, pp. 643–51 (649–51). Cf. C. Hauri, *Das pentatematische Schema der altpersischen Inschriften*, Wiesbaden, 1973, p. 57.

[163] For these details, see Schweiger, *Kritische Neuedition* (n. 161 above), II, pp. 505–17 (his abbreviations are adopted here). Cf. also Schmitt, *Beiträge* (n. 11 above), 80–81. Important also Scheil, *Inscriptions des Achéménides. Supplément* (n. 161 above), p. 128, and F. Vallat, 'Les Inscriptions du Palais d'Artaxerxès', *Cahiers de la Délégation Archéologique Française en Iran*, 10, 1979, p. 149 and fig. 47, 7.

[164] Vallat, 'Fragments de Suse' (n. 159 above); Vallat, 'Epigraphie Achéménide', *Cahiers de la Délégation Archéologique Française en Iran*, 2, 1972, pp. 201–2, figs 203–17, 253–6; see again Schweiger, *Kritische Neuedition* (n. 161 above), II, pp. 507–11.

3) h^{a}-a : i-m^{a}-a-m^{a} : h^{a}-d^{i}-i-š : t^{a}-y^{a} : j^{i}-v^{a}-d^{i}-i-y^{a} : p^{a}-r^{a}-d^{a}-y^{a}-d^{a}-a-m^{a} : a-d^{a}-m^{a} : a-k^{u}-u-n^{a}-v^{a}-m^{a} : AM_1 : a-n^{a}-h^{a}-i-t^{a}
4) <:> /u\-t^{a}-a: m^{i}-t^{a}-r^{a}: m^{a}-a-m^{a} p^{a}-a-t^{u}-u-v^{a}: h^{a}-c^{a}-a: v^{i}-i-s^{a}-p^{a}-a: g^{a}-s^{a}-t^{a}-a: u-t^{a}-m^{a}-i-y^{a}: k^{a}-r^{a}-t^{a}-m^{a}:

1) adam R̥taxšaçā XŠ vazr̥ka XŠ XŠ-yānām XŠ DH-yūnām XŠ ahyāyā
2) BU-yā Dārayava.uš XŠ-ahyā puça Haxāmanišiya // θātiy R̥taxšaçā XŠ vašnā AM-
3) hā imā[m] hadiš taya jīvadiy pardai̯dām adam akunavam AM anahita
4) utā mitra mām pātuv hacā vispā gastā utamaiy kr̥tam.

The two extant sentences are sufficiently clear and do not need detailed discussion, with the exception of the sequence which is relevant for the present investigation about 'paradise': *imā[m] hadiš taya jīvadiy pardai̯dām adam akunavam*.

Although *pardai̯dām* has been generally, but not unanimously, considered as the unique attestation of the Old Persian word for 'paradise', it was Lecoq[165] who recently tried to demonstrate that such an occurrence has nothing to do with it.[166] The French scholar has actually explained this form as an aorist, to be properly read *pariyadām* [<p^{a}-r^{a}-i-y^{a}-d^{a}-a-m^{a}>], meaning 'I consecrated'. But such a solution is based on a strong emendation, while the Old Persian syntax of the whole passage results, notwithstanding this risky correction,[167] in a very peculiar construction ('this palace which [I] consecrated I made' [*imām hadiš jīvadiy *pariyadām adam akunavām*]). Furthermore, it is not clear why Artaxerxes should have consecrated the palace before its construction. The logical ritual sequence should be just the opposite.

We must recall that already Hauri[168] had tried to explain p^{a}-r^{a}-d^{a}-y^{a}-d^{a}-a-m^{a} [**par^{a}dayadām*] as the final result of a (miswritten) contracted verbal form like **parididāyam* (< **pariyadidāyam*), i.e. as an imperfect, first pers. sing., of an unattested present stem **didāya-* of the verbal root *dā* (cf. Gr. περιτειχίζω). Hauri,[169] thus, translated 'nach dem Willen Ahuramazdas machte ich diesen Sitz, den ich im Leben ummauerte'. Thus, as in the case of Lecoq's solution, the syntax is doubtful, while the introduction of an unknown (and isolated) present stem of a very well attested Iranian root makes such an interpretation unconvincing, to say nothing of the bold emendation of the form as it was engraved on all the surviving fragments.[170]

[165] Lecoq, 'Paradis en vieux-perse?' (n. 76 above), pp. 210–211; Lecoq, *Les inscriptions* (n. 107 above): 116, 274. See the opposite opinion expressed by Uchitel, 'Persian Paradise' (n. 82 above), p. 140, n. 15.

[166] Lecoq's ingenious interpretation has been accepted by Gignoux, 'Quelques réflexions' (n. 19 above, p. 4, n. 3) and Tuplin, *Achaemenid Studies* (n. 2 above, p. 93, n. 49) without any discussion of the passage.

[167] The introduction in a so late text of an aorist form seems to me far-fetched, because the category was already collapsing in the older phases of Old Persian. See also Werba, '(Indo-) Iranische Rekonstructe' (n. 6 above), p. 282, n. 103.

[168] Hauri, *Das pentatematische Schema* (n. 162 above), pp. 57–8.

[169] Ibid., p. 55.

[170] Cf. Schmitt, *Beiträge* (n. 11 above), p. 84.

It is more prudent to start again from the basic interpretation given by Kent:[171] 'this is the palace which I built in my life as a pleasant retreat' (*imām hadiš taya j*^{i}*vadiy paradayadām adam akunavām*). It would also be possible to read the passage, as suggested by Steve,[172] **paradidām* [<p^{a}-r^{a}-d^{i}-i-d^{a}-a-m^{a}>] 'beyond the wall' ('en dehors des murs'); but, in this case, we would have to accept a radical collapse of the Old Persian orthography (not to mention another strong emendation) and a number of syntactic problems, already remarked upon by Schmitt.[173] Steve, in addition, implicitly suggested that this new (emended!) form would syntactically work like an adverb, but such a solution is unnecessary. In fact, the proposed <p^{a}-r^{a}-d^{i}-i-d^{a}-a-m^{a}> can be an accusative sing. n.[?] (the way in which the ending is written remains in any case a problem; see below), referring to *imām hadiš taya*. Furthermore, although his explanation is not convincing, Lecoq[174] is absolutely right when he notes that the place of this form between *jīvadiy* (as nom. sing. m. referring to *adam*) and *adam* itself, would be peculiar. His interpretation of that word, however, does not explain the fact that *-diy*, if one accepts the solution suggested by Kent,[175] could be an enclitic (IE. **ǵ*^{h}*i*), which – like Vedic *-hí*, Avestan *zī*, Greek (*ναί*)-*χι* (etc.), Latin (**ne*)-*gi*, *nec* (etc.), Bulgarian, Serbian, etc. *-zi* (after pronouns) – systematically takes the second position in a sentence, i.e. it must occur after the first word.[176] While Lecoq has not broached this syntactical problem, Benveniste,[177] who clearly rejected Kent's solution, did not offer, in his turn, a definitive analysis of the sentence. He simply considered the passage as 'barbaric', the production of an ignorant scribe, probably not Persian (this hypothesis, in my opinion, is doubtful to say the least), then translated 'cette résidence (appellée) *jīvadiy paradayadām*'.[178] In any case, we cannot deny that the mistakes attested in this short inscription are very many and of various kinds (as has been carefully noted by all the scholars who have dealt in more detail with

[171] Kent, *Old Persian Grammar* (n. 162 above), p. 155. Cf. also Kent, 'The Recently Published Old Persian Inscriptions' (n. 161 above), p. 228.

[172] Steve, *Nouvelle Mélanges Epigraphiques* (n. 161 above), pp. 96–8.

[173] Schmitt, *Beiträge* (n. 11 above), pp. 82–4.

[174] Lecoq, 'Paradis en vieux-perse?' (n. 76 above), p. 210.

[175] Kent, 'The Recently Published Old Persian Inscriptions' (n. 161 above), p. 229; *Old Persian Grammar* (n. 160 above), p. 191; the existence of *-diy* was accepted by Mayrhofer in Brandenstein and Mayrhofer, *Antiguo Persa* (n. 12 above), p. 105.

[176] See Pokorny, *Indogermanisches Etymologiches Wörterbuch* (n. 8 above), I: 417–418; M. Mayrhofer, *Kurzgefaßtes etymologisches Wörterbuch des Altindischen. A Concise Etymological Sanskrit Dictionary*, III, *'Y-H'*, Heidelberg, 1976 (KEWA III), p. 592; *Etymologisches Wörterbuch des Altindoarischen*, II/20, Heidelberg, 1996 (EWA II/20), pp. 814–815 (here, Mayrhofer explicitly denies the existence of O.P. *-diy* in DB 4, 69, but without any reference to the present inscription). Cf. also E. Schwyzer and A. Debrunner, *Griechische Grammatik, auf der Grundlage von Karl Brugmanns Griechischer Grammatik*, II, *Syntax und syntaktische Stilistik*, Munich, 1950, p. 577; Frisk, *Griechisches Etymologisches Wörterbuch* (n. 4 above), II, p. 1099; Chantraine, *Dictionnaire étymologique* (n. 3 above), p. 1259; Kellens and Pirart, *Les Textes vieil-avestiques* (n. 128 above), pp. 182–6, 322. The reader will find a remarkable presentation of the Vedic data in H. Hettrich, *Untersuchungen zur Hypothaxe im Vedischen*, Berlin and New York, 1988, pp. 171–89.

[177] See E. Benveniste, 'Persica', *Bulletin de la Société de Linguistique de Paris*, 30/1, 1930, pp. 58–74 (67); idem, in A. Meillet and E. Benveniste, *Grammaire du Vieux-Perse*, 2nd edn, Paris, 1931, p. 25.

[178] Benveniste, in Meillet and Benveniste, *Grammaire* (n. 177 above), p. 222.

the source):[179] they are of an orthographic, grammatical and syntactical nature, to say nothing of the defective writings and the frequent inaccuracies in engraving the same cuneiform characters (as in j^{i}-v^{a}-d^{i}-i-y^{a}, where the sign for *j^{i}-* presents only two final parallel horizontal wedges instead of three,[180] and the expected sign for -i- between j^{i}- and v^{a}- was written over it or simply omitted).[181]

Schmitt, whose interpretation of the passage is at present the most reliable, had focused on this text in its many aspects. He reads and explains the same passage in his latest edition of the Old Persian inscriptions as follows:

(A) *θāti R̥taxšaça xšāyaθiya*:	Es kündet Artaxerxes, der König:
(B) *vašna A.uramazdāhā*, –	Nach dem Willen Ahuramazdās, –
(C) *ima hadiš*,	dies (ist) der Palast,
(D) *taya jīvadi paradaịdām akunavam*	den ich noch zu Lebeszeiten als Lust-/Jagdschloß errichtet habe.[182]

In a footnote the same scholar remarks that the writing of p-r-d-y-d-a-m is an 'inverse Schreibung von **pardēd* (O.P. **paridaịda-*)'.[183] But Schmitt had already dealt in detail with the same problematic passage, which he translated as follows:

> Nach dem Willen des Auramazdā, dies ist der Palast, den ich wirklich noch zu Lebzeiten (?) als Lust-/Jagdschloss (als Schloss im Park, im Grünen) errichtet (vollendet) habe.[184]

Schweiger, however, interpreted it as follows: 'Nach (gemäß) dem Willen des Ahuramazdā [!] baute ich gar während des Lebens (für das Leben) diesen [!] Hadiš als Paradies/Wohlbehütetes'.[185]

In his translation, Schweiger openly followed with close regard to *jīvadi paradaịdām* a solution already proposed by Weissbach,[186] but the actual meaning is not grammatically clear. Unfortunately both Weissbach and Schweiger have not offered any supporting explanation for this translation, while the function of *-diy* also remains unexplained in

[179] See, for instance, Scheil, *Inscriptions des Achéménides* (n. 159 above), p. 93; Scheil, *Inscriptions des Achéménides. Supplément* (n. 161 above), p. 128; Kent, 'The Recently Published Old Persian Inscriptions' (n. 161 above), pp. 191, 228–9; Schmitt, *Beiträge* (n. 11 above), pp. 81–3.

[180] Kent, 'The Recently Published Old Persian Inscriptions' (n. 161 above), pp. 191, 228; W. Hinz, *Altpersischer Wortschatz*, Leipzig, 1941, p. 153. Only Scheil, *Inscriptions des Achéménides* (n. 159 above), p. 93, n. 3, wonders whether the form of the sign is the result of a mistake, or a later variant. In the case of ji-va-di-i-ya, Benveniste, 'Persica' (n. 177 above), p. 67, assumed the presence of a third mistake, the short *-a-* in -va-, because he tried to explain the word as *jiva + adiy*, but this solution is completely out of date.

[181] See again Schmitt, *Beiträge* (n. 11 above), p. 81.

[182] Schmitt, *Die altpersischen Inschriften* (n. 8 above), pp. 194–5.

[183] Ibid., p. 195.

[184] Schmitt, *Beiträge* (n. 11 above), pp. 82, 84. 'According to the will of Ahuramazdā, this is the Palace, which I really erected (completed) during my lifetime (?) as (a castle in a park, in the green) for pleasure/hunting'.

[185] Cf. Schweiger, *Kritische Neuedition* (n. 161 above), I, p. 145.

[186] See again Weissbach, 'Achämenidisches' (n. 162 above), pp. 649–51, and F. H. Weissbach, 'Rezension an V. Scheil 1929', *Archiv für Orientforschung*, 7, 1931–32, pp. 37–45 (45).

their interpretation. It is also important to recall that Werba has offered in his turn an alternative approach to this passage, suggesting that, if p-r-d-y-d-a-m were interpreted as an adjectival *éntheos*-compound meaning 'eine(n) Wall/Mauer (um sich) herum [**pari*] habend' or 'mit einer/m Wall/Mauer um[geben]', referring to *hadiš* through the correct neuter *taya*, the tantalizing *j*^{i}*vd*^{i}*iy* could assume a 'lokale/modale Bestimmung' like 'allenthalben'.[187] This way, the attested °*diy* might, then, be either an adverbial suffix or a locative singular of a stem in °*d*-. Of course, this explanation presupposes a different (and again emended) reading of the two first signs of *j*^{i}*vd*^{i}*iy*, which makes such a solution certainly interesting and worthy of consideration, but not compelling or, so to speak, definitive.

An alternative possibility would be that of reading *jīvadiy* as *jīva*^{n}*diy* (< **jīva*^{m}*diy*), if we consider that, although only in some circumstances, nasals were not expressed in writing before stops and spirants.[188] This argument could be theoretically rejected on the grounds that *-m-* is normally preserved before enclitics.[189] However, this phenomenon occurs only with *-ca*, *-ciy* and *-šim*,[190] while we do not have consistent examples with *-diy* or similar cases, in particular where an internal *-m-* should have become *-n-*. Furthermore, *-diy*, if it is a so-called 'sentence clitic' (normally used in Wackernagel's position),[191] it behaves very differently from *-ca*, which is a classical 'word clitic', a fact that implies a much stronger juncture between it and the word it follows. In any case, some attested facts remain as supporting evidence for such a phenomenon: *Ka*^{m}*pa*^{n}*da*, *Ka*^{m}*būjiya* and in particular *ha*^{n}*duga-* (< *ha*^{m}*duga-*), etc., where *-m-* and *-n-* before *-d*, *-p* and *-b*, were never written.

On the other hand, as R. Schmitt very kindly remarks,[192] this interpretation seems difficult in the light of DNb 51 and 51f, where we find *ciyākaram-tai̯* (after the new reading offered by Sims-Williams,[193] and accepted by Schmitt himself in his new edition of the Old Persian inscriptions).[194] To this pertinent criticism I raise two objections: 1) *ciyākaram-tai̯* (in reality c^a-i-y^a-a-k^a-r^a-m^a-m^a-t^a-i-y^a) is so written (twice) after an occurrence of the same word alone (*ciyākaram ahi*); 2) My suspicion is that the scribe was so afraid to miss the *-m* before *-tai̯*, that he wrote it twice, and this happened in the

[187] Werba, '(Indo-) Iranische Rekonstructe' (n. 6 above), p. 285, n. 118; cf. also pp. 282–3, n. 103; see again Werba, pp. 279–80, n. 90.

[188] Kent, *Old Persian Grammar* (n. 160 above), p. 39, par. 111.

[189] See R. Schmitt, 'Zur Enklise im Altpersischen', in *Verba et Structurae. Festschrift für Klaus Strunk zum 65. Geburtstag*, Innsbruck, 1995, pp. 205–301.

[190] Kent, *Old Persian Grammar* (n. 160 above), p. 18, par. 39.

[191] Cf. J. Wackernagel, 'Über ein Gesetz der indogermanischen Wortstellung', *Indogermanische Forschungen*, 1, 1892, pp. 333–436, repr. in Wackernagel, *Kleine Schriften*, I, Göttingen, 1953, pp. 1–103.

[192] Letter dated 10 October 2010.

[193] N. Sims-Williams, 'The Final Paragraph of the Tomb-Inscription of Darius I (DNb, 50–60): The Old Persian Text in the Light of an Aramaic Version,' *Bulletin of the School of Oriental and African Studies*, 44/1, 1981, pp. 1–7 (3–4).

[194] R. Schmitt, *Epigraphisch-exegetische Noten zu Dareios' Bisutun-Inschriften*, Vienna, 1990, p. 110.

two following clauses (D and E) of DNb § 11. This dittography, doubtless, needs an explanation: I wonder whether it could be due to an excess of precision, which turned out to be a *verschlimmte Verbesserung*. But against this solution we must consider that it is quite possible that the persons charged with engraving the inscription, as R. Schmitt kindly reminds me,[195] did not properly understand what they were writing, and that they were just copying their portion of text in a very schematic way and, furthermore, from right to left. In any case, also accepting this pertinent remark, the possibility that the double -m- (in two following cases) was due not only to a simple *lapsus*, but also to a misunderstanding of the orthographic rules, remains open. Another possibility is that the mistake was already attested in the *Vorlage*, i.e. in the working copy (probably shorter) that the *lapicida* was using for his activity. Furthermore, the example given does not involve a case with a final *-m* in front of *-d*. In this case, my alternative proposal in the light of the other, although few, cases (but in good texts) where an expected final *-m* was dropped out, cannot be altogether excluded.

Furthermore, we must consider two other facts:

1) We must note with regard to *-diy* [*-di*] that, although Kent's explanation[196] of it as an emphatic enclitic particle is to be rejected in the case of the old reading *ufraštā-diy* (DB 4, 69, rightly corrected by Schmitt[197] in *ufraštā diš*), such an occurrence still needs to be investigated. In fact, Schmitt has taken, in his turn, the enclitic *-diy* , which he fittingly translated as 'wirklich',[198] simply as an emphatic particle, referring only to the word *jīva-* ('wirklich noch zu Lebzeiten') and not to the whole sentence, as we should expect in the case of the *Wackernagels Gesetz*. I do not a priori rule out such an explanation, but it has to be said that this also is an *ad hoc* solution, although such a limit turns out to be inevitable because this is, at present, the unique attestation of such an enclitic in the whole Old Persian corpus. On the other hand, we could still maintain the earlier solution offered by Kent, but on the condition that we ascribe to *taya* only very weak syntactic force, rather like an *eżāfe*, in the structure of the whole sentence, a condition which would not imply the immediate insertion of the enclitic after the relative pronoun.

2) The grammar of the passage is very disturbed, probably not only because of the work of sloppy scribes (given the number of mistakes in all the fragments, we should assume the presence of a whole band of unqualified *lapicidae*, even if this remotely plausible explanation is not the only possible one), but also because of a certain corruption/evolution of the Old Persian language towards a new linguistic stage between a sort of Later Old Persian and Old Middle Persian. The mixture of a decline in standard

[195] Letter dated 18th November 2010.

[196] Kent, *Old Persian Grammar* (n. 160 above), p. 191.

[197] Schmitt, 'Epigraphisch-exegetische Noten' (n. 194 above), pp. 48–9 (with literature).

[198] Schmitt, *Beiträge* (n. 11 above), p. 82. He omitted "wirklich" in his later edition of the text, but it would be better to maintain it.

for the selection of the scribes and the new status of the real language might produce, as a final result, the highly corrupt texts we can presently read in the attested materials belonging to the times of Artaxerxes.

For these reasons, even if the omission of the *-m* in the written *jīvadiy* were not due to a (correct) orthographic rule, such a mistake might easily occur in that text.

Anyway, I offer such a new interpretation of the passage as an alternative with respect to the 'basic' solution given by Schmitt (and already by Kent, although with some minor but pertinent differences), just remarking that the main aim of the present note was that of preserving the existence of the word for 'paradise' in the Old Persian lexicon, so that I will be not discontent with the fact that the *hadiš* of Artaxerxes was not a 'living place of pleasure', but that it was simply built up as a 'paradise' during the life of that king.

My problem with an uncritical acceptance of that 'basic' translation is due to the fact that the syntactic location of *jīvadiy*, referred to Artaxerxes (as subject of the verb *akunavam*), is at least strange, between *taya* and *paradaidām*. We should expect *jīva* after *akunavam*, as a fitting determination to the order given by the king (and carefully executed by his functionaries) *during his life*, and not when he was already *mr̥ta*. Unfortunately, there were only two parallel attestations of *jīva* and *mr̥ta* (XPh 48; 54) and this material does not allow a more useful syntactic comparison.

Thus, in the modest framework of such a tentative alternative, the postulated *jīva*$^{n/m}$*diy* could more easily refer to the following *paradaidām*. If so, we may translate the whole passage as 'this is the palace [which] I built indeed as a living place of pleasure' or 'this is the palace (named) indeed "Living Paradise" [which] I built up'.[199]

The earlier explanation advanced by Scheil,[200] who, in fact, thought that the very name[201] of the palace was *jīvadiy paradayadām* (just like the name[202] of Babylon was that of a 'residence of life' [Sum. TIN.TIR.RA = Akk. *šubat balāṭi*]),[203] would be now supported with a fresh argument, if we read {*jīva*^{m}*diy paradaidām*}. The *-ā-* in the ending of *paradaidām* remains a problem, but the orthography of this sentence is so corrupt that such a bizarre occurrence cannot be considered as an overwhelming problem, and, perhaps, as Werba supposes,[204] we have to do with a new orthographic rule, 'wobei das

[199] With regard to the semantic role of the *-diy*, if we accept Schmitt's interpretation of it, the enclitic would simply and only mark **jīvam ... paradaidām*. Then, we could also translate the sentence as: 'this is the palace [which] I built as a living place, indeed, of pleasure' or 'this is the palace (named) "Living Paradise", indeed, [which] I built up'.

[200] Scheil, *Inscriptions des Achéménides* (n. 159 above), pp. 91–3. About Susa, see Fauth, 'Der königliche Gärtner' (n. 17 above), p. 6.

[201] Scheil, *Inscriptions des Achéménides* (n. 159 above), p. 92: 'ce palais *Paradis de vie* (?) j'ai fait'.

[202] Ibid., p. 93 made also reference to the fact that already Xerxes had given the name of '*Visadahyum*' to the Portico of 'all the nations'.

[203] See *Chicago Assyrian Dictionary*, XVII, '*Š*', part III, ed. E. Reiner, Chicago IL, 1992, pp. 179–80, sub *šubtu* A3a. Cf. also Fauth, 'Der königliche Gärtner' (n. 17 above), p. 20.

[204] Werba, '(Indo-) Iranische Rekonstructe' (n. 6 above), p. 283.

Morphem -am nach dem auslautenden Stamm-Konsonanten eben wie anlautendes <*am*°> geschrieben wurde'.

Incidentally, I do not exclude that, in the linguistic competence of the scribe(s) a certain (incorrect) relation between *imām* (as, in fact, it was engraved in all the attested copies) and *paradai̯dām* was established, in particular if **jīva*^{m}*diy paradai̯dām* was the name of the *hadiš*. I must observe that presently, for instance, many ordinary Italian people (but increasingly also journalists and public persons especially in television) are no longer able to use the relative pronouns properly, and frequently introduce feminine forms (considered more difficult and polite) instead of the masculine ones: e.g., 'il libro della quale [*sic*] ti ho parlato' instead of the right sentence 'il libro del quale/di cui ti ho parlato'; or the same people prefer a generic *che*, which is not declined, thus producing such monstrosities as 'il libro che ti ho parlato'. If the problems emerged from a grammatical and syntactical incongruence of a similar nature, the later Old Persian scribes might have taken *hadiš*, in its turn, no longer as a regular neuter stem, but as a feminine one in *-i-*, making an improper *Kreuzung* between a nominative in *-iš* and the regular neutral accusative in *-iš* of the stem *hadiš-*[205] (cf. Av. *hadiš-*, n., name of a divinity protective of the native place;[206] Ved. *sádas-*, n.). In this regard, we must not forget that *imām hadiš* occurs also in A2Sc.[207] Such a mistake would imply that the word had become very rare and that the grammatical opposition between nom./acc. was collapsing, at least in a number of lesser-known stems. That this can be the case is shown from the evidence that an emended reading such as *imā[m] hadiš* also presents a problem in the case of the final long *-ā* of the neuter pronoun, which in the nom./acc. singular should be short (see, e.g., Inscription Xerxes, Persepolis, XPd 16–17: *ima hadiš adam akunavam*).

Furthermore, we can also observe, for instance, that the correct spelling of the verb <a-k^{u}-u-n^{a}-v^{a}-m^{a}> (line 3; so attested only in a fragment identified by Vallat [A^2Sd^{V2}= A^2Sd^{7}])[208] was written as *akunavām* (i.e with an irregular *-ām* instead of *-ăm*) in A^2Sd^{b} (in A^2Sd^{a} the reading *aku[navam]* is only fruit of an integration,[209] in A^2Sd^{d} the word is completely missing) while, in A^2Sd^{c}, the text shows <a-k^{u}-u-v^{a}-n^{a}-š^{a}-a-š^{a}>.[210] These mis-spellings are probably also due to the progressive decay/evolution of the Old Persian current language in such a late period and the increasing uncertainty in how to write a formal and archaic linguistic register that did not correspond to the real current speech, as, perhaps, we can see also in the case of the wrong (but interesting) agreement between *imām* (acc. sg. f.) and *hadiš* (nom./acc. sg. n.). However, we cannot exclude that the

[205] Brandenstein and Mayrhofer, *Handbuch* (n. 8 above), p. 122.

[206] Bartholomae, *Altiranisches Wörterbuch* (n. 8 above), p. 1759.

[207] See Schmitt, *Die altpersischen Inschriften* (n. 8 above), p. 193.

[208] Vallat, 'Epigraphie' (n. 164 above); Schmitt, *Beiträge* (n. 11 above), p. 81 (= A2Sd7); Schweiger, *Kritische Neuedition* (n. 161 above), II, p. 515.

[209] Scheil, *Inscriptions des Achéménides* (n. 159 above), pp. 91–2.

[210] See also Steve, *Nouvelle Mélanges Epigraphiques* (n. 161 above), p. 96.

peculiar writing <a-k^{u}-u-v^{a}-n^{a}-š^{a}-a-š^{a}> originated with a scribe who had only a distant (and imprecise) memory of forms like a-k^{u}-u-n^{a}-u-š^{a} (*akunauš*) and a-k^{u}-u-n^{a}-v^{a}-š$^{a(n)}$ (*akunavaša*, which Gershevitch and I tried to explain as *akunava^{n}š*),[211] and so invented another form, completely meaningless.

Thus, coming back to the word for 'paradise': although it would be theoretically possible to suppose a feminine stem, *paradaidā-* on the model of O.P. *didā-*, f. 'fortress' (as apparently supposed by Benveniste),[212] we should, if my hypothesis is sound, expect a form like *jīvā$^{m/n}$diy*. In addition, the corresponding Avestan form (*pairi.daēza-*, m.) – although Werba may be right in his sharp distinction, at least in principle, between two different compounds – does not support this solution, and we have good reasons to suppose that the actual pronunciation of the word was, at that time, more or less like **pardēda-/*pardēdə-*, or simply *pardēd* (< **paridaida-*), as Schmitt has also suggested.[213]

Before concluding, I should like to remark that the attested writing could also be the result of a mistake just repeated in some copies, although one of them actually (A^{2}Sd^{b}, line 3) permits us, perhaps, to suppose an alternative situation. Schweiger,[214] in fact, there transliterates p^{a}-r^{a}-d^{a}[-..-..-..]-m^{a}, but cautiously he remarks, on the basis of Scheil's drawing,[215] that in such an occurrence there is empty space only for two signs, and not for three! This evidence could mean (of course, if the drawing is correct)[216] that the engraved word might have been, at least once, exactly written as p^{a}-r^{a}-d^{a}[y^{a}-d^{a}]-m^{a}, but it is difficult to speculate too much on this isolated and unclear attestation, although the already quoted case of *akunavam* shows that just one fragment can still preserve the correct expected form.

Finally, it is interesting to note that, in the Elamite and Akkadian versions of the same inscription, the palace is named *tacara-*, instead of *hadiš*.[217] This additional evidence could support the idea that such a word was no longer *à la page*, and that some scribes might be not so conversant with it and its correct gender. We must also remark that the location of the παράδεισοι in proximity of a royal palace is frequently mentioned in ancient

[211] See Allegri and Panaino, 'On the *š*-ending' (n. 53 above), pp. 15–21.

[212] Benveniste, 'Persica' (n. 177 above), p. 67.

[213] We must recall, according to I. Gershevitch ('Zoroaster's Own Contribution', *Journal of Near Eastern Studies*, 23, 1964, pp. 12–38, in particular p. 34), that we have here 'a mere transposition into cuneiform characters of the Aramaic spelling **PRDYD'* of O.P. **paridaidā-*', but Schmitt, *Beiträge* (n. 11 above), pp. 82–3, is right when he notes that this is an *ad hoc* explanation.

[214] Schweiger, *Kritische Neuedition* (n. 161 above), II, p. 512, n. 28.

[215] Scheil, *Inscriptions des Achéménides. Supplément* (n. 161 above), p. 128, line 3 of the first of two published drawings.

[216] I must thank A. V. Rossi, who kindly remarked, during the discussion of my paper, that he actually found in many cases Scheil's drawings very inaccurate. For this reason we cannot be sure of the real status of the inscription.

[217] See again Steve, *Nouvelle Mélanges Epigraphiques* (n. 161 above), pp. 97, 98.

literatures,[218] and this kind of combination seems to have had an ideological function.[219] Anyway, I do not find compelling reasons to delete the entry *paradaịda-* from the attested *Sprachgut* of Old Persian.

[218] See Bremmer, 'The Birth of Paradise' (n. 1 above), p. 50, and passim. Cf. also P. Briant, 'À propos du roi-jardinier: remarques sur l'histoire d'un dossier documentaire', in *A Persian Perspective. Essays in memory of Heleen Sancisi-Weerdenburg*, eds W. Henkelman and A. Kuhrt, Leiden, 2003, pp. 33–49, about the king as 'gardener'.

[219] It is interesting, as again Bremmer, 'The Birth of Paradise' (n. 1 above), p. 45, that the 'royal *parádeisos*' was chosen by the Phoenicians as their first target in their revolt of 351 BC. This fact shows that such a place was 'emblematic of Persian authority'.

Enoch, Eden, and the Beginnings of Jewish Cosmography

Annette Yoshiko Reed (University of Pennsylvania)

Where does our world end and the otherworld begin? Can one cross the boundaries between them, whether in life or after death? Is it possible to describe, with any certainty, the spaces above and beyond the visible cosmos? Today, we are accustomed to thinking of the traditional Western answers to such questions as among the ideas bequeathed to us by Judaism and Christianity. Striking, however, is how few of these answers can be found in the Hebrew Bible. Biblical references to creation, the cosmos, and Eden (especially Genesis 2–3) have garnered a disproportionate amount of attention in the history of Christian biblical interpretation and in modern scholarship. Yet such references are surprisingly few and often quite elusive.

In the surviving literature of ancient Israel[1] one finds little systematic concern for primeval time, cosmic spaces, or eschatological horizons. If this reticence reflects an era in which Israel's literary elite may have eschewed speculation into the unknown for a focus on familiar spaces, human life, and historical time,[2] then it also challenges us to grapple with the dramatic shift in the centuries thereafter. As with otherworldly spirits, the afterlife, and eschatology, so too with the structure of the cosmos and the place of paradise therein: topics treated only in passing, terse, or allusive fashion in biblical materials became the focus for intensive, explicit discussion in the Jewish literature of the Second Temple period (especially Hellenistic and Roman eras; 333 BCE–CE 70).[3]

The reasons for this shift remain puzzling, and perhaps all the more so now, with the maturation of scholarship on Second Temple Judaism in the wake of the discovery of the Dead Sea Scrolls. In past research this period was largely relegated to the status of a

Author's Note: Earlier versions of this piece were presented at the Warburg Institute conference on 'The Cosmography of Paradise' (5 March 2009) and as the Plenary Address, entitled 'Cosmology, Epistemology, and Paradise', of the Mid-Atlantic Region Society of Biblical Literature Annual Meeting (Baltimore MD, 26 March 2009).

Abbreviations used in this essay:

Gr^{Pan} = Codex Panopolitanus

MT = Masoretic

LXX = Septuagint

[1] I.e., as reflected in biblical texts and traditions from before, during, or in the wake of, the Babylonian Exile (586–538 BCE) – or, in other words, in materials that are generally accepted to preserve pre-exilic and exilic perspectives, even if their 'final' forms might be later. Inasmuch as my focus here falls on the contrast between these 'early' materials and the Jewish literature that took form during and after the 3rd century BCE, I shall leave it for others to debate the precise dating of Genesis 2–3, the redacted Pentateuch, etc.

[2] So, e.g., A. Segal, *Life after Death: A History of the Afterlife in Western Religion*, New York, 2004, pp. 120–24, emphasizing the reaction against Canaanite practices and the polemic against local expressions of Near Eastern traditions among the Israelite populace.

[3] On angels, see S. Olyan, *A Thousand Thousands Served Him: Exegesis and the Naming of Angels in Ancient Judaism*, Tübingen, 1993; on ouranology, otherworldly realms, eschatology, and the fate of the wicked and righteous dead, see J. J. Collins, *The Apocalyptic Imagination*, 2nd edn, Grand Rapids MI, 1998, and discussion below.

transitional age (e.g., 'inter-testamental'), and its literary products were interpreted as a preface to the origins of Christianity, rather than as part of the history of Judaism. Accordingly, the rise of interest in topics such as angels, the afterlife, and the end of time was often explained in apologetic terms – whether as a result of corrupting, foreign 'influence' upon Judaism (e.g., Persian, Hellenistic) or as a sign of the supposed distancing of the divine in the centuries preceding Jesus.[4] As Second Temple Judaism has been increasingly studied for its own sake and on its own terms, however, it has become clear that the continuities with ancient Israelite, early Christian, and rabbinic Jewish traditions are many and profound, and the patterns of discontinuity far more complex than any simple trajectory permits.[5]

We may be now in a position to attend to this complexity and to see what it might disclose about the pre-histories of many ideas now common in Western cultures. In the case of paradise, the gap between the early biblical and Second Temple evidence cautions us against assuming that it was always and everywhere a topic of pressing concern; interest in the topic may be frequent and widespread, but it is hardly universal or constant.[6] Hence, to tell the pre-history of paradise in the West is not merely to pinpoint moments of 'origin' or 'invention' for ideas and motifs that later became significant. Rather, it is also to ask why interest in particular questions arose when and where it did, and why and how it was given literary expression.

Towards this aim, the present essay considers the place of paradise in our earliest surviving examples of systematic Jewish cosmography, namely, the Enochic *Astronomical Book* and *Book of the Watchers* (*c.* third century BCE). Through a focus on paradise I explore some of the epistemological ramifications of the practice of cosmography, particularly in relation to evolving ideas about revelation and writing in Second Temple Judaism. In the process I hope to shed light on the transformation of ancient Israelite traditions about the cosmos and protology, as forged in the Judaean scribal cultures of the early Hellenistic age and as informing enduring beliefs in Judaism and Christianity alike.

[4] See further Olyan, *A Thousand Thousands* (n. 3 above), pp. 3–9.

[5] So, e.g., J. D. Levenson, *Creation and the Persistence of Evil*, San Francisco CA, 1988, pp. 31–46 – noting the continued vitality of the 'combat myth' in the First and Second Temple periods and beyond, but also the significant changes in the articulation thereof, e.g., as when '[i]n the latter half of the Second Temple period, the paucity of references to eschatological combat between God and the sea monster in the Hebrew Bible gives way to the almost psychedelic world of Jewish apocalyptic' (pp. 32–3).

[6] So, e.g., W. Brueggemann, *Genesis*, Atlanta GA, 1982, p. 41; cf. T. Stordalen, *Echoes of Eden: Genesis 2–3 and the Symbolism of the Eden Garden in Biblical Hebrew Literature*, Leuven, 2000, pp. 21–6. Although Stordalen may well be correct to attenuate past judgments of Genesis 2–3 as 'exceedingly marginal' in the Hebrew Bible, some of his arguments for doing so are not wholly convincing. His appeal to the theories of Mircea Eliade, et al, about cosmogony as a privileged site for culture-making, for instance, can be readily countered by recent critiques of these very theories (see, e.g., T. Masuzawa, *In Search of Dreamtime: The Quest for the Origin of Religion*, Chicago IL, 1993). Also questionable is his emphasis on the popularity of Genesis 2–3 within later Jewish and Christian literature to posit its significance earlier as well; such an argument seems to miss precisely what makes the lack of early biblical discussion so fascinating, namely, the contrast with the later explosion of explicit, sustained, and systematic interest in Genesis 2–3 (especially in materials dating from, during, and after the 1st century CE).

1. Writing the Cosmos before Cosmography

Some information about the structure of the cosmos, of course, can be found already in the Pentateuch, conveyed in the course of Genesis's account of how God formed 'the heavens and the earth'. In the first chapter of Genesis the reader/hearer learns that the world as we know it was formed through the separation of 'waters from waters' (1:6), and in the second, that the first dwelling-place of humankind was a garden, *gan Eden* (2:15, cf. 2:8; Septuagint Genesis 2:8: παράδεισος ἐν Εδεμ) planted at the source of four rivers.[7] When combined within the redacted Pentateuch, the result is an image of the inhabited world as situated in a space of water that is controlled – albeit by acts of divine order that can readily be un-done, as when 'gates in heaven' were opened and 'fountains of the Great Deep' gushed forth to flood the earth (7:11; 8:2).[8]

Nevertheless, the narrative thrust of the redacted Pentateuch is marked by a progressive shift away from cosmological concerns. That heaven – and knowledge of heaven – belongs solely to God and His angels is asserted when pre-history passes into history and when the limited parameters of human space and historical time become drawn and set. The reader/hearer is told, in Genesis, that the boundaries between heaven and earth were crossed by the 'sons of God', right before the Flood (6:1–5), and challenged by the builders of the Tower of Babel, right after (11:4). Both had disastrous results. Thereafter, a sharp epistemological divide marks off human from divine. God and His messengers are said still to roam the earth – sometimes surprising unwary humans through appearances that momentarily blur the line between supernatural and natural.[9] Yet humankind is described as knowing the realm of the divine only from the voices that resound from above, and from blessings and punishments sent down from the skies.[10] God's heavenly abode is depicted as a place so distant from earth that it is only in dreams that a man like Jacob might see the ladder that connects them, only angels who can climb this ladder, and only from the most sacred of mountains at the most sacred of times that a man like Moses might catch a fleeting glimpse of God's feet, resting on the sapphire floor of the firmament (Genesis 28:12–17; Exodus 24:9–10).

[7] On the relevant terminology and associated imagery, see Stordalen, *Echoes of Eden* (n. 6 above), pp. 81–186, 250–60; J. N. Bremmer, 'Paradise: From Persia, Via Greece, Into the Septuagint', in *Paradise Interpreted*, ed. G. P. Luttikhuizen, Leiden, 1999, pp. 1–20. On the four rivers, see n. 38 below.

[8] Compare Psalms 74:12–17; 104:6–9; Job 38:8–11. On the contrast between the image of creation as completed in Genesis 1:1–2:4 (i.e. cosmogony, as determinative for the Pentateuch in its present form) and the image of creation as part of an ongoing battle with the Sea, Leviathan, etc., found elsewhere in the Hebrew Bible, see Levenson, *Creation* (n. 5 above), especially pp. 14–25. That the 'combat myth' was not wholly supplanted or suppressed is suggested by its appearance even in rabbinic and later materials, on which see also M. Fishbane, *Biblical Myth and Rabbinic Mythmaking*, Oxford, 2003, pp. 37–70, 112–31.

[9] On the latter, see now J. L. Kugel, *The God of Old: Inside the Lost World of the Bible*, New York, 2003.

[10] For voices from above, see Genesis 21:17; 22:11, 15; Exodus 20:22; Deuteronomy 4:36. For heaven as the source of divine gifts of rain, dew, and bread, see Genesis 27:28; Exodus 16:4; Deuteronomy 28:12; 33:28; and, as the source of divine punishments of fire and drought, see Genesis 19:24; Deuteronomy 11:7.

Eventually, Jewish and Christian exegetes would delight in re-reading the protology of the redacted Pentateuch through the lens of the belief that *Endzeit gleicht Urzeit*[11] – with a divine judgment upon all of the earth, as in the Flood (cf. Genesis 6–9), the granting of paradisiacal bliss to the righteous in reversal of the expulsion from Eden (cf. 3:8–24), and/or a re-creation akin to God's first cosmogonic acts (cf. 1:1–31), whereby the purity of primeval space and time might be restored.[12] Significantly, for our purposes, such reading practices seem to have spread in the Second Temple period, together with the rise of literary interest in a range of other topics raised only tersely or tacitly, if at all, in earlier writings: the names and functions of angels, the origins and purpose of demons, the postmortem fates of the human wicked and righteous, the timetables of world-history, the topographies of heaven and the ends of the earth.

The exegetical derivation of eschatology from pentateuchal protology is now largely naturalized, particularly due to the prominence of this approach in Christian interpretations of Genesis.[13] Its development in Second Temple times, however, arguably necessitated reading against the grain of the very texts that were then coming to be elevated as 'Scripture'.[14] In the Pentateuch one finds no such nostalgia for Eden, nor any sustained concern for postmortem, otherworldly, or eschatological realities.[15] Rather, through its narratives, its authors/redactors situate the aim of life in our own space and

[11] H. Gunkel, *Schöpfung und Chaos in Urzeit und Endzeit, eine religionsgeschichtliche Untersuchung über Gen 1 und Ap Joh 12*, Göttingen, 1895. See further M. E. Stone, 'Three Transformations in Judaism: Scripture, History, and Redemption', *Numen*, 32, 2, 1985, pp. 218–35 (223–8). Notably, this seems to form part of a broader pattern whereby – as John J. Collins observes – '[t]he sages and prophets of the Hellenistic Near East looked not only to the great leaders of the past but also to the ancient myths to provide meaningful paradigms for the present, and especially to the myths of kingship and creation' ('Jewish Apocalyptic against Its Hellenistic Near Eastern Environment', *Bulletin of the American Schools of Oriental Research*, 220, 1975, pp. 27–36 [30]).

[12] In Second Temple Jewish materials it is possible to distinguish between a focus on the Flood as protological precedent for the Eschaton and a focus on the story of Adam, Eve, and Eden – as shown by M. E. Stone, 'The Axis of History at Qumran', in *Pseudepigraphical Perspectives: The Apocrypha and Pseudepigrapha in Light of the Dead Sea Scrolls,* eds E. G. Chazon and M. E. Stone, Leiden, 1999, pp. 133–50. Sustained interest of this sort in the Flood (as well as in Enoch, the 'sons of God', Noah, etc.) appears to have preceded sustained interest in the story of Adam, Eve, and Eden – indeed, note already Isaiah 24–7. See further below, as well as my discussion in *Fallen Angels and the History of Judaism and Christianity*, Cambridge, 2005, pp. 51–3, 110–16, 220–21.

[13] Note, especially, Christian readings of Genesis 2–3 as a 'Fall of Man' rectified by redemption through Jesus. The view of Christ as a second Adam, for instance, is found already in the letters of Paul, and extended soon thereafter in early Christian discussions of Mary as a second Eve; see further G. A. Anderson, *The Genesis of Perfection: Adam and Eve in the Jewish and Christian Imagination*, Louisville KY, 2001 – also on the contrast with Jewish views of creation as completed and culminating in the Tabernacle, Temple, and Torah.

[14] On the post-exilic elevation of the Pentateuch and its exegetical effects, see J. L. Kugel, *Traditions of the Bible: A Guide to the Bible as it was at the Start of the Common Era*, Cambridge MA, 1998, pp. 1–42.

[15] I here focus on the Pentateuch due to its prominence in the Second Temple period, especially as being privileged in shaping scribal pedagogies as well as the nostalgic ideal of ancient Israelite religiosity. It is perhaps in prophetic and psalmic materials, however, that we find early expressions of hope for vindication, cosmic restoration, etc., later concretized and systemized in the historical time-tables, eschatological predictions, mythic geographies, etc., of the apocalyptic literature of Second Temple times. For a poignant example pertaining to Eden, see Isaiah 51:3, where the post-exilic Deutero-Isaiah yearns for God to make Zion 'like Eden, her desert like the garden of the Lord'. For a parallel – and related – pattern with respect to the emergence of the doctrine of resurrection, see now J. D. Levenson, *Resurrection and Restoration of Israel,* New Haven CT, 2006.

time: the proper re-connection of human and divine is decisively located within the Promised Land of Israel and its Temple, on the one hand, and within genealogical and historical time, on the other.[16] Lest one imagine otherwise, it is stressed that one need not look beyond our own known world to gain the knowledge needed for life. This point is made explicit, for instance, in words attributed to Moses in Deuteronomy:

> It is not in the heavens (לא בשמים הוא) that you should say, 'Who will go up to the heavens for us, and get it for us so that we may hear it and do it?' And it is not beyond the sea, that you should say, 'Who will cross to the other side of the sea for us, and get it for us, so that we may hear it and do it?' (30:12–13)

Although heaven is deemed the primary dwelling-place of Israel's God, Israelites are thus dissuaded from contemplating heaven, the sun, the moon, and stars, lest they be tempted to 'be led astray, and bow down to them, and serve them' (Deuteronomy 4:19).[17] So, too, with Sheol: even as allusions to the realm attest the belief that it lies below the earth, as a destination for the dead (e.g. Genesis 37:35; 42:38; 44:29, 31), the redacted Pentateuch discourages further inquiry through the association of the dead and their places with practices condemned as 'foreign' and abhorrent (e.g. Deuteronomy 14:1; 18:10–11; cf. Numbers 16:20–33).[18]

A similar reticence can be glimpsed in other early materials now preserved in the Hebrew Bible. Even though one finds a number of references to God's heavenly abode, for instance, their function is rarely cosmographical *per se*. To reconstruct a cohesive cosmology, in fact, the modern scholar must mine and gather myriad scattered references, which – in their own contexts – serve aims other than the mapping of the structures of the seen and unseen world.[19] Just as many references to the cosmos are oriented towards praise and petition of God as Creator, so references to the contents and inhabitants of heaven most often function as means to telegraph His power and majesty; likewise, references to Sheol are often geared towards lamenting the limitations of the human condition and/or proclaiming the power of God to rescue and redeem.[20] Furthermore, much to the frustration of the modern scholar, many such references simply resist

[16] In Levenson's words: 'In the Torah, the future beatitude is conceived as serene and prosperous life in the Land of Israel. In these later literatures [e.g., Second Temple apocalyptic, early Christian], although the Land is sometimes important, the blessed future is conceived as a new age or a new world' (*Creation* [n. 5 above], 44). Note also, e.g., descriptions of the Tabernacle as a new creation, the Jerusalem Temple as a new Eden, and the Torah as Eden and/or the Tree of Life, on which see pp. 78–89; idem, *Resurrection* (n. 15 above), pp. 82–107; as well as further references below.

[17] See also II Kings 23:4; Isaiah 47:13; Amos 5:26–7.

[18] On Sheol, see sources cited above, as well as P. S. Johnston, *Shades of Sheol: Death and Afterlife in the Old Testament*, Downer's Grove IL, 2002.

[19] See, e.g., J. E. Wright, *An Early History of Heaven*, Oxford, 1999, pp. 52–97; Fishbane, *Biblical Myth* (n. 8 above), pp. 70–80.

[20] Fishbane further notes how protological events are often cited in Psalms (e.g. 74; 89) as 'paradigmatic acts of divine power whose like [the speaker] hopes to evoke in his own time' (*Biblical Myth* [n. 8 above], p. 41).

certainty, floating at the level of prophetic metaphor, sapiential rhetoric, or poetic allusion.

It may not be coincidental that the most extensive and detailed of these references occur precisely in the context of asserting what the human mind simply cannot know. This tendency is perhaps most pronounced in the sapiential tradition – as most famously in the book of Job, where cosmogony and geography are marshalled towards the aim of evoking the enormity of the divine knowledge that lies completely beyond human comprehension.[21] Likewise is Eden placed beyond the bounds of access to humans in historical time, not just in the Pentateuch, but also in prophetic traditions (especially Ezekiel 28). Biblical references to Eden are surprisingly few, and they serve, as Edward Noort has shown, primarily 'to describe the transformation from the Golden Age into the real world or to announce the fall of a mighty ruler', articulating what has been lost that cannot be re-gained again (cf. Isaiah 51:3).[22] Whether in a garden, as in Genesis, or on a mountain, as in Ezekiel, Eden is a place from which to be expelled, in a moment of transition definitive of the nature and limits of life thereafter. The precise location of the place is secondary.[23]

To be sure, some impulse towards cosmography can be glimpsed already in the celebration of God as Creator in late biblical prophecy (esp. Deutero-Isaiah; e.g. Isaiah 40:25–8).[24] Furthermore, metaphorical, poetic, and liturgical uses of cosmological motifs would not have been possible, nor resistance to further speculation even meaningful, unless some traditions of this sort were already circulating in ancient Israel – whether orally or in writing, whether as self-consciously 'Israelite' traditions or as part of an understanding of the world shared with other Near Eastern cultures. And, of course, the Hebrew Bible preserves only a small fragment of ancient Israelite religiosity, selected and shaped to promote certain practices as 'traditional' and normative (e.g. centralized worship at Jerusalem, Israelite devotion to a single deity) and to dismiss others as 'foreign' and deviant (e.g. worship at local shrines, commerce with the dead, Israelite devotion to multiple

[21] See Job 11:5–6; 28; 38–40; L. Perdue, 'Wisdom in the Book of Job', in *In Search of Wisdom: Essays in Memory of John G. Gammie*, eds L. Perdue, B. Scott, and W. Wiseman, Louisville KY, 1993, pp. 73–98 (92–8). The emphasis on the limits of human knowledge might seem, at first sight, to stand in tension with the didactic aims of the sapiential tradition as well as the attitude towards Nature conveyed by Proverbs 8. Nevertheless, expressions of this sort can be found already in Proverbs 30:1–4; compare Qohelet (esp. 3:21) and the *Wisdom of the ben Sira* (on which see below), and contrast the Wisdom of Solomon (especially 7:17–21). On similar sentiments in biblical prophecy (e.g. Isaiah 40:12–14; Jeremiah 31:37), see M. Himmelfarb, *Ascent to Heaven in Jewish and Christian Apocalypses*, Oxford, 1993, pp. 76–7.

[22] E. Noort, 'Gan-Eden in the Context of the Mythology of the Hebrew Bible', in *Paradise Interpreted* (n. 7 above), pp. 21–36 (35).

[23] Ibid., pp. 27–34.

[24] Most influential for later images of heaven, of course, are Isaiah 6 and Ezekiel 1; 10. Note also the place of Ezekiel 40–48 as a model for the 'tour' narratives by which later apocalypses commonly communicate their views of the structure and contents of the cosmos; see M. Himmelfarb, *Tours of Hell: An Apocalyptic Form in Jewish and Christian Literature*, Philadelphia PA, 1983, esp. pp. 50–67. Sapiential precedents are emphasized by Philip S. Alexander, who points particularly to Proverbs 8 ('Enoch and the Beginnings of Jewish Interest in the Natural Sciences', in *The Wisdom Texts from Qumran and the Development of Sapiential Thought*, eds C. Hempel, A. Lange, and H. Lichtenberger, Leuven, 2002, pp. 223–44 [237–8]).

deities).[25] Of the notions of creation, the cosmos, and paradise in the latter, we can only speculate.[26] Accordingly, any conclusions about broader trajectories in ancient Israel and Second Temple Judaism must come with the caveat that we speak specifically of shifts within their literary cultures; for, indeed, what we know of both eras reflects what scribes – and their patrons and pedagogies – have deemed apt for preservation in writing.[27]

Nevertheless, it remains that there is nothing in the surviving literary heritage of ancient Israel that prepares us for the extensive cosmographical writings that emerge within Judaism in the early Hellenistic age.[28] Perhaps oldest among them is the *Astronomical Book*, which describes Enoch's tours of heaven, guided by the angel Uriel, and outlines the names, paths, number, times, gates, laws, and angelic leaders of the sun, moon, wind, stars, and seasons.[29] What the *Astronomical Book* does for heaven, the *Book of the Watchers* does for earth. This work claims to report what Enoch saw during and after his ascent to heaven, when he travelled with angels to and beyond the edges of the inhabited world. In effect, Enoch is here described as achieving precisely what Deuteronomy had deemed it improper even to desire: he brings back the knowledge that 'is in heaven' and from 'beyond the sea'. In the process, the works attributed to him claim to preserve eyewitness accounts of many of the phenomena cited in Job (especially 28; 37–8) as exemplary of what humankind cannot know.[30]

What proves interesting, for our purposes, is that these works seem to have been penned by scribes who deliberately transgressed the epistemological divides outlined in the pentateuchal and sapiential traditions from which they drew and to which they were

[25] For this point with respect to cosmology, see Wright, *An Early History* (n. 19 above), pp. 88–95.

[26] The scholarly tendency has been to reconstruct these and other 'popular' Israelite beliefs on the basis of the beliefs and practices critiqued in the Pentateuch, Deuteronomistic History, etc., as correlated with parallel traditions in other Near Eastern cultures. Although the value of this approach is unassailable, my caution here is based on the uncertainty involved in any such reconstruction – particularly insofar as it often involves following the biblical authors'/redactors' polemical construction of local, etc., religiosity as 'foreign' (and, conversely, only pentateuchal, Deuteronomistic, etc., ideas as distinctively 'Israelite') by progressing on the assumption that the beliefs and practices of most Israelites would be indistinguishable from those of surrounding peoples.

[27] I draw here and below on recent attempts to reorient our understanding of the formation, redaction, collection, and meanings of early biblical writings in light of what we know about scribal practices in ancient Israel and the Near East more broadly; especially D. M. Carr, *Writing on the Tablet of the Heart: Origins of Scripture and Literature*, Oxford, 2005; K. van der Toorn, *Scribal Culture and the Making of the Hebrew Bible*, Cambridge MA, 2007.

[28] This point is similarly stressed by Alexander, 'Enoch and the Beginnings' (n. 24 above), pp. 231–2, 240; J. C. VanderKam, 'Chapters 72–82: The Book of the Luminaries', in G. W. E. Nickelsburg and J. C. VanderKam, *1 Enoch 2: A Critical Commentary on the Book of 1 Enoch, Chapters 37–82*, Minneapolis MN, 2011, 334–569 (367). See also J. Ben-Dov, *Head of All Years: Astronomy and Calendars at Qumran in their Ancient Context*, Leiden, 2008.

[29] See further VanderKam, 'Chapters 72–82' (n. 28 above); O. Neugebauer, 'The "Astronomical Chapters" of the Ethiopic Book of Enoch (72 to 82)," in M. Black, *The Book of Enoch or 1 Enoch: A New English Edition*, Leiden, 1985, pp. 386–414; M. Albani, *Astronomie und Schöpfungsglaube: Untersuchungen zum astronomischen Henochbuch*, Neukirchen, 1994; H. Drawnel, *The Aramaic Astronomical Book from Qumran: Text, Translation, and Commentary*, Oxford, 2011.

[30] M. E. Stone, 'Lists of Revealed Things in the Apocalyptic Literature', in *Magnalia Dei: The Mighty Acts of God: Essays on the Bible and Archaeology in Memory of G. Ernest Wright*, eds F. M. Cross, W. Lemke, and P. D. Miller, Garden City NJ, 1976, pp. 414–52 (435–9); K. Coblentz Bautch, *A Study of the Geography of 1 Enoch 17–19: 'No One Has Seen What I Have Seen'*, Leiden, 2003, pp. 223–7 and Table 9.

otherwise deeply indebted.[31] In what follows, then, I would like to reflect upon this moment of transgression, asking what it might tell us about the beginnings of Jewish cosmography, the transformation of ancient Israelite traditions in the Second Temple period, and the pre-history of Western ideas about paradise.

2. The Cosmos in the Astronomical Book *and* Book of the Watchers

Written in Aramaic and attributed to the antediluvian sage Enoch, the *Astronomical Book* and the *Book of the Watchers* are among the earliest extant Jewish works outside the Hebrew Bible (third or early second century BCE), predating and likely even influencing the latest book therein (i.e. the book of Daniel, mid-second century BCE).[32] These works are also our oldest examples of the literary genre of the apocalypse.[33] Furthermore, as we shall see, they preserve the first known attestations of a number of traditions pivotal for the development of Western ideas about paradise.

The earlier of the two, the *Astronomical Book*, includes only one chapter that focuses on earth (i.e. *1 Enoch* 77). Here, one finds a four-fold division of the earth into west, south, east, and north, which corresponds to the four-fold division of the heavens. This is followed by another, three-fold division,[34] described as follows:

[31] Particularly in the case of the *Book of the Watchers*. On its connections with the Pentateuch, see M. Himmelfarb, *A Kingdom of Priests: Ancestry and Merit in Ancient Judaism*, Philadelphia PA, 2006, especially pp. 39–41, and with Wisdom, see G. W. E. Nickelsburg, *1 Enoch 1: A Commentary on the Book of 1 Enoch, Chapters 1–36; 81–108,* Hermeneia, Minneapolis MN, 2001, pp. 38–9, 50–3, 58–61; idem, 'Enochic Wisdom and its Relationship to Mosaic Torah', in *The Early Enoch Literature*, eds G. Boccaccini and J. J. Collins, Leiden, 2007, pp. 81–94. See also J. C. VanderKam, 'Scripture in the Astronomical Book of Enoch', in *Things Revealed: Studies in Early Jewish and Christian Literature in Honor of Michael E. Stone*, eds E. G. Chazon, D. Satran, and R. A. Clements, Leiden and Boston MA, 2004, pp. 89–103.

[32] The *Astronomical Book* and *Book of the Watchers* are both preserved in Ge'ez in the Ethiopian collection *Maṣḥafa Henok Nabiy* ('Book of Enoch the Prophet', typically called '*1 Enoch*' by Western scholars), as chapters 72–82 and 1–36 respectively. Portions of both are also extant in Aramaic fragments from Qumran (4Q201–2, 4Q204–5, 4Q208–11), the earliest of which date from around 200 BCE (i.e. 4Q201, 4Q208). The *Book of the Watchers* is also partially extant in a Greek MS (c. 5th century CE, Gr^{Pan}) as well as in citations from the Byzantine chronographer George Syncellus. The Aramaic, Greek, and Ge'ez versions of the *Book of the Watchers* correspond relatively closely, while the extant Aramaic and Ge'ez of the *Astronomical Book* are notably divergent. For further details, references to the relevant editions, etc., see Nickelsburg, *1 Enoch 1* (n. 31 above); Nickelsburg and VanderKam, *1 Enoch 2* (n. 28 above); M. A. Knibb, *The Ethiopic Book of Enoch: A New Edition in Light of Aramaic Dead Sea Fragments*, Oxford, 1978, 2 vols, II, pp. 1–46; idem, 'The Book of Enoch or Books of Enoch? The Textual Evidence for 1 Enoch', in *The Early Enoch Literature* (n. 31 above), pp. 21–40; A. Y. Reed, 'The Textual Identity, Literary History, and Social Setting of 1 Enoch', *Archiv für Religionsgeschichte*, 5, 1, 2003, pp. 279–96 and Chart 1.

[33] See further J. J. Collins, 'Towards the Morphology of Genre', *Semeia*, 14, 1979, pp. 1–20; M. E. Stone, 'Enoch and Apocalyptic Origins', in *Visionaries and their Apocalypses*, ed. P. Hanson, Philadelphia PA, 1983, pp. 92–100; as well as further references in A. Y. Reed, 'The Origins of the *Book of the Watchers* as "Apocalypse" and its Reception as "Apocrypha"', *Henoch*, 30, 1, 2008, pp. 55–9.

[34] Translations here follow E. J. C. Tigchelaar, 'Eden and Paradise: The Garden Motif in Some Early Jewish Texts (1 Enoch and Other Texts Found at Qumran)', in *Paradise Interpreted* (n. 7 above), pp. 37–62 (62); Coblentz Bautch, *A Study* (n. 30 above), pp. 200–201; VanderKam, 'Chapters 72–82' (n. 28 above), p. 482; cf. J. T. Milik, *Books of Enoch*, Oxford, 1976, p. 290. Note that in the Ge'ez, this split is depicted as a three-fold division of the northern quadrant of the earth, while the Aramaic fragments from Qumran show that the original intention here was to describe a three-fold division of the entire world (J. C. VanderKam, '1 Enoch 77, 3 and a Babylonian Map of the World', *Revue de Qumran*, 42, 1983, pp. 271–8 [274–5]).

Aramaic, 4Q209 fr. 23 lines 8–10 and 4Q210 fr. 1 col. ii lines 19–20	Ge'ez, *I Enoch* 77:3b
… one of them for the dwelling of the sons of men in it; and one of them […]	one of them is the dwelling-place of men (*māxdar la-sab'*).
[and one of them] for the deserts and for …	And another: seas of water, and the deeps, and forests, and rivers, and darkness, and mist.
[… and for the *parde*]*s* of righteousness.[35] …	And another part: the garden of righteousness (*gannata ṣedq*).

The *Astronomical Book* then goes on to list what appear to be places at the edge of the earth: seven great mountains, higher than all the mountains of the earth, from which snow comes (77.4), seven great rivers, larger than any earthly rivers, which flow into the Red Sea, the Great Ocean, and the Wilderness (77.5–7), and seven great islands (77.8).[36] Although the fragmentary character of the extant Aramaic and the divergences in the Ge'ez make it difficult to know for certain,[37] this mapping of earthly spaces seems to move from known centre to unknown periphery – from worldly to other-worldly. If so, the '*pardes* of righteousness' is poignantly set at the very crossing-point between them.[38]

The *Astronomical Book* also proves significant, for our purposes, due to its rhetorical and epistemological stance. Its presentation of information about these and other features of the cosmos is couched in claims of certainty and totality (e.g. 74.2; 76.14; 81.1; 82.1). Enoch is here said to have learned 'everything' about the celestial cycles by virtue of his travels with angels, his instruction by them, and his access to heavenly books, and he is said to have transmitted 'all of it' to his son Methuselah. Indeed, from one perspective, the work can be read as a defence of the very practice of Jewish cosmography: far from exhibiting any reticence about investigating 'the sun, the moon, and the stars, all the host of heaven', it purports that trustworthy knowledge about them

[35] 4Q209 fr. 23 line 9: [… ולפרד]ס קושטא. I transliterate *pardes* here and below, so as to mark it as a technical term; indeed, as Nickelsburg notes (*1 Enoch 1* [n. 31 above], p. 327), this word-choice may be significant, as one would expect Aramaic גנתא here for Hebrew גן.

[36] On the geography here, and its points of contact with ancient Greek and Mesopotamian traditions, see P. Grelot, 'La Géographie mythique d'Hénoch et ses sources orientales', *Revue Biblique*, 65, 1958, pp. 33–69 (34–6, 47–69); cf. VanderKam, 'Chapters 72–82' (n. 28 above), pp. 493–5.

[37] VanderKam, '1 Enoch 77' (n. 34 above), pp. 274–5; Tigchelaar, 'Eden' (n. 34 above), p. 47.

[38] If so, this is comparable to what Noort observes of the description of the location of Eden in Genesis 2–3, namely, that the list of rivers reflects move from 'from well-known to unknown, from geography to myth, from … the familiar Euphrates to a broad description of a land of gold', whereby 'geography… demonstrate[s] the reality of Paradise' even as the 'mystified location' does not permit it to be 'an accessible and locatable place' ('Gan-Eden' [n. 22 above], pp. 28, 33). On Gihon in relation to the Temple and Jerusalem, see also Stordalen, *Echoes of Eden* (n. 6 above), pp. 280–1, 357–71; Levenson, *Resurrection* (n. 15 above), pp. 87–95.

has circulated amongst humankind since before the Flood, in written records of revealed wisdom.[39]

This stance is enabled by the emphasis on the power of writing to transcend the boundaries of ordinary space and time.[40] Not only is Enoch presented as the antediluvian 'author' of the *Astronomical Book*, but the book itself describes how he stopped to chronicle what he saw in the course of his journeys, so that his son could preserve and transmit the revelations:

> All this, Uriel the holy angel... showed me. I [i.e. Enoch] wrote down their positions as he showed me... (*I Enoch* 74:2)
>
> All their laws ... I have shown you everything, my son Methuselah. (76:14)
>
> Now, my son Methuselah, I am telling you these things and am writing [them] down. I have revealed all of them and have given you the books about these things. My son, keep the book written by the hand of your father so that you may give it to the generations of the world. (82:1)

In addition, Enoch's scribal acts include the reading of heavenly writings:

> The book about the motions of the heavenly luminaries... which Uriel showed me... (72:1)
>
> He [i.e. Uriel] said to me: 'Enoch, look at the heavenly tablets, read what is written on them...' I looked at all the heavenly tablets, read everything that was written... (81:1)

The knowledge revealed to the reader/hearer of the book is thus doubly authorized: it is presented as the record of human sight, speech, and memory but also as a written emblem of the transfer – and the very possibility of an *exact* transfer – of heavenly knowledge down to earth.

Similar ideas about cosmology, knowledge, and writing can be found in the *Book of the Watchers*, together with even more references relevant to paradise. The present form of the book is composite, and much of the scholarship about it has focused on delineating and investigating its constituent parts.[41] A focus on cosmography, by contrast, draws our

[39] The implication, more specifically, is that written records of revealed wisdom about such matters resides in Israel; on the significance of such a claim, in light of the 'competitive historiographies' of the early Hellenistic age, see below.

[40] See further A.Y. Reed, 'Pseudepigraphy and/as Prophecy: Continuity and Transformation in the Formation and Reception of Early Enochic Writings', in *Revelation, Literature, and Community in Late Antiquity*, eds P. Townsend and M. Vidas, Tübingen, 2011, pp. 25–42.

[41] It is common to distinguish five sub-units in the *Book of the Watchers*: *I Enoch* 1–5, 6–11, 12–16, 17–19, and 20–36 – although some of these are themselves composite. Cosmographical materials cluster mostly in the last two, with 20–36 often posited to be a later reflection or expansion on 17–19; see discussion and references in Coblentz Bautch, *A Study* (n. 30 above), pp. 13–23. On the redaction history of this material, particularly in relation to its combination with the sub-units concerning the fallen angels (6–11, 12–16), see also C. Newsom, 'The Development of *I Enoch* 6–19: Cosmology and Judgment', *Catholic Biblical Quarterly*, 42, 1980, pp. 310–29. For an attempt to make sense of the redacted form of chapters 17–36, as three journeys with a 'coherent structure', see J. Stock-Hesketh, 'Circles and Mirrors: Understanding 1 Enoch 21–32', *Journal for the Study of the Pseudepigrapha*, 11, 2000, pp. 27–58.

attention to the orchestrating logics and anthological structures of its redacted form – the form in which, and by which, the significance of paradise is here expressed.[42]

The redacted form of the *Book of the Watchers* is shaped by an overarching concern for charting the proper scope, source, and purpose of human knowledge. It begins with an exhortation by Enoch, concerning what he knows and how. As in the *Astronomical Book*, claims of revelation enable a stance of certainty, and the rhetoric is totalizing:

> From the words of the Watchers and holy ones, I heard everything,[43] and as I heard everything from them, so I also understand what I saw. (*I Enoch* 1:2)

The ramifications of this angelic knowledge for human life are then proclaimed with reference to divine judgment: Enoch warns of an age in which God will descend from His abode in heaven, treading down upon Mount Sinai with His angels, to judge all the creatures of the earth (1:4–9).[44]

With this, the *Book of the Watchers* turns to explore the twin themes of knowledge and judgment. It does so first from the perspective of the inhabited world, beginning with an appeal to the elements and cycles of Nature, as experienced by humankind in daily life and historical time: the rising and setting of the sun, moon, and stars, the shifts of the seasons and the blossoming of trees, the flow of the seas and rivers (2:1–3:3).[45] Through the voice of Enoch, the text calls for contemplation of them, and it stresses the ethical aims of proper human inquiry into Nature. The created world is here described as providing models for obedience to God's laws – even convicting, by comparison, people who stray from the paths of righteousness (3:4).[46]

From the positive *exempla* of Nature obedient, the text then turns to negative *exempla*, namely, the angelic Watchers who descended to earth in the days before the Flood (6–16). In comparison with the terse description of the deeds of the 'sons of God' in Genesis

[42] The discussion below extends my earlier efforts to consider the *Book of the Watchers* as a redacted whole, especially in 'Heavenly Ascent, Angelic Descent, and the Transmission of Knowledge in 1 Enoch 6–16', in *Heavenly Realms and Earthly Realities in Late Antique Religions*, eds R. S. Boustan and A. Y. Reed, Cambridge, 2004, pp. 47–66. Here, as there, I am indebted to John Collins's methodological insights, e.g., in 'The Apocalyptic Technique: Setting and Function in the Book of the Watchers', *Catholic Biblical Quarterly*, 44, 1982, pp. 91–111.

[43] See 4Q201 col. i line 3, and textual notes in Nickelsburg, *1 Enoch 1* (n. 31 above), p. 137. Note the lack of totalizing rhetoric in Balaam's oracle in Numbers 24:15–17, which seems to have served as a model for this opening proclamation.

[44] On this passage and its parallels with Deuteronomy 33 and other earlier theophanies, see L. Hartman, *Asking for a Meaning: A Study of 1 Enoch 1–5*, Uppsala, 1979, pp. 23–4.

[45] See further Hartman, *Asking for a Meaning* (n. 44 above), pp. 17–21.

[46] Below, we shall see how much the same argument is made with reference to natural elements and cosmic cycles that lie beyond the inhabited world and everyday human experience. For similar notions in later literature, see *Wisdom of ben Sira* 16:26–8; *I Enoch* 100:10–11; 101:1–8 (*Epistle of Enoch*); *I Enoch* 41:5–8 (*Parables of Enoch*); *II Baruch* 19:1–4; 48:9–10; *Testament of Naphtali* 3:2–4:1; Nickelsburg, *1 Enoch 1* (n. 31 above), pp. 152–5; M. E. Stone, 'The Parabolic Use of Natural Order in Judaism of the Second Temple Age', in *Gilgul: Essays on Transformation, Revolution, and Permanence in the History of Religions*, eds S. Shaked et al, Leiden, 1987, pp. 298–308.

6:1–4, the account here is lengthy and detailed. Most notable, for our purposes, is the treatment of the transgression of the boundaries between heaven and earth. Not only are these angels accused of abandoning their appointed homes in heaven, seeking immortality through fleshly reproduction, defiling human women, and siring monstrous Giants (e.g. 6:1–7:2; 9:8–9; 12:4; 15:3–16:1), but they are said to have revealed heavenly secrets on earth and to have taught corrupting knowledge to humankind (e.g. 7:1; 8:1–4; 9:6–8; 13:2; 16:3).

The topics of their teachings include the workings of the metals for the sake of weapons and jewellery, the manipulation of plants and chemicals for spells and cosmetics, and divination from the sun, moon, stars, and meteorological phenomena.[47] Within the redacted *Book of the Watchers* the result is a striking contrast: these practices of extracting gain from the earth and knowledge from the skies are a perversion of what was promoted, at the outset, as proper inquiry into the cosmos. Whereas Enoch's speech in *I Enoch* 2–5 showed how proper contemplation of the earth and the skies can foster righteousness, the account of angelic descent in chapters 6–11 describes how the knowledge revealed by the Watchers resulted in the corruption of humankind and the spread of earthly evils. Their tale thus serves as a warning about the corrupting power of knowledge about the cosmos wrongly gained, as well as providing a paradigm for the improper crossing of the physical and epistemological divide between heaven and earth.

Furthermore, the rectification of the disorder caused by angelic descent occasions the revelation of the mechanisms of divine justice through which Nature, humankind, and angels are rightly connected. According to the *Book of the Watchers* the antediluvian earth was ravaged by the corruption of humankind and bloodied by the hunger of the Giants, and it cried out to heaven, together with the voices of the souls/spirits of the creatures slain (7:3–6; 8:4). With this, the perspective of the text shifts away from the inhabited world, following these voices upwards to the 'gates of heaven', as their pleas reach the archangels who 'look down upon the earth from the sanctuary of heaven' (9:1–2). Upon hearing the pleas, the archangels petition God. He, in turn, commissions them to mete out punishment and to purify the earth (9:4–11:2).

The archangelic intervention into worldly affairs is here said to encompass, not just the sending of the Flood and the salvation of Noah (cf. Genesis 6–9), but also the binding of the fallen angels in darkness and desert, the destruction of the bodies of the Giants through sword, and the punishment of angelic and human sinners alike (*I Enoch* 10:1–11:2). The account of the divine instructions to the archangels, moreover, includes the promise that the cleansing of the earth from wickedness shall be accompanied by its renewal: the righteous shall be free from wrath 'for all the generations of eternity' (10:22), dwelling on a pure earth 'tilled with righteousness', filled with 'all the trees of joy' (10:18), and nourished by the opening of 'the storehouses of blessing in heaven' (11:1). Protology

[47] On the teachings of the Watchers, see further Reed, *Fallen Angels* (n. 12 above), pp. 24–57.

and eschatology are thus brought into correlation, with Flood and final judgment seamlessly joined, as if in the timelessness of divine will.

Up until this point in the text (i.e. 1–11), the *Book of the Watchers* departs only slightly from the cautious stance towards cosmological inquiry conveyed by Deuteronomy, Job, and other early biblical writings (see above). To celebrate proper human knowledge of the divinely-created world, it ethicizes the act of observing the elements and cycles of Nature, and it condemns the use of such knowledge for improper aims like divination. Indeed, if anything, the boundaries between flesh and spirit are sharply underscored – together with the epistemological divide between heaven and earth – by the *Book of the Watchers*'s detailed and dramatic descriptions of the disastrous results of angelic descent.

The chapters that follow, however, open up the possibility that human bodies and knowledge can cross these very boundaries. Having told the tale of the fallen angels from the perspective of earth (6–8) and from the perspective of heaven (9–11), the text retells this tale from the perspective of Enoch (12:1ff).[48] The intermediary status of this 'scribe of righteousness' is signalled by the accounts of his earthly encounters with angels. Heavenly ones instruct him to rebuke the fallen ones (12:3–13:1), who then plead with him to write a petition on their behalf and to deliver it to God in heaven (13:3–6). The power of writing – even to cross the boundaries between earth and heaven – is then narratively confirmed, as Enoch's penning of the petition occasions his rapture.[49]

The *Book of the Watchers* subsequently purports to record Enoch's own memory of his ascent to heaven, as told to the fallen angels after his return. He recounts how clouds, shooting stars, mists, wind, and lightning carried him up to a three-chambered heavenly palace (13:7–14:9). Even as God's heavenly abode is here described in terms that evoke the Jerusalem Temple,[50] Enoch himself is made to emphasize that the place is incomprehensibly and frighteningly otherworldly. It is presented as a realm of coexisting opposites that could never be joined on earth, such as hailstones and tongues of fire, water and flame, and a divine presence that is flanked by countless hosts of angels yet cannot be seen or approached by any (14:9–23). The sight of all this is said to cause Enoch to collapse, falling down 'on my face, prostrate, trembling' (14:24).

This chapter, *I Enoch* 14, has been much discussed by scholars as the oldest surviving account, in Jewish literature, of a heavenly ascent.[51] Within the *Book of the Watchers* it also serves an important literary function: the account of Enoch's heavenly journey marks

[48] On the shifts of perspective that structure the patterns of repetition and expansion in *I Enoch* 6–16 – enabling the interweaving of strands of earlier materials into a new whole – see further Reed, 'Heavenly Ascent' (n. 42 above), pp. 53–65.

[49] It is significant, as Himmelfarb notes, that Enoch's upward journey is a case of rapture, rather than the result of any ritual of self-induced ascent (*Ascent* [n. 21 above], pp. 109–10). Compare Elijah's rapture in II Kings 2:1–14, and contrast the understanding of the *yored ha-merkavah* in Hekhalot literature.

[50] Himmelfarb, *Ascent* (n. 21 above), pp. 14–28.

[51] See, e.g., ibid., pp. 9–28; I. Gruenwald, *Apocalyptic and Merkavah Mysticism,* Leiden, 1980, pp. 32–47; P. Schäfer, *The Origins of Jewish Mysticism*, Tübingen, 2009, pp. 55–67.

a pivot in the work, whereby the focus turns from the descent of the fallen angels and their corrupting teachings on earth, to the elevation of Enoch and his access to salvific knowledge from heaven.[52]

What immediately follows (i.e. 14:24–16:4) is an account of the scribe's encounter with God, from whom he receives his commission to rebuke the Watchers. The last nineteen chapters of the *Book of the Watchers* (i.e. 17–36) are then taken up by a detailed description of Enoch's travels to the ends of the earth, guided and instructed by angels.[53] Although this 'tour' material has garnered far less scholarly attention than the chapters concerning the fallen angels and Enoch's heavenly ascent (i.e. 6–16), it makes up more than half of the work and significantly shapes the meaning and message. This material also proves central for our understanding of the emergence of Jewish cosmography and the place of paradise therein.

The description of Enoch's travels in *1 Enoch* 17–36 begins with a brief, ostensibly first-hand, account of Enoch's north-westward journeys around the earth's edges, from the 'places of the luminaries and the treasures of the stars' to the 'living waters and fires of the west, which provides all sunsets', to the 'great sea of the west' and the 'place where no flesh walks', 'the mouth of all the rivers of the earth and the mouth of the abyss' (17:2–8).[54] Significantly, for our purposes, the subsequent summary of what Enoch saw (18:1–5) includes features that recall the *Book of the Watchers*'s introductory appeal to the obedience of elements and cycles of Nature (see above).[55] Exemplary, in this regard, is his

[52] See further Reed, 'Heavenly Ascent' (n. 42 above), pp. 61–5.

[53] For attempts to map the geography here presupposed, see Grelot, 'La Géographie mythique' (n. 36 above), pp. 46–7; Milik, *Books of Enoch* (n. 34 above), pp. 22–41; Coblentz Bautch, *A Study* (n. 30 above), pp. 160–90. Grelot's map draws on a range of Enochic and aligned sources, both early and late; that of Milik focuses on the *Astronomical Book* and *Book of the Watchers*; and that of Coblentz Bautch focuses on *1 Enoch* 17–19; compare the map of chapters 21–32 in Stock-Hesketh, 'Circles and Mirrors' (n. 41 above), p. 30.

[54] For a detailed analysis of *1 Enoch* 17–19, with a comprehensive survey of the history of research, see now Coblentz Bautch, *A Study* (n. 30 above). As noted already by Grelot ('La Géographie mythique' [n. 36 above], pp. 41–4), the geography presupposed in this and other accounts of Enoch's travels to and around the ends of the earth include features familiar from both Near Eastern and Greek traditions, as attested in the *Epic of Gilgamesh* and Babylonian world maps on the one hand, and Odyssey's journey in Homer, *Od.* 11 and other Greek *nekyiai*, on the other. Particularly insofar as attempts to isolate a single dominant model have proven unsuccessful (see, e.g., the critique of Grelot in VanderKam, '1 Enoch 77' [n. 34 above]), it seems best to conclude, with Nickelsburg, that what is reflected in the *Book of the Watchers* is most likely 'a popular geography that was a composite of Mesopotamian and Greek ideas', even as it simultaneously gives us 'a hint of much broader Jewish "scientific" and apocalyptic speculation' (*1 Enoch 1* [n. 31 above], p. 280).

[55] As is often noted, these verses break the flow of *1 Enoch* 17–19 and are puzzling when read in a strict chronological fashion. For a survey of scholarly attempts to account for this and other repetitions, etc., through reconstructions of their sources and redaction history, see Coblentz Bautch, *A Study* (n. 30 above), pp. 17–23. Here, by contrast, I focus on the progression of themes, particularly as they relate to the rest of the *Book of the Watchers*; this approach reflects my concern with the meanings generated by the interweaving of received materials and, hence, with the ways in which repetition, expansion, etc., functions in the redacted whole. Indeed, as is clear already from the structure of its accounts of the fallen angels (6–16), the main redactional principle governing the combination of materials in the *Book of the Watchers* was not chronology nor what modern readers might call 'narrative logic'. Nevertheless, its patterns of repetition serve to generate new levels of meaning, which can be overlooked when its hypothetical sources are examined in isolation. Throughout chapters 17–36, as we shall see,

sight of the 'storehouses of all the winds'[56] with which God 'arranged all that is created' (18:1): not only do these winds blow upon the inhabited earth, but they are revealed to be the forces that bear the earth and support the firmament, stretch out the skies, and move the sun, stars, and clouds across the heavens (18:2–5).[57]

Enoch is then said to head southwards, past seven great mountains and then beyond (18:6–10). There, he sees the prisons, not just of the fallen angels (19:1–2), but also of seven stars, in a great chasm of heavenly fire at 'the edge of the great earth, where the heavens come to an end' (18:10).[58] The reader/hearer thus learns that the commonalities between humans, angels, and Nature go beyond the shared ideal of their obedience: natural phenomena can stray from their appointed tasks, and they too suffer punishment when they do.[59]

The thematic oscillation between obedience and disobedience continues with a list of exemplars of angelic faithfulness. To the fallen angels and the seven wayward stars (6–16; 18–19) are contrasted seven angels who watch over the inhabited world (20:1–8). The domains of five of them are connected to the judgment of the wicked and the protection of the righteous. The other two watch over specific places: Uriel is in charge of the inhabited world and Tartarus (20:2 in Gr.$^{Pan.}$), while Gabriel is in charge of paradise, cherubim, and serpents (20:7). With this, the focus of the text shifts back to the spatial concretization of its earlier assertions about the inevitable punishment of all creatures who transgress divine law: it returns to wayward angels and stars – this time, separating the two and using Enoch's travels to reveal the fearsome places of their final punishment (21:1–10).[60]

The significance of these various distinctions becomes clear with the subsequent descriptions of the places that Enoch visits when he journeys back into the parts of the cosmos bounded by earth below and firmament above. From this point in the text

the threads of received traditions seem to have been coaxed into harmony through a focus on thematic concerns (especially knowledge and judgment); taken together, they progressively build up an argument for the human choice of a righteous path for life (e.g. through the further establishment of Enoch's unique authority as a source for information about the cosmos; the extension of the correlation of natural order and human righteousness; the assertion of the totality of divine justice through the focus on the punishment of the Watchers and other negative *exempla*; the exhortation to righteousness with appeal to the realities of post-mortem and eschatological judgment, as 'proven' through purportedly first-hand, eye-witness, testimony to the places set and prepared for it).

[56] Gr.^{Pan} τοὺς θησαυροὺς τῶν ἀνέμων πάντων; Ethiopic *mazāgebta kʷellu nafāsāt*. On such 'storehouses' compare Deuteronomy 28:12 and Job 38:22.

[57] Cf. Job 26:7, 11; 36:29; 37:16; 38:6; also Jeremiah 31:37. Even though Enoch is here described as learning of matters deemed unknowable in earlier literature, it remains that these verses – as Nickelsburg notes – display many of the 'well-known trappings of biblical cosmology'; see *1 Enoch 1* (n. 31 above, pp. 284–5) for references to further biblical parallels.

[58] The description of this paradigmatically otherworldly space is, notably, punctuated by a notice emphasizing the totality and uniqueness of Enoch's eye-witness knowledge: 'And I Enoch alone saw the sights, the extremities of all things (Gr.$^{Pan.}$: τὰ πέρατα πάντων) and no one has seen what I saw' (*1 Enoch* 19:3).

[59] Compare the reference to the end-time disruption of orderly celestial cycles in *1 Enoch* 80 in the *Astronomical Book* (although possibly a secondary addition to that work; see J. C. VanderKam, *Enoch and the Growth of an Apocalyptic Tradition*, Washington DC, 1984, pp. 76–9).

[60] For an analysis of Enoch's movements in *1 Enoch* 21–32, see Stock-Hesketh, 'Circles and Mirrors' (n. 41 above).

(22:1ff) the selection and arrangement of sights is shaped by a concern for the judgment of the wicked and righteous among humankind, both directly after death and in the eschatological future. These chapters of the *Book of the Watchers* are well-known for including what George Nickelsburg has termed 'the earliest detailed treatment of the fate of the dead' in Jewish literature (esp. 22; 24:2–27:2), and begin with 'one of the two earliest extant Jewish testimonies to a belief in a post-resurrection judgment' (i.e. 22:1–4).[61] Here, Enoch visits a mountain in the Far West of the world with four 'hollows' (Gr.[Pan.] κοιλώματα; 22:1–3), which – he learns – are where the 'souls of all the sons of men' (נפש]ת כל בני אנשא) are gathered:

> ... the pits for the place of their confinement (פחתיא לבית עגנן)... until the day on which they will be judged... the time of the day of the end of the Great Judgment (זמן יום קצא ד[י] דינא רבא), which will be exacted upon them... (*I Enoch* 22:3–4; 4Q206 fr. 2 col. ii lines 1–3)

Three of the hollows are dark, and these are said to be reserved for different types of wicked souls. The souls of the righteous, by contrast, dwell in the single bright hollow, which is filled with light and watered by a fountain.[62] Whereas Enoch had responded to the sights of the places of the punishments of stars and angels with pure terror (21:8–9), his response here shifts to blessing (22:14),[63] as if in a moment of recognition that what he has seen attests the justice that pervades even the far edges of the cosmos.

To reach the next major destination[64] Enoch travels past a ceaseless fire to the west of the end of the earth (23:1–4). Beyond this are the seven massive mountains at the very edge of the world (24:2–3). Although all are impressive, his sights are said to fix on the largest among them – a throne-like mountain on which stands a glorious tree. When Enoch questions his angelic guide, he is told that the mountain 'is the seat ... where the Great Holy One ... will sit, when [He] descends to visit the earth in goodness' (25:3).[65] The tree is described in terms that evoke the Tree of Life in Genesis's account of Eden.[66]

[61] I.e. building on Isaiah 26:14–19 and possibly influencing Daniel 12:2; so Nickelsburg, *1 Enoch 1* (n. 31 above), p. 304. See further M.-T. Wacker, *Weltordnung und Gericht: Studien zu 1 Henoch 22*, Würzburg, 1982.

[62] Interestingly, Enoch is here said to hear the voice of Abel, calling out for the punishment of Cain (*I Enoch* 22:6–7), thus recalling the description in *I Enoch* 9 of the pleas of the earth and slain creatures travelling up to heaven to cry out for vengeance against the violence of the Giants.

[63] 4Q205 fr. 1, col. i lines 2–3: 'Blessed is the judgment of righteousness (ברוך דין קושט[א]) and blessed are you, O Lord of majesty and righteousness, who are Lord of eternity'.

[64] Throughout *I Enoch* 17–36 major destinations are marked by question-and-answer dialogue between Enoch and his accompanying *angelus interpres*. The first site to be so marked is the prison of the fallen angels (18:12–19:2); the rest are locations relevant to the judgment of humankind.

[65] I.e. thereby echoing the opening proclamation concerning God's descent upon Mount Sinai to judge the earth, on which see above.

[66] See, e.g., Stock-Hesketh, 'Circles and Mirrors' (n. 41 above), pp. 46–8. This resonance represents one of several thematic affinities with Genesis 2–3 – although *I Enoch* 24–25 is perhaps less concretely connected to this pentateuchal tale than is implied by Stock-Hesketh's parallel chart and comparative analysis (pp. 49–52, conflating traditions related to trees, garden, mountain, etc., in *I Enoch* 24–5 and 32); the text, moreover, gives no basis for his suggestion that the Tree of Wisdom is imagined to be transplanted, like the Tree of Life, to the centre of the earth at the end of time (p. 52).

The archangel Michael explains that no flesh can touch this tree until after the final Judgment, when it will be transplanted to the 'holy place, the house of God' (25:5). Imagery of the primeval Eden is thus intertwined with imagery of the Temple (cf. Ezekiel 47:7–12) to evoke an eschatological Eden as well.[67] The fruits of the Tree of Life, in fact, are here said to be saved for the eschatological future, when they will be eaten by the righteous, who will thus live long lives, without illness, as in primeval times (25:5–6).

Here too, the scribe responds with blessing (25:7), as he does in response to the next major destination on his tour (26:1–27:5). This is a mountain at the 'centre of the earth' (26:1), in a fertile land, with a dry valley stretching beneath it. It is also, as scholars have often noted, the only stop on the tour that bears features that clearly correspond to known geography: what is here described appears to be the topography surrounding what is – in Enoch's future and the reader's present – the city of Jerusalem.[68] It is made clear, moreover, that in the valley below it, the human wicked shall suffer their final fate (27:1–4; i.e. *Gehinnom*).

With Enoch's final major destination we come at last to the '*pardes* of righteousness' (4Q206 fr. 3 line 21: [א]פרדס קשט; Gr.[Pan.] παράδεισος τῆς δικαιοσύνης; Eth. *gannata ṣedq*). According to the *Book of the Watchers* he encounters this garden when travelling far eastwards,[69] past many ranges of mountains, and past deserts and places of spices (28:1–32:2).[70] When he sees the place from afar (32:3) he stops, as if transfixed. What attracts his gaze is a tree in its midst, which is 'very large and beautiful and glorious and magnificent' and from which 'the holy ones eat and learn great wisdom' (32:4; cf. Gr.[Pan]).[71] When Enoch inquires about this tree, his angelic guide answers as follows:

> This is the Tree of Wisdom (Gr.[Pan.] τὸ δένδρον τῆς φρονήσεως)[72] from which your father of old and your mother of old, who were before you, ate and learned wisdom; and their eyes were

[67] Himmelfarb, for instance, likens this transplantation of the Tree of Life to the use of Edenic imagery (e.g., plentiful trees, life-giving stream) in the description of the restored Temple in Ezekiel 40–48, which she interprets – following Levenson – as evoking an image of the Temple as the 'eschatological Eden at Zion' that even 'surpasses the Eden of the past' ('The Temple and the Garden of Eden in Ezekiel, the Book of the Watchers, and the Wisdom of ben Sira', in *Sacred Places and Profane Spaces*, eds J. Scott and P. Simpson-Housley, New York, 1991, pp. 63–80 [64–9, especially 65]). Compare Grelot, 'La Géographie mythique' (n. 36 above), pp. 41–4; and see discussion below.

[68] See further Milik, *Books of Enoch* (n. 34 above), 36–7; Nickelsburg, *1 Enoch 1* (n. 31 above), p. 318. For Jerusalem as the centre of the earth, see already Ezekiel 5:5; 38:12.

[69] I.e. consistent with the description of Eden in MT Genesis 2:8 as planted מקדם ('in the east' and/or 'in ancient times', the former interpretation of which is reflected in LXX κατὰ ἀνατολάς). For the '*pardes* of righteousness' of *I Enoch* 32 as located in the northeast, in particular, see Grelot, 'La Géographie mythique' (n. 36 above), pp. 62–3; Milik, *Books of Enoch* (n. 34 above), p. 37. In *Jubilees* (8:16) and *II Enoch* (42:3, 65:10) Eden is similarly located in the east. For other traditions about its location, see Kugel, *Traditions of the Bible* (n. 14 above), pp. 103–7, 139–40.

[70] On these chapters, see J. T. Milik, 'Hénoch au pays des aromates (ch. xxvii à xxxii): Fragments araméens de la grotte 4 de Qumran', *Revue Biblique*, 58, 1965, pp. 70–77. Milik's attempt to connect this journey to specific trade routes has been widely dismissed as implausible.

[71] On this reading, see Nickelsburg, *1 Enoch 1* (n. 31 above), p. 322; Tigchelaar, 'Eden' (n. 34 above), p. 42.

[72] Compare the connection of Wisdom (MT חכמה; LXX σοφία) to a 'tree of life' in Proverbs 3:18 (MT עץ החיים; LXX ξύλον ζωῆς; as in Genesis 2:9; 3:22, 24; cf. δένδρον ζωῆς in LXX Proverbs 11:30; 13:12; 15:4), as well as the description of Wisdom in Edenic terms in *Wisdom of ben Sira* 24. See discussion below.

> opened, and they knew that they were naked, and they were driven from the garden. (*I Enoch* 32:6)[73]

To this, Enoch is silent. Conspicuously absent are the blessings that punctuate his visits to other locales associated with the judgment of humankind (22:14; 25:7; 27:5). Instead, the text veers quickly to its conclusion (33–6), with a brief list of other sights that Enoch saw,[74] followed by a coda of praise for the God who made all that he was shown. Whether as an afterthought, or as the culmination of the tour, the '*pardes* of righteousness' is thus distinguished from the rest of the major sights that Enoch is said to have seen.

3. Paradise and Apocalyptic Epistemology

As noted above, the *Astronomical Book* and the *Book of the Watchers* are our oldest surviving examples of an extensive literary effort within Judaism to map the structure of the cosmos. In the *Book of the Watchers*, we also find gathered nearly all of the elements that would eventually constitute Western ideas about paradise. These include a concern for the location of the garden of Adam and Eve, the Tree of Knowledge/Wisdom, and the Tree of Life; a description of the architecture, contents, and inhabitants of God's heavenly abode as visited by a righteous visionary; an account of the places where the souls of the righteous reside, divided from the wicked after death; and the correlation of protology and eschatology, with the promise of a restored world in the far future, to which shall be returned the purity of primeval space and time.

But what – we might ask – is the function of the '*pardes* of righteousness' within the texts themselves? Above, we noted how the reference to this locale in the *Astronomical Book* likely serves to mark the border between the inhabited world and what lies beyond. More puzzling is the place of paradise within the *Book of the Watchers*. Among the major destinations on Enoch's tours, this site is the only one not granted any significance for the future. It has no bearing on the post-mortem fate of humankind or any function in end-time events. It is neither the present nor future location of the Tree of Life. Nor is it associated with the eventual restoration of the righteous. It may not even be imagined as an intended stop on Enoch's tour; after all, the scribe is said to have glimpsed it from afar during the course of his journeys, rather than having been led to it by the angels. In addition, the explanation of its significance is somewhat laconic, with surprisingly little attention lavished on the Tree of Wisdom,[75] and the structure of the passage about this place deviates from the common pattern in this section of the *Book of the Watchers*.[76] Furthermore, even though *I Enoch* 32 contains the earliest unequivocal reference to the

[73] Unfortunately, the Aramaic here (4Q206 fr. 4 lines 1–11) is very fragmentary.

[74] Many of these sights correspond to those described in greater detail in the *Astronomical Book*; see VanderKam, *Enoch and the Growth* (n. 59 above), pp. 114, 136.

[75] See further Himmelfarb, 'The Temple' (n. 67 above), p. 70; eadem, *Ascent* (n. 21 above), p. 74

[76] See further Tigchelaar, 'Eden' (n. 34 above), pp. 41–2.

story of Adam and Eve outside Genesis,[77] one finds little sense of the later prominence of Genesis 2–3 as a nexus for protological and eschatological speculation or as a focus for the mapping of the topographies of the ends of the earth.

Why, then, is this site included at all? In an influential 1958 article, Pierre Grelot posited an exegetical motive, namely, the desire to reconcile Genesis's description of Eden as a garden located in the east with Ezekiel's description of Eden on a mountain, perhaps also with reference to the northern 'mountain of God' in Isaiah 14. Grelot speculated about a mythic geography in which God dwelt in the mountain-paradise in the northwest, then planted an earthly replica of its garden in the northeast after forming humankind, to which He temporarily transported the Tree of Life; after the expulsion of Adam and Eden, the Tree was moved back to the mountain-paradise, to remain until the Eschaton, when another earthly replica will be established at the centre of the earth in Jerusalem. Grelot further posits the mountain-paradise as the Eden in which Enoch is said to dwell in *Jubilees* (see below).[78] In his view the *Book of the Watchers* thus reflects a cosmology with two paradises: [1] a northeastern garden associated with Adam and Eve, containing the Tree of Knowledge/Wisdom (i.e. in *I Enoch* 32), and [2] a northwestern site, the mountain-throne of God, containing the Tree of Life that will be transplanted to the centre of the earth at the end of time (i.e. in *I Enoch* 25).[79]

Inasmuch as later traditions commonly double Genesis's garden into a heavenly paradise and a terrestrial Eden, it is tempting to read such doubling already in the *Book of the Watchers*. More recently, however, Eibert Tigchelaar has questioned this reading, on the grounds of Grelot's selective use of traditions from early and late sources to reconstruct his Enochic world map; particularly misleading, in his view, is Grelot's conflation of the mythic geographies of third- and second-century BCE writings with that of the first-century BCE/CE *Parables of Enoch* (*I Enoch* 37–71).[80] When read on its own terms, by contrast, the *Book of the Watchers* reveals only a single paradise, and the place is granted relatively little significance. In fact, in Tigchelaar's estimation, the inclusion of this site is best understood in terms of an impulse towards comprehensiveness: the

[77] I should here stress 'unequivocal', inasmuch as scholars have certainly posited that earlier references to gardens, trees, rivers, etc., might encode references specifically to the Eden of Genesis 2–3. For a maximalist example, see Stordalen, *Echoes of Eden* (n. 6 above), pp. 305–456, on what he deems Eden 'metaphors', 'similes', and 'allegories'. Contrast Tigchelaar, 'Eden' (n. 34 above), pp. 37–8, 45–6, who stresses that '[t]he narrative of Adam and Eve in the Garden of Eden receives no attention in the Hebrew Bible outside of Gen 2–3' (p. 37) and cautions – even in the case of the *Book of the Watchers* – that we cannot simply assume that early examples of traditions featuring imagery or themes similar to those of Genesis 2–3 all reflect the influence of this particular passage, let alone its exegesis; some may attest other streams in the broader complex of traditions surrounding Eden, gardens, trees, rivers, etc., in ancient Israelite and Near Eastern literatures in relation to the themes of creation, life, and renewal.

[78] Grelot, 'La Géographie mythique'(n. 36 above), p. 43.

[79] Those who follow Grelot in this – in whole in part – include Milik, *Books of Enoch* (n. 34 above), p. 40; Himmelfarb, 'The Temple' (n. 67 above), p. 70; Stock-Hesketh, 'Circles and Mirrors' (n. 41 above), p. 47.

[80] Tigchelaar, 'Eden' (n. 34 above), pp. 39, 44–9. See also Coblentz Bautch, *A Study* (n. 30 above), pp. 207–20, who similarly concludes that 'it is inadvisable… to maintain that there is a second paradise or Eden in the Book of the Watchers' (p. 220).

garden was 'placed at the end of Enoch's journey to the east in order to complete the description of the East, and not because of some special importance of this location'.[81]

Martha Himmelfarb also notes the passing treatment of the garden of Adam and Eve in *I Enoch* 32 but offers a different explanation.[82] In the *Book of the Watchers*, as is well known, it is the descent of the Watchers before the Flood that serves as the primary explanation for earthly sin and suffering. Accordingly, it may not be coincidental that Eden is here integrated, even as its significance for salvation history is downplayed. For instance, *I Enoch* 32:6 includes none of the elements that permit Genesis 2–3 to be read as an account of the origins of evil. In addition, the *Book of the Watchers* severs any possible connection of the primeval garden with future restoration, by virtue of the association of post-mortem and eschatological reward – as well as the Tree of Life – with other locales. Himmelfarb proposes, therefore, that the authors/redactors of the *Book of the Watchers* may have been forced to include the story of Adam and Eve, due to its fame from Genesis, but did so in a manner that defused its significance, both for the past genealogy of error and for the promise of future redemption.[83]

What these various theories share is the assumption that the treatment of the '*pardes* of righteousness' in the *Book of the Watchers* must have been shaped primarily by concerns outside the text, whether rooted in biblical exegesis, common notions of eastern geography, or competing aetiologies of evil. What I would like to propose here, by contrast, is that Enoch's trip to this locale – the final major destination of the tour in the *Book of the Watchers* – serves an important literary function within the text and that it may, moreover, help us to grasp something of the poignantly transitional character of this very early cosmographical work. When one situates the passage about the '*pardes* of righteousness' within the apocalypses as a redacted whole, and in relation to its over-arching concerns, what is striking is the emphasis on wisdom. In *I Enoch* 32:6, Enoch is said to have been transfixed by precisely those features of the Tree of Wisdom/Knowledge by which Eve is said to be tempted in Genesis 3:6: its beauty and the promise of wisdom. He then notices that this wisdom is accessible only to the 'holy ones' (Gr. ἅγιοι, presumably rendering Aramaic קדישין),[84] the very highest class of the angels in the schema of the *Book of the Watchers*. Yet, even despite the text's repeated celebration of Enoch's angel-like status and his access to knowledge from the angels, he is not permitted entry. He does not cross into the bounds of Eden. Nor is he said to taste of the Tree. This is the one line that is drawn and maintained between Enoch and the angels, and – according

[81] Tigchelaar, 'Eden' (n. 34 above), p. 41.

[82] Himmelfarb, 'The Temple' (n. 67 above), pp. 70–1; see also Reed, *Fallen Angels* (n. 12 above), p. 51.

[83] Note, however, Himmelfarb's caution in raising this possibility. Indeed, although it is tempting to read the *Book of the Watchers* against the background of competing ideas about the 'origins of evil' in the sins of Adam and Eve, and to see its authors/redactors as reacting against positions similar to the interpretations of Genesis 2–3 attested so richly in later literature (especially during and after the 1st century CE), we have no evidence that such interpretations were current already in the Judaism of the 3rd century BCE.

[84] I.e. following the emendation of *I Enoch* 32:3 suggested by Nickelsburg, on which see above.

to the *Book of the Watchers* – it shall remain uncrossed by humankind even in the eschatological future.

The *Astronomical Book* and the *Book of the Watchers* would open the way for a flurry of Jewish (and, later, Christian) speculation about the topographies of realms outside the inhabited world – above, beyond, and below – and about the possibility that the righteous can experience those realms, whether in visionary journeys and/or in blessed life in eternal or eschatological spaces after death.[85] Eventually, some Jewish and Christian writers would let Enoch into Eden, and others would grant him all the knowledge and power of the angels. Many would also make a paradisiacal place for the righteous dead in heaven or at the ends of the earth.[86]

Later reflection on Enoch's post-ascent abode, in fact, appears to have developed in tandem with reflection on the location(s) of paradise. In most materials from the second century BCE, for instance, Enoch and Eden are both located on the horizontal plane, and thus situated within a geography similar to that of *Book of the Watchers*. In a unit about the birth of Noah affixed to the end of the *Epistle of Enoch*, for instance, Methusaleh is said to visit Enoch at 'the ends of the earth' (*1 Enoch* 106–7; cf. the Genesis Apocryphon from Qumran, 1QapGen 2:23), and in the *Book of the Giants*, the Giant Mahaway must cross 'the Great Desert' to reach him (4Q530 2 ii 5).[87] Similarly, *Jubilees* describes Enoch as 'writ[ing] down the judgment and condemnation and witness of humankind' in Eden where the Flood did not reach (4.23–6). At the same time, however, *Jubilees* enhances earlier connections between Eden, Temple, and heaven – not least by describing both Eden and heaven in terms that pointedly recall the Temple.[88] By the first century BCE, these various threads have become interwoven: the *Parables of Enoch* (*1 Enoch* 37–71) assert that Noah walks to the 'ends of the earth' to see Enoch (65:1–5) but also that Enoch, after his ascent, was 'set ... between two winds, between the north and the west, where the angels took

[85] See further Himmelfarb, *Tours of Hell* (n. 24 above); eadem, *Ascent* (n. 21 above); R. Bauckham, 'Visiting the Places of the Dead in the Extra-Canonical Apocalypses', in *The Fate of the Dead: Studies on the Jewish and Christian Apocalypses*, Leiden, 1998, pp. 81–96. On trajectories in 'ancient Jewish sciences' in the wake of these works, see Ben-Dov, *Head of All Years* (n. 28 above); idem, 'Scientific Writings in Aramaic and Hebrew at Qumran: Translation and Concealment', in *Aramaic Qumranica: The Aix-en-Provence Colloquium on the Aramaic Dead Sea Scrolls*, eds Katell Berthelot and Daniel Stökl ben Ezra, Leiden, 2010, pp. 379–402; Mladen Popović, 'The Emergence of Aramaic and Hebrew Scholarly Texts: Transmission and Translation of Alien Wisdom', in *The Dead Sea Scrolls: Transmission of Traditions and Production of Texts*, eds S. Metso, H. Najman, and E. Schuller, Leiden, 2010, pp. 81–114.

[86] See sources cited above as well as Wright, *An Early History* (n. 19 above), pp. 117–214; A. Scafi, *Mapping Paradise: A History of Heaven on Earth*, London and Chicago IL, 2006, pp. 32–43.

[87] L. Stuckenbruck, *The Book of the Giants from Qumran*, Tübingen, 1997, pp. 133–4; cf. Tigchelaar, 'Eden' (n. 34 above), p. 47.

[88] Gary Anderson, 'Celibacy or Consummation in the Garden? Reflections on Early Jewish and Christian Interpretations of the Garden of Eden', *Harvard Theological Review*, 82, 1989, pp. 121–48 (128–31); Martha Himmelfarb, 'Earthly sacrifice and Heavenly Incense: The Law of the Priesthood in *Aramaic Levi* and *Jubilees*', in *Heavenly Realms and Earthly Realities in Late Antique Religions* (n. 42 above), pp. 103–22. On the precedents, see nn. 16 and 67 above.

cords to measure for me the place of the chosen and the righteous ... and [where] I saw the first fathers and the righteous, who were dwelling in that place from of old' (70:3).[89] Likewise, in the following centuries, Enoch is increasingly moved into heaven. In the shorter recension of the *Testament of Abraham* (11:5), as in *II Enoch* and *Sefer Hekhalot/III Enoch*, he sits in heaven writing the deeds of the living.[90] Heaven, moreover, is no longer a locale limited to God and His angels. It begins to bustle with the souls of the dead, whether waiting to be judged or taking up their homes in paradise (e.g., *Testament of Abraham B* 8–12; *II Enoch* 7–10, cf. 42; *III Enoch* 41–48).

In the *Book of the Watchers*, by contrast, Enoch may see heaven, but it is still a place so otherworldly that his terror of it matches his terror of the places far beyond the edges of earth and sky. Likewise, he may see the '*pardes* of righteousness', but he remains outside its bounds, far from the fruits of wisdom by which he is so tempted. It is here – I suggest – that we might glimpse an enduring trace of the older reticence towards cosmological speculation that we noted at the very outset. In the *Book of the Watchers* the garden of Adam and Eve and Tree of Wisdom are featured as the final major sights of a tour in which Enoch might otherwise seem privy to all the information about the cosmos known to God and His angels. They stand, in effect, as a reminder that there are limits to human knowledge, even for the wisest and most righteous of scribes.

4. Conclusions

Traditionally, the development of 'mainstream' Judaism has been reconstructed primarily on the basis of the Hebrew Bible and classical rabbinic literature, while apocalypses and other so-called 'inter-testamental' texts have been placed in a trajectory leading to Jesus, the New Testament, and Christianity – and, at times, to 'mystical' varieties of Judaism. Among the results has been the enduring assumption that works like the *Astronomical Book* and the *Book of the Watchers* must represent a radical departure from the 'normative', 'this-worldly' Judaism exemplified by the Pentateuch and/or its later rabbinic interpreters.

At first sight, our inquiry might appear to confirm this impression. Inasmuch as the *Astronomical Book* and *Book of the Watchers* transgress the bounds of pentateuchal and sapiential epistemologies, these cosmographical writings do herald the rise of an 'other-worldly' perspective within Second Temple Judaism. Some tension with 'this-worldly' perspectives, moreover, is suggested by our evidence for their early reception. A number of scholars have proposed that a reaction against Enochic texts and traditions can be glimpsed already in the *Wisdom of ben Sira*. Writing in Hebrew text in the second-

[89] One wonders whether the 'ends of the earth' are imagined here – as in later midrashim – as a place where the dome of the firmament comes close enough so as almost to meet the ground. See above on early ideas about 'where the heavens come to an end'.

[90] Himmelfarb, *Ascent* (n. 21 above), pp. 29–46.

century BCE, ben Sira seems to reassert the values of the Pentateuch and the biblical Wisdom tradition in the face of the apocalyptic innovations of the early Hellenistic age.[91] In contrast to the cosmological concerns of the *Astronomical Book* and *Book of the Watchers*, for instance, he cautions against inquiry into what lies beyond the inhabited world:

> Seek not what is too difficult for you, nor investigate what is beyond your power. Reflect upon what has been assigned to you, for you do not need hidden things (Heb. נסתרות; Gr. τῶν κρυπτῶν). (*Wisdom of ben Sira* 3:21–2)[92]

Significantly, for our purposes, ben Sira's critique is coupled with a presentist twist on the promise of paradise: not only does he proclaim that 'the fear of the Lord is like blessed Eden' (Heb. כעדן ברכה; cf. Gr. ὡς παράδεισος εὐλογίας; 40:27), but he equates first-created Wisdom with the Pentateuch itself (24:23).[93] The result is a pointed confirmation of the capacity of humankind to gain knowledge, life, and blessing here on earth, within the bounds of the inhabited world, in historical time, and without any ancient visionary or otherworldly tourist as revelatory mediator.[94]

The contrast with the *Astronomical Book* and *Book of the Watchers* is striking, and it is indeed tempting to imagine the contrast as presaging a later conflict – whether between 'this-worldly' Judaism and 'other-worldly' Christianity,[95] or between 'mainstream' and

[91] See further R. Argall, *1 Enoch and Sirach: A Comparative Literary and Conceptual Analysis of the Themes of Revelation, Creation, and Judgment*, Atlanta GA, 1995, especially pp. 74–6, 250; B. G. Wright, 'Ben Sira and the *Book of the Watchers* on the Legitimate Priesthood', in *Intertextual Studies in Ben Sira and Tobit*, eds J. Corley and V. Skemp, Washington DC, 2005, pp. 241–54; idem, 'Wisdom, Instruction and Social Location in Ben Sira and 1 Enoch', in *Things Revealed* (n. 31 above), pp. 105–21; idem, 'Sirach and 1 Enoch: Some Further Considerations', in *The Origins of Enochic Judaism: Proceedings of the First Enoch Seminar*, ed. G. Boccaccini, Turin, 2002, pp. 179–87.

[92] See also *Wisdom of ben Sira* 20:30; 34:1–8. Lest we be tempted to draw too sharp a contrast with the *Book of the Watchers*, etc., it should be noted that ben Sira also describes the ideal scribe as one who 'studies the wisdom of all of the ancients and occupies himself with prophecies' (39:1–3); see sources cited in the previous note, as well as discussion below.

[93] Compare the use of the imagery of a 'tree of life' to describe wisdom, the fruit of righteousness, etc., already in the book of Proverbs (3:18; 11:30; 13:12; 15:4), and note also the description of the wisdom, instruction, etc., flowing from the Torah/Pentateuch as akin to the rivers Tigris, Euphrates, Pishon, and Gihon in *Wisdom of ben Sira* 24:22–8. See further Stordalen, *Echoes of Eden* (n. 6 above), pp. 372–6; Argall, *1 Enoch and Sirach* (n. 91 above), pp. 93–4.

[94] See further Himmelfarb, 'The Temple' (n. 67 above), pp. 72–5. Interestingly, what the two share is an association of the promise of paradise with Jerusalem and its Temple.

[95] Note, e.g., R. H. Charles's contrast between the alleged 'legalism' of Pharisaism and post-Christian Judaism, on the one hand, and the 'apocalyptic' and ethical concerns of 'Old Testament pseudepigrapha' and Christianity, on the other, in *The Apocrypha and Pseudepigrapha of the Old Testament*, Oxford, 1913, II, pp. viii–xi. One recent iteration of this theme is the proposed distinction between 'Mosaic'/'Zadokite' and 'Enochic' Judaisms in Second Temple times, whereby the former is linked to rabbinic Judaism and the latter to Christianity; see, e.g., G. Boccaccini, *Roots of Rabbinic Judaism*, Grand Rapids MI, 2002, especially pp. 89–102, 134–50. Cf. A. Y. Reed, 'Interrogating "Enochic Judaism": 1 Enoch as Evidence for Intellectual History, Social Reality, and Literary Tradition', in *Enoch and Qumran Origins: New Light on a Forgotten Connection*, ed. G. Boccaccini, Grand Rapids MI, 2005, pp. 336–44.

'mystical' ideologies within Judaism itself.[96] After all, some sectors of Judaism saw a flowering of speculation into the cosmos, otherworldly spirits, the afterlife, the end-times, etc., in the wake of the *Astronomical Book* and *Book of the Watchers*, but their literary products would be preserved and extended primarily by Christians. Late antique rabbis seem to have rejected such works, even as they embraced the words of ben Sira.[97]

Are the beginnings of Jewish cosmography, then, marginal to the history of Jewish thought, and meaningful mainly for our understanding of Christianity? To be sure, ben Sira couched his critiques in explicit claims to traditionalism, as did rabbinic sages after him, and their claims pivoted on their elevation of the Pentateuch: these self-claimed Torah-experts granted such scriptures an epistemological monopoly, to which all knowledge from Nature and 'sciences' was theoretically subordinated.[98] Much like the *Wisdom of ben Sira*, moreover, the classical rabbinic literature features pragmatic concerns that we might term 'this-worldly' (e.g. ethics, halakhah), alongside sayings critiquing those who dwell on the unknown and otherworldly. *Mishnah Ḥagigah* 2.1, for instance, attests tannaitic efforts to curtail the exposition of *ma'aseh bereshit* ('work of creation') and to condemn speculation into 'what is above and what is below, what is before and what is after' (i.e. heaven, netherworld, pre-creation, eschatology).[99] And, just as ben Sira's famous critique (3:21) would be quoted favourably by amoraim commenting on this very mishnah (*y. Ḥagigah* 2.1/77c; *b. Ḥagigah* 13a), so Deuteronomy's assertion that 'it is not in heaven' (30:12a; see above) would be marshalled for the articulation of a rabbinic epistemology that privileges consensus wrought from learned discourse on earth, over

[96] See, e.g., Gruenwald, *Apocalyptic* (n. 51 above), especially pp. 17–18; N. Séd, *La Mystique cosmologique juive*, Paris, 1981. For an extension of the supposed rivalry between 'Mosaic' and 'Enochic' perspectives (see above) to post-70 Judaism, see Alexander, 'Enoch and the Beginnings' (n. 24 above), pp. 233–4; idem, 'From Son of Adam to Second God: Transformations of the Biblical Enoch', in *Biblical Figures Outside the Bible*, eds M. E. Stone and T. A. Bergren, Harrisburg PA, 1998, pp. 87–122. A similar contrast, albeit with a different reading of the *Wisdom of ben Sira*, seems to lie at the heart of Rachel Elior's construction of a unified world-view of a Jewish 'successionist priesthood' from Enochic, Qumranic, Hekhalot, etc., sources, as posited to stand in opposition to the 'mainstream' perspectives defended by what she deems the Jewish 'leadership' (e.g., Hasmonean, Pharisaic, rabbinic); for the contrasting positions on ouranology, heavenly secrets, cosmic cycles, natural phenomena, angels, etc., that she credits to them, see *The Three Temples: On the Emergence of Jewish Mysticism*, transl. D. Louvish, Portland OR, 2004, especially pp. 201–22. Cf. Schäfer, *The Origins* (n. 51 above), especially pp. 9–17; M. Himmelfarb, 'Merkavah Mysticism since Scholem: Rachel Elior's *The Three Temples*', in *Wege mystischer Gotteserfahrung: Judentum, Christentum und Islam*, eds P. Schäfer and E. Müller-Luckner, Munich, 2006, pp. 19–36.

[97] On the fates of Enochic and other 'outside books' in post-70 Judaism, see references and discussion in Reed, *Fallen Angels* (n. 12 above), pp. 136–47, 206–17, 233–72. Contrast the rich *Nachleben* on these materials in early Christianity, for which see also my essay, 'Beyond Revealed Wisdom and Apocalyptic Epistemology: Early Christian Transformations of Enochic Traditions about Knowledge', in *Early Christian Literature and Intertextuality*, eds C. A. Evans and H. D. Zacharias, London, 2009, I, pp. 138–64.

[98] See further P. Schäfer, 'Bereshit bara Elohim: Bereshit Rabba, Parasha 1, Reconsidered', in *Empsychoi Logoi: Religious Innovations in Antiquity: Studies in Honour of Pieter Willem van der Horst*, eds A. Houtman, A. de Jong, and M. Misset-van de Weg, Leiden, 2008, pp. 267–89.

[99] On this mishnah and the tradition surrounding it, see Schäfer, *The Origins* (n. 51 above), pp. 180–85, 207–10, 233–4; D. J. Halperin, *The Merkabah in Rabbinic Literature*, New Haven CT, 1980, pp. 19–63; A. Y. Reed, 'Was there Science in Ancient Judaism? Historical and Cross-Cultural Perspectives on Religion and Science', *Studies in Religion/Sciences Religieuses*, 36, 2007, pp. 461–96 (476–82).

claims of revelation or charismatic power from heaven (esp. *b. Bava Metzia* 59b; cf. Romans 10:6–8).

Closer analysis, however, suggests that these positions may lie in a continuum more variegated than any simple dichotomy suffices to convey. Assumptions about Judaism's 'this-worldly' orientation may have led scholars to highlight those portions of the classical rabbinic literature which focus on legal and ethical concerns and/or those which ignore or curtail cosmological and eschatological speculation.[100] Yet it remains that this rich and massive corpus also includes traditions expressing an interest in cosmic cycles for calendrical and 'scientific' aims, in a manner not unlike the *Astronomical Book*, as well as traditions expressing an interest in mythic geography and the correlation of *Urzeit* and *Endzeit*, in a manner not unlike the *Book of the Watchers*.[101] And, just as the authors/redactors of these early cosmographies enthusiastically broached the limitations of human knowledge asserted in older works such as Deuteronomy and Job, so the constraints on cosmogony and cosmology in the Mishnah were tested and transgressed by the rabbis responsible for works such as *Genesis Rabbah* (esp. 1–8), the Babylonian Talmud (*b. Ḥagigah* 12b–13a), and *Pirke de-Rabbi Eliezer* (esp. 3–11).[102] In addition, paradise continues to emblematize the crossing between known and unknown, in all its danger and power: much as this place marked the border of the visible world in the *Astronomical Book* and the limits of human knowledge in the *Book of the Watchers*, so it serves as a nexus for exploring the profits and perils of secret knowledge and visionary experience in tales about the 'four who entered *pardes*' in the Tosefta and Talmudim (*t. Ḥagigah* 2.3–4; *y. Ḥagigah* 2.1/77b; *b. Ḥagigah* 14b–15b).[103] As with parallel Christian debates about Nature, Revelation, Science, and Scripture, rabbinic Jewish discussions may thus reflect something of the generative tension between the cosmographical

[100] Levenson, for instance, has noted how assumptions about the 'this-worldly' focus of rabbinic Judaism have been fostered by the traditional focus on halakhic materials. He proposes that midrashim, by contrast, reveal vibrant 'mythic' concerns as well – including creative developments of ancient Near Eastern traditions in continuity with apocalyptic literature, Dead Sea Scrolls, and other Second Temple traditions, no less than with later Kabbalah (*Creation* [n. 5 above], pp. 33–4). See also Fishbane, *Biblical Myth* (n. 8 above), pp. 1–23, on traditional assumptions about the incompatibility of monotheism and 'myth', as they have shaped notions of Judaism more broadly.

[101] On the former, see S. Fraade, *From Tradition to Commentary: Torah and Its Interpretation in the Midrash Sifre to Deuteronomy*, Albany NY, 1991, pp. 153–62; and sources cited in n. 45 above. On the latter, see Levenson, *Creation* (n. 5 above), pp. 32–46; Fishbane, *Biblical Myth* (n. 8 above), 95–252.

[102] See further Schäfer, 'Bereshit bara Elohim' (n. 98 above); idem, 'From Cosmology to Theology: The Rabbinic Appropriation of Apocalyptic Cosmology', in *Creation and Re-Creation in Jewish Thought: Festschrift in Honor of Joseph Dan on the Occasion of his Seventieth Birthday*, eds R. Elior and P. Schäfer, Tübingen, 2005, pp. 39–58; P. S. Alexander, 'Pre-Emptive Exegesis: Genesis Rabba's Reading of the Story of Creation', *Journal of Jewish Studies*, 43, 1992, pp. 230–45; A. Y. Reed, 'Who Can Recount the Mighty Acts of the Lord? Cosmology and Authority in *Pirqei de Rabbi Eliezer* 1–3', *Hebrew Union College Annual,* 80, 2009, pp. 115–41. Note also the para-rabbinic sources discussed in Séd, *La Mystique cosmologique juive* (n. 96 above); P. Schäfer, 'In Heaven as it is in Hell: The Cosmology of *Seder Rabbah de-Bereshit*', in *Heavenly Realms and Earthly Realities* (n. 88 above), pp. 234–74.

[103] I.e., with *pardes* initially serving as a metaphor for esoteric wisdom and only later reinterpreted as metonym for ascent to heaven in the Babylonian Talmud and parallels in the Hekhalot literature, as shown by Schäfer, *The Origins* (n. 51 above), pp. 196–206, 231–8.

reticence of the Pentateuch and the cosmological curiosity resultant from its own claims that the God of Israel is also the Creator.

We have seen how the creative power of this point of tension can be glimpsed already at the beginnings of Jewish cosmography – both within and between early writings. Not only does the *Book of the Watchers* express ambivalence towards celestial wisdom, but its elevation of Enoch's expertise is tempered by his exclusion from Eden and the wisdom of the fruits of its famed Tree. Accordingly, if ben Sira does indeed critique those who cultivate Enochic traditions, his concern might be best understood as part of the same conversation about how best to negotiate the proper scope and sources of Jewish knowledge in the early Hellenistic age.[104]

It is in this context – I suggest – that we might best situate the beginnings of Jewish cosmography. Whatever their points of contrast, the *Astronomical Book*, *Book of the Watchers*, and *Wisdom of ben Sira* express many of the same priestly, pedagogical, and sapiential concerns, and they bear marks of having emerged from similar scribal settings.[105] Taken together these works may thus bespeak a culturally fertile moment and milieu in pre-Maccabean Palestine, marked by competing scribal pedagogies in conversation concerning the practices of learning and writing about the cosmos.[106] The intensification of such debates may reflect the growing prestige of the scribe and the valorization of writing in the early Hellenistic age, as extending ideas in prophetic works like Ezekiel and Jeremiah, and as transforming the understanding of knowledge, revelation, and 'authorship' throughout the Second Temple period.[107] Such shifts may also owe something to the emergence of new patterns of patronage, particularly with the further expansion of imperial, administrative, and mercantile demands for literary skills under Ptolemaic and Seleucidic rule. In any case, it remains that works like the *Astronomical Book*, *Book of the Watchers*, and *Wisdom of ben Sira* attest a surprisingly sharp self-consciousness about scribalism and the power of writing. They share, moreover, a concern to re-map the traditional scope of scribal expertise and to extend the authority of Jewish writing into new domains – astral 'science' in the case of the *Astronomical Book*, mythic geography and post-mortem realities in the case of the *Book of the Watchers*, the training

[104] See now B. Wright, '1 Enoch and Ben Sira: Wisdom and Apocalyptic in Relationship', in *The Early Enoch Literature* (n. 31 above), pp. 159–76.

[105] See further J. Z. Smith, 'Wisdom and Apocalyptic', in *Visionaries and their Apocalypses* (n. 33 above), pp. 101–20; P. Davies, 'The Social World of Apocalyptic Writings', in *The World of Ancient Israel,* ed. R. Clements, Cambridge, 1989, pp. 251–71; J. J. Collins, 'Wisdom, Apocalyptic, and Generic Compatibility', in *In Search of Wisdom* (n. 21 above), pp. 165–86.

[106] See further Reed, *Fallen Angels* (n. 12 above), pp. 58–83; Himmelfarb, *A Kingdom of Priests* (n. 31 above), pp. 42–5. To be sure, similar epistemological concerns are present already in earlier sapiential and prophetic traditions; they appear to have become more pressing, however, during the periods of Persian and Hellenistic rule, concurrent with shifts in the status of scribes within Judaean society.

[107] See further Himmelfarb, *Ascent* (n. 21 above), pp. 95–114; E. F. David, *Swallowing the Scroll: Textuality and the Dynamics of Discourse in Ezekiel's Prophecy*, Sheffield, 1989; J. Wyrick, *The Ascension of Authorship: Attribution and Canon Formation in Jewish, Hellenistic, and Christian Traditions*, Cambridge MA, 2004.

of secular aristocracy in the case of the *Wisdom of ben Sira*, and in all three, the location of paradise and its Tree(s).

If so, then the catalysts for Jewish cosmography cannot be reduced simply to external 'influence'. But neither can we extricate 'inner-scribal' and 'inner-Jewish' developments from their broader, cross-cultural contexts. The elevation of the pre-exilic past and the development of new models of Jewish pedagogy, for instance, may have been spurred by encounters with 'classical' Greek wisdom and Hellenistic education, in a manner parallel to (and perhaps connected with) the discourse about the cosmos and the past among Egyptians and Babylonians in the Ptolemaic and Seleucidic Empires.[108] This shared discourse, in turn, seems to have contributed to the emergence of a new, cross-cultural *koine* among native elite in the Hellenistic Near East. It is perhaps not surprising, then, that the geography of the *Astronomical Book* and *Book of the Watchers* interweaves elements from ancient Babylonian and Greek traditions – following no one model but rather appearing to reflect common cosmological ideas.[109]

New claims concerning Jewish scribal expertise, however, may have also formed part of a defence of the ancient literary heritage of Israel, answering the challenges posed by the intellectual prestige of ancient Greek culture and the cosmopolitan 'scientific' discourse forged by learned elite in the Hellenistic Near East. In this regard it is notable that the astronomy of the *Astronomical Book* reflects none of the Hellenistic innovations of its time, drawing instead on ancient Babylonian precedents.[110] Like 'Pseudo-Eupolemus's soon after them,[111] the authors/redactors of the *Astronomical Book* and the *Book of the Watchers* seem to appeal to the antediluvian Enoch to argue that Jewish learning encompassed the same questions that were being asked and answered by the wise men, philosophers, and astronomers among their neighbours – albeit with a claim to

108 I.e. far from simply adopting or rejecting Greek ideas, some Jewish scribes may have been spurred to articulate their own pedagogies in response – and, significantly for our purposes, to do so in part by recovering ancient Near Eastern mythic and 'scientific' traditions. See further Carr, *Writing* (n. 27 above), pp. 202–12, on ben Sira and early Enochic apocalypses – and, more broadly, pp. 177–272, on Jewish sources of the period in the context of a broader pattern in the Hellenistic Near East, marked by the creative hybridization of native and Greek models of textuality, memory, authority, etc., and the emergence of 'Hellenistic-style anti-Hellenistic' pedagogical practices.

109 See Nickelsburg, *1 Enoch 1* (n. 31 above), p. 280, and discussion and references in nn. 36 and 53 above.

110 See further Albani, *Astronomie* (n. 29 above); Neugebauer, 'The Astronomical Chapters' (n. 29 above), pp. 386–414; Reed, 'Was there Science?' (n. 99 above), pp. 469–70; M. E. Stone, 'Enoch, Aramaic Levi and Sectarian Origins', *Journal for the Study of Judaism*, 19, 1988, pp. 159–70 (164–8); cf. Alexander, 'Enoch and the Beginnings' (n. 24 above), pp. 231–6.

111 See Eusebius, *Praeparatio evangelica* 9.17.3–9, preserving Alexander Polyhistor's account of a (Jewish? Samaritan?) source that claims that Abraham 'discovered astronomy/astrology and the Chaldean [art]' (ἀστρολογίον καὶ Χαλδαικήν [sc. τέχνην] εὑρεῖν; 9.17.3), taught the Phoenicians 'the movements of the sun and moon' (9.17.5), and introduced the Egyptians 'to astronomy/astrology and the rest' – stressing, moreover, that the 'original discovery' should be credited to Enoch and that 'Enoch, not the Egyptians, discovered astrology first' (9.17.8). On the 'competitive historiography' of this and other Greek Jewish authors concerning Israel's possession of astral wisdom prior to the Greeks, see my discussion in 'Abraham as Chaldean Scientist and Father of the Jews: Josephus, *Antiquities of the Jews*, 1.154–168, and the Greco-Roman Discourse about Astronomy/Astrology', *Journal for the Study of Judaism*, 35, 2, 2004, pp. 119–58 (136–42).

extreme antiquity far greater than that of the Greeks; a claim to an unbroken tradition of written transmission rivaling even those of the Babylonians and Egyptians; and a claim to a unique source in angelic revelation, heavenly books, and eye-witness accounts of otherworldly realms.

Our inquiry into cosmography thus points us to the transitional character of the *Astronomical Book* and *Book of the Watchers*, as products of the early Hellenistic age that laid the groundwork for later developments in Judaism, Christianity, and Western culture. At the same time, however, our consideration of paradise has served to remind us of the poignant continuities that connect these traditions with the Hebrew Bible and its ancient Israelite roots. The earliest extant examples of Jewish cosmography certainly transgress the bounds of older views about human knowledge and the proper focus of human concern. But they maintain the same ethical aim and eschatological horizon. In the *Astronomical Book*, as we have seen, cosmography is marshalled towards a task familiar from psalmic, sapiential, and other early biblical traditions – namely, the celebration of Israel's God as the same deity who created and rules the cosmos. What Enoch here sees and transmits, in fact, could not be further from the types of knowledge about the sun, moon, and the stars that might encourage someone to 'bow down to them, and serve them' (cf. Deutoronomy 4:19); the celestial bodies, rather, are revealed to be part of an orderly and systematic heavenly hierarchy, guided by angels, who are overseen by leaders, who in turn are all governed by the God of Israel and by the laws specified for them. Similarly, in the *Book of the Watchers*, Enoch may bring back knowledge from 'in heaven' and 'from across the sea' (cf. Deutoronomy 30:12–13), but the ramifications of this knowledge remain surprisingly similar to the ramifications of the narratives and commandments in Deuteronomy and the rest of the Pentateuch. What Enoch learns of the unseen spaces of heaven and earth speaks to the necessity of obedience to God's law – albeit not just for humankind, but also for winds, stars, and angels. Furthermore, when he visits the edges of the earth, he learns that it is the Land of Israel that is the place that proves most significant for the fate of humankind – albeit not just only for Jews. The Land is lauded as the place at the centre of the entire earth, prepared for the punishment of the wicked and the reward of the righteous alike. It is presented as the only place, moreover, in which the righteous can hope to experience some measure of a return to paradise, eating at the end of time from the Tree of Life that once rested in the primeval garden.

Gnostic Paradises

Einar Thomassen

Heirs to the traditions of the Bible, however much against the grain, the so-called Gnostics[1] also adopted the idea of paradise. Naturally, they reinterpreted it, as they did with so many other biblical ideas. Since the god who designed the Garden of Eden as a dwelling-place for the human being is not the ultimate god, but is regarded by Gnostics as an inferior (and often decidedly mean-spirited) being, paradise, too, is frequently given a negative interpretation in gnostic texts. Paradise may, for example, be interpreted as a trap designed to lead the human being astray and to keep him in the grip of the world ruler's power. That notion is well attested in gnostic texts, but several other interpretations of paradise can be found in them as well. The purpose of this contribution is to show some of their variety, thereby filling an omission in previous treatments of the subject.[2]

Paradises can be located on several levels in the gnostic ontological hierarchy. Most frequently, paradise is a theme found within accounts of the creation of the world and of the first earthly human, with explicit reference to the narrative of the book of Genesis. These accounts – of a paradise situated somewhere in the cosmos and which are surprisingly diverse – will provide the main material for this paper. At the end, however, we shall see that the theme of paradise is also sometimes used to describe realities which lie beyond the cosmos and which have an eschatological and soteriological significance.

Abbreviations used in this essay:
NHC = Nag Hammadi Codices.
BG = Berlin Gnostic Codex.

[1] Although the term 'gnostic' has become controversial in recent years, it will be used here as a shorthand reference to that general area of groups and texts in Antiquity which relates to the Judaeo-Christian tradition while professing a theological dualism, and holding that salvation is attained by knowledge about the origins of the inner self in a supramundane realm. The fact that the edges of this general area are fuzzy need worry only those with an essentialist approach to historical categories.

[2] See in particular W. C. van Unnik, 'Der Neid in der Paradiesgeschichte nach einigen gnostischen Texten', in *Essays on the Nag Hammadi Texts in Honour of Alexander Böhlig*, ed. M. Krause, Leiden, 1972, pp. 120–32; P. S. Alexander, 'The Fall into Knowledge: The Garden of Eden/Paradise in Gnostic Literature', in *A Walk in the Garden*, eds P. Morris and D. Sawyer, Sheffield, 1992, pp. 91–104; J. Magne, *From Christianity to Gnosis and From Gnosis to Christianity: An Itinerary through the Texts to and from the Tree of Paradise*, Atlanta GA, 1993, especially pp. 59–103 (see n. 5 below); G. Lettieri, 'The Ambiguity of Eden and the Enigma of Adam', in *The Earthly Paradise: The Garden of Eden from Antiquity to Modernity*, ed. F. R. Psaki, Binghampton NY, 2002, pp. 23–54, especially 39–54. G. P. Luttikhuizen, 'A Resistant Interpretation of the Paradise Narrative', in *Paradise Interpreted*, ed. idem, Leiden, 1999, pp. 140–52, deals primarily with the Nag Hammadi text *The Testimony of Truth*; T. Rasimus, *Paradise Reconsidered in Gnostic Mythmaking: Rethinking Sethianism in Light of the Ophite Evidence*, Leiden, 2009, focuses on the figure of the serpent.

The Cosmography of Paradise: The Other World from Ancient Mesopotamia to Medieval Europe, Warburg Institute Colloquia 27, 2016

The Cosmic Paradise: Paradise as a Trap

It is appropriate to begin with what is probably the most representative of all preserved gnostic texts, the *Apocryphon of John*.[3] This tractate, generally considered as promoting 'Sethian' gnosticism, presents a very negative view of the creator god and world ruler Yaldabaoth, as well as of the cosmos itself over which he presides together with his subordinate planetary archons. Yaldabaoth was conceived and brought forth by the aeon Sophia in her misguided attempt to produce offspring all by herself, without the collaboration of her partner aeon or the assent of the supreme deity, the Invisible Spirit. As a result, she bore a monster. Yaldabaoth inherited, however, a portion of his mother's spiritual nature, and the main theme of the story as it subsequently unfolds thereby becomes the retrieval of this spiritual element and how it can be restored to the Fullness. Humankind is the battleground where the struggle over the spirit is fought out. Yaldabaoth and his partners first create the cosmos for themselves. The transcendent deity then decides to manifest himself, and the archons see his image reflected from above in the cosmic waters. Using their own soul-stuff as clay, they mould a human being, Adam, after the image. Their creature is unable to move, however, until the higher powers put into the mind of Yaldabaoth the idea of breathing into it the spirit obtained from his mother Sophia. As a result, not only does Adam begin to move, but the rulers perceive that he has become superior to themselves. In reaction, they make for him a material body which will serve to confine him.

It is at this point, in direct connection with the creation of Adam's material body, that paradise is introduced into the story. Adam is placed in the garden and invited to eat from its trees, in order that he may exercise his new corporeal faculties:

> The chief archon took and placed him in paradise, telling him to enjoy its delights. This was, however, in order to deceive him, because the food of those delights is bitter and their beauty perverse. The food is a trap and the trees a sacrilege: their fruit is deadly poison and their promise is death.[4]

As constructed in the *Apocryphon*, then, paradise represents all the pleasures of the body. These pleasures are deceptive, however, because the soul which gives itself over to them must by necessity share the fate of the body itself: corruption and death.

The Tree of Knowledge, too, is given a fresh interpretation as part of this exegetical reversal of the biblical narrative. If the other trees of the garden – those from which the rulers want Adam to eat – are harmful, then the Tree of Knowledge, which they forbid

[3] Three different copies of the *Apocryphon of John* are found among the NHC; another copy (of the short recension) is part of the BG; and an early version of the text is reported by Irenaeus in his *Adversus Haereses*, 1.29. For an accessible, recent study of the text, see K. L. King, *The Secret Revelation of John*, Cambridge MA, 2006. In this article, except where stated otherwise, I have used the translations in *The Nag Hammadi Scriptures*, ed. M. Meyer, New York, 2006, modifying them when necessary. Another excellent resource is *Écrits gnostiques*, eds J.-P. Mahé and P.-H. Poirier, Paris, 2007.

[4] BG, pp. 55–6; cf. NHC II, p. 21. In the NHC II version the archons collectively are the agents who place Adam in paradise.

him to touch, must be beneficial. Just as the rulers intend to keep Adam inside the prison of the body by distracting him with sensual pleasures, they also intend to prevent him attaining the knowledge which would reveal to him their deceptive schemes. Moreover, the one who encouraged Adam to eat from the Tree of Knowledge, the *Apocryphon* tells us, was not the serpent but Christ himself, in the form of an eagle.[5]

What, then, are we to make of the Tree of Knowledge? As the story develops, the allegorical nature of the exegesis becomes particularly evident. The Tree of Knowledge is not, in fact, a tree at all. It represents an interior capacity for attaining knowledge, a mental faculty called the '*epinoia* of light', or 'enlightened insight', which the highest deity has given to Adam and which acts as his 'helper'. This faculty is also called life, *zoē*, which suggests that it is identified with the figure of Eve (cf. Genesis 2:18, 3:20).[6]

It is not necessary here to dwell on all the further details of the *Apocryphon*'s version of the paradise narrative. Suffice it to say that Yaldabaoth next creates Eve, the first human female, in an attempt to take away Adam's *epinoia,* insight (rather than his rib[7]) after sinking him in oblivious sleep. This physical Eve, however, retains the insight from which she has been made, and when Adam awakes and sees the female next to him, he recognizes that part of himself in her, and becomes aware of what had been deposited within him.[8] They both then eat from the 'tree', thus activating their latent insight, and acquire knowledge. Frustrated, Yaldabaoth throws them out of paradise and cloaks them in 'thick darkness'.[9]

Remarkable in this rewriting of the paradise story is not only the consistent reversal of the values of the original story, but also the extent of its allegorization. The *Apocryphon* has no interest in paradise as a place; it is not located in a cosmography. Rather, paradise and its trees represent the temptations which the cosmic rulers engineer in order to induce the human being – all human beings, in fact, not just Adam as an individual[10] – to forget

[5] Thus NHC II, p. 23: 'As for me, I appeared in the form of an eagle upon the Tree of Knowledge, which is the Insight of the pure enlightened Forethought [i.e. it is sent by the personified providence of the supreme deity], that I might teach the human beings and awaken them from the depth of sleep.' The short recension of BG and NHC III says that Insight herself (see below) was the eagle encouraging the humans to eat. The identification of Insight with Christ in the longer recension is evidently a secondary addition. This situation creates problems for Magne's reconstruction (as in n. 2 above) of the origins of gnostic mythology in the identification of the serpent of paradise with Jesus the Instructor, based on the reversal of the Genesis narrative which Magne claims underlies the Emmaus episode in Luke 24. See Rasimus's general assessment (n. 2 above), p. 65, n. 2, of Magne's work. As Rasimus notes, 'The serpent of *paradise* is not usually depicted as a manifestation of Jesus in Gnostic texts.'

[6] BG, pp. 52–4; NHC II, p. 20.

[7] The first ruler 'put the part he had taken from the power of the human being into the female creature. It did not happen, however, the way Moses said: "Adam's rib"' (NHC II, pp. 22–3).

[8] 'Adam saw the woman beside him. At once enlightened insight appeared and removed the veil that covered his mind. He sobered up from the drunkenness of darkness. He recognised his counterpart and said, "This is now bone from my bones and flesh from my flesh"' (NHC II, p. 23).

[9] BG, pp. 61–3; NHC II, pp. 23–4. Cf. Genesis 3:21–4.

[10] The story of the first pair of humans is generalised in the following passage: 'For this reason a man will leave his father and his mother [i.e. the archons who made him] and will join himself to his wife [i.e. insight], and the two of them will become one flesh. For his partner will be sent to him, and he will leave his father and mother' (NHC II, p. 23).

about the spirit they have inside them and succumb to the passions of the irrational soul created by the rulers. The trees *are* these passions, just as the Tree of Knowledge is nothing other than the capacity for knowledge which is hidden within humans and which Christ teaches them to discover.[11]

Paradise as a Lesson

A different interpretation of the story is offered by the Valentinian text from Nag Hammadi known as the *Tripartite Tractate.* Unlike the Sethian Gnostics of the *Apocryphon of John,* the Valentinians did not consider the creator and ruler of the physical world to be evil; rather, though unbeknown to himself, he is a servant in a plan of salvation which has ultimately been designed by the Father of the All. This perspective has implications for the Valentinian interpretation of paradise, which is created by the demiurge. In its retelling of the anthropogenic narrative, the *Tripartite Tractate* describes how the human protoplast was made by the demiurge and his archons as a mixture of matter and soul. Spirit was then secretly infused into Adam by the higher powers. The tractate next explains paradise:

> … it is also said that a paradise was planted for him, that he might eat from the fruit of three kinds of trees; it was a garden of the threefold order, and one that provided pleasures. The noble and distinctive substance in him was much superior to <the> creation, and <was> a blow to them. For that reason, they issued a commandment, threatening him, and they brought over him great danger: death. Only the food of the bad trees did he allow him to eat and enjoy, whereas from that tree which had the double character [he] was not allowed to eat – and much less from that of life – in order that he might not acquire an honour [equal to] themselves …[12]

The three kinds of trees in this version of paradise correspond to the anthropological tripartition into matter, soul and spirit. The ordinary, bad trees belong to the material sphere; as in the *Apocryphon* they represent the corrupting 'delights' of the garden which are best avoided. The tree with the double character represents soul and its inclination to both good and evil, characteristic of the intermediate position of the soul (a Platonic idea which was appropriated by the Valentinians).[13] The psychic nature of this, the Tree of Knowledge, also follows from its association for the Valentinians with commandment and the Law, characteristics of the god of the Old Testament.[14] The Tree of Knowledge

[11] Among the Nag Hammadi texts the *Hypostasis of the Archons* contains an account (NHC II, pp. 88–91) which resembles that of the *Apocryphon of John*, except that it is less interested in paradise as such, and focuses on the prohibition of eating from the Tree of Knowledge; see I. S. Gilhus, *The Nature of the Archons*, Wiesbaden, 1985, especially pp. 52–72; and Alexander, 'The Fall into Knowledge' (n. 2 above). The *Testimony of Truth* also comments on the paradise narrative (NHC IX, pp. 45–8), but only in order to demonstrate that the god of the Law is a 'malicious grudger'. For a further similar account, see Irenaeus, *Adversus Haereses*, 1.30.6–8.

[12] NHC I, pp. 106–7. I quote my translation in *The Nag Hammadi Scriptures*.

[13] Cf. *Tripartite Tractate* NHC I, p. 106; Irenaeus, *Adversus Haereses,* 1.6.1, etc.

[14] This is especially clearly stated in the Gospel of Philip: 'That tree was the law. It can give knowledge of good and evil, but it neither freed Adam from evil nor made him good, and it brought death to those who ate of it. For when it was said, "Eat this and do not eat that", death began' (NHC II, p. 74).

is thus given a quite different interpretation from that found in the *Apocryphon of John* and other Sethian texts, where the knowledge given by the tree is understood as salvific *gnosis*. Instead, the Tree of Life, which in the *Apocryphon* forms part of the archons's deception,[15] is given the highest value in the *Tripartite Tractate*, being identified with the spirit. The valuation of the trees of knowledge and life is thus reversed by the Valentinians *vis-à-vis* the *Apocryphon*.

The archons's jealousy over Adam's spiritual and superior nature is a traditional gnostic motif; but this idea too has been modified. The Valentinian demiurge and his associates are not really malevolent, as are their Sethian counterparts: their kind of jealousy is that of the Old Testament lawgiver, who operates by means of commandments, threats and punishment, on the basis of a limited idea of justice. Evil, on the other hand, is the serpent: in stark contrast to the Sethian accounts, the serpent deceived the first human and made him disobey the commandment not to eat from the Tree of Knowledge, with the fatal consequence that the human was expelled from paradise and humanity thenceforth became mortal.[16]

Most importantly, however, all this happened as part of a preordained plan of salvation designed by the highest god himself. The first human was never meant to remain in paradise. His visit there and his subsequent expulsion instead served a pedagogical purpose:

> This is the expulsion he was made to suffer, when he was expelled from the pleasures of those who belong to the imitation and those who belong to the likeness. It is, however, a work of providence, in order that it may be realised that the enjoyment the human being may have for such pleasures is short compared with the eternal existence of the place of rest. It was a work the Spirit had ordained, because it had planned in advance that the human should experience that great evil which is death – that is, the complete ignorance of all things – and also experience all the evils that arise from that, so that he, after the cravings and anxieties that result from these, might partake of the greatest good, namely, eternal life, which is the complete knowledge of the All and the partaking of all good things.[17]

The expression 'those who belong to the imitation and those who belong to the likeness' refers, in the special terminology of this text, to the respective powers of matter and soul. The brief experience of paradise thus taught humanity that the pleasures of the body and the soul are short-lived and worth nothing compared to the delights of eternal life. Moreover, the experience of ignorance, suffering and ultimately death – in other words the experience of living in a body – is a necessary lesson for humanity which enables us to appreciate the true value of the opposite state which the Father has planned for us

[15] 'They put their Tree of Life in the middle of paradise' (NHC II, p. 21).

[16] 'Through the evil power called "the serpent" – for he is more crafty than all the other evil powers – the human <was> deceived ... It made him transgress the commandment, so that he should die. And he was expelled from that place and all its pleasures' (NHC I, p. 107).

[17] NHC I, pp. 107–8.

to enjoy. In this way, the story of the first human's short stay in paradise becomes an image of humanity's general condition in a world of corruption.

Paradise as a Failed Route of Escape

Different again is the account of paradise made in the untitled tractate from the Nag Hammadi Library referred to by modern scholars as *The Origin of the World.*[18] The account of creation here has much in common with that of the *Apocryphon of John.* The sinister world creator Yaldabaoth emerges out of the chaos fatally brought into being by the aeon Pistis Sophia. Yaldabaoth then spawns seven sons and makes a realm for himself in the chaos consisting of seven heavens. There follow the traditional boast of the world ruler – 'I am God, and there is no other but me' – and the quick and sudden response of his mother who calls out that he is wrong, because the immortal human exists above him, while she reveals her reflection in the waters.

At this point, however, the story diverges from that of the *Apocryphon,* because Sabaoth, one of Yaldabaoth's sons, realizes upon the manifestation of Pistis that there exists a superior realm and revolts against his father. Sabaoth ascends above the spheres of his father and takes up residence in the eighth heaven. Representing the ideas of conversion and justice, Sabaoth occupies a position similar to that of the psychic demiurge in the hierarchy of the Valentinian system and can be identified with the god of Israel. His introduction into the story entails a different interpretation of paradise from that of the *Apocryphon.*[19] Paradise in *The Origin of the World* is invented not by the world ruler as a trap for humans, but by Sabaoth, apparently under the guidance of Pistis Sophia,[20] as a means for human souls to escape from the world of matter and ignorance. This is how it is described:

> Then Justice created a beautiful paradise beyond the sphere of the moon and the sphere of the sun in the land of delights, which is in the east amidst the stones. And desire dwells in the middle of the beautiful and inviting trees. And the tree of eternal life, having appeared by the will of God, is placed in the north of paradise in order to give immortality to the souls of the saints, who will leave behind their poor modelled bodies at the end of the age.
>
> Now, the Tree of Life shines like the sun; its branches are lovely, its leaves are like those of the cypress, its fruit glistens like a cluster of grapes and its height reaches the sky. And next to it is found the Tree of Knowledge, which is endowed with the power of God. Its splendour is like the shining full moon; its branches are lovely, its leaves are like those of the fig tree and its fruit is like good, delicious dates. This tree is placed in the north of paradise in order to arouse the

[18] This is a rather difficult text, due to later reworking which resulted in inconsistencies. The credit for sorting out the layers of the text belongs above all to Louis Painchaud, *L'Écrit sans titre*, Sainte-Foy, Louvain and Paris, 1995. See also Painchaud's annotated translation in *Écrits gnostiques* (n. 3 above), pp. 401–65.

[19] The paradise theme in this text is discussed by M. Tardieu, *Trois mythes gnostiques,* Paris, 1974, pp. 175–214. Several of Tardieu's interpretations are, however, criticised by Painchaud in *L'Écrit sans titre* (n. 18 above), pp. 93–4, 339 n. 147, 347–8, 367, 379.

[20] Although this is not said explicitly in connection with the creation of paradise (NHC II, pp. 110–11), it can be assumed from the general description of Pistis's guidance of Sabaoth (pp. 104–6).

> souls from the oblivion of the demons, so that they may advance to the Tree of Life, eat its fruit and condemn the authorities and their angels.[21]

It may first be noted that the passage combines several traditional themes. The location of paradise in the third planetary sphere is an idea known above all from Paul (II Corinthians 12:1–4); it belongs to Jewish tradition and is also attested elsewhere.[22] A pre-Ptolemaic sequence of the planets seems to be assumed, with the moon and sun set in the two lowest spheres.[23] The 'delight' (*tryphē*) of paradise, its situation in the east and the precious stones associated with it are of course biblical notions and equally traditional.[24]

The statement 'desire (*epithymia*) dwells in the middle of the beautiful and inviting trees' is problematic. This could suggest a negative understanding of the trees as representing carnal desires, similar to the interpretation of the trees in the *Apocryphon of John*. Unlike the *Apocryphon,* however, paradise as such is not negatively understood in *The Origin of the World*: it is not a trap designed to activate the passions, so it seems that *epithymia* must here be given a more neutral interpretation, probably as referring to the natural appetites of the human being.[25]

The really important features of paradise are in any case the two trees which lead to salvation, representing two distinct stages in that process: the fruit of the Tree of Knowledge arouses humans from their state of oblivion; and the Tree of Life grants immortality. The trees grow in the north of the garden, which is a singular idea. It must be in the north that the exit from paradise can be found – a passage which leads those who follow it out of the planetary spheres and into the heavenly realm beyond.[26] It thus appears that paradise is understood as a place inserted by the higher powers into the region of the 'authorities' – that is, the planetary rulers – in order to provide a route of escape

[21] NHC II, p. 110 (my translation).

[22] Cf. *II Enoch* 8:1; *Apocalypse of Moses* 37:5; *Life of Adam and Eve* 37, 40; *Apocalypse of Paul* 20. In *III Baruch* 4:1–10, paradise actually seems to be located in the fourth heaven. According to Irenaeus, *Adversus Haereses*, 1.5.2, reporting a Valentinian source, 'paradise is above the third heaven, and is a fourth Archangel with power'. The *Excerpts from Theodotus* (51) similarly locate paradise in the fourth heaven. See I. de Vuippens, *Le Paradis terrestre au troisième ciel*, Paris, 1925; P. R. Gooder, *Only the Third Heaven?*, London and New York, 2006.

[23] An attractive alternative interpretation is suggested by Tardieu, *Trois mythes gnostiques* (n. 19 above), pp. 176–7: the location of paradise beyond the moon and the sun means that it is above all the planetary spheres and thus 'dans un au-delà ou un en-deçà du temps marqué par la faute et le malheur'.

[24] Genesis 2:8, 2:12, 3:23; Ezekiel 28:13–14; *I Enoch* 18:6. See Tardieu, *Trois mythes gnostiques* (n. 19 above), pp. 177–90.

[25] In his annotated translation in *Écrits gnostiques* (n. 3 above), p. 438, Painchaud takes the word to mean the desire for knowledge, although in his *L'Écrit sans titre* (n. 18 above), pp. 369–71, he is more circumspect on the issue. Such an interpretation is, however, nowhere suggested in the text itself, and the fact that the archons later invite the first humans to eat from those trees (p. 118) argues against it. A neutral meaning is therefore more likely in this case.

[26] In his annotated translation in *Écrits gnostiques* (n. 3 above), p. 438, Painchaud, refers to *I Enoch* 34:1–2, where the gates of heaven are said to be in the north. See also Painchaud's discussion in *L'Écrit sans titre* (n. 18 above), p. 372.

for humans trapped in the cosmos. The first humans were nevertheless unable to avail themselves of this route. A very complicated anthropogony describes the creation of the earthly Adam by the archons; how Pistis sent him her daughter Life-Eve to instruct him; how the archons wanted to violate Eve; and how she escaped them by entering into the Tree of Knowledge, leaving behind an image of herself which became the earthly Eve and which the archons violated instead, resulting in the birth of Abel and the rest of Eve's children.[27] Placed in paradise by the archons, at any rate, Adam and Eve disobeyed their command not to eat from the Tree of Knowledge; instead they listened to the advice of the 'instructor', another agent sent by Pistis, and ate from the tree. Then 'their minds opened' and 'the light of knowledge shone upon them'.[28] Dismayed, the archons threw Adam and Eve out of paradise and down to earth, setting cherubim with a flaming sword to guard the Tree of Life, to make sure no earthly being should gain access to it.[29] Thus Adam and Eve acquired knowledge by eating from the first tree, but were unable to complete the second stage in the process of redemption which paradise was intended to provide: the attainment of immortality by eating from the Tree of Life.[30]

After this, paradise no longer plays a role in the text as a location giving access to salvation. In a curious episode it is told that Sophia took revenge on the archons who had blocked access to paradise, by throwing them down to earth;[31] then she brought forth the phoenix as a symbol of paradise lost.[32] The phoenix, a bird which continuously dies and revives, became a reminder of mankind's loss of paradise as a way to eternal life.[33] A reviser of the text has added a section here, which expands on the notion of the phoenix and also suggests that access to the Tree of Life has been reopened in a new form:

> There are in the world and until the end of the world three human beings and their descendants: the spiritual, the psychic and the earthly. In the same way there are three *phoinikes* in paradise: the first is immortal, the second lives for a thousand years and the third is consumed, as is written in the *Holy Book*. Likewise, there are three baptisms: the first is spiritual, the second is fire, the third is water.[34]

[27] NHC II, pp. 112–18.

[28] NHC II, p. 119.

[29] Ibid., pp. 120–21.

[30] Cf. the interpretation of Alexander, 'The Fall into Knowledge' (n. 2 above), p. 101: 'The Garden of Eden/paradise becomes for [the author], in effect, an image of the spiritual world from which man has fallen, and to which his spirit would hope to return in the *eschaton*'. As we shall see, this eschatological return is in fact considered to be effected by means of a ritual of baptism.

[31] This alludes to the fall of the giants in Genesis 6.

[32] NHC II, pp. 121–2.

[33] For a translation of this passage, see Painchaud, *L'Écrit sans titre* (n. 18), pp. 459–60. For the traditional association of the phoenix with paradise, see R. van den Broek, *The Myth of the Phoenix According to Classical and Early Christian Traditions*, Leiden, 1972, especially pp. 310–332. On this particular text, see also M. Tardieu, 'Pour un phénix gnostique', *Revue de l'histoire des religions*, 183, 1973, pp. 117–42; and *Trois mythes gnostiques* (n. 19 above), pp. 231–62. Tardieu's analysis is, however, sharply questioned by Painchaud, *L'Écrit sans titre* (n. 18 above).

[34] NHC II, p. 122.

The word *phoinix* is used in a play on words: it refers on the one hand to the phoenix bird which was mentioned just before, and on the other hand to the date palm, *phoenix dactylifera*, three varieties of which are here said to constitute the trees of paradise.[35] The *phoinix* which is 'consumed' (another double meaning) represents the ordinary trees of paradise, corresponding to earthly, i.e. material, humans and baptism by water; the *phoinix* which lives for a thousand years means the Tree of Knowledge, psychic humans and baptism by fire; and the immortal *phoinix* symbolises the Tree of Life, spiritual humans and spiritual baptism. It follows from this that access to the paradisiacal Tree of Life, barred by the archons at the beginning of history, has been reopened by the introduction of spiritual baptism. The paradise theme has thus been relocated from cosmology to the realm of the sacraments. The association of the Tree of Life with sacraments is attested elsewhere, and not only in gnostic sources. *The Gospel of Philip,* for example, states that

> The Tree of Life is in the middle of the garden. It is an olive tree, and from it comes chrism, and from chrism comes resurrection.[36]

The lost paradise can be regained, and eating from the Tree of Life, which Adam and Eve were never allowed to do in paradise, is eventually made possible in the form of a ritual of redemption.

Paradise as the World

The idea that paradise symbolizes the situation of humanity in the cosmos at large figures implicitly, as was shown above, in the background of both the *Apocryphon of John* and the *Tripartite Tractate.* An explicit formulation of this idea appears in that most peculiar example of allegorical ingenuity known as the *Book of Baruch*, written by a certain Justin and preserved for posterity by Hippolytus of Rome in his *Refutatio omnium haeresium* (5.26–7).[37] In this text, Edem/Eden is imagined as a female cosmic principle, the upper half of her body being that of a young woman and her lower part having the shape of a serpent.[38] She is one of three primordial principles, the other two being the transcendent

[35] Painchaud, *L'Écrit sans titre* (n. 18 above), p. 467; and his annotations in *Écrits gnostiques* (n. 3 above), p. 454.

[36] NHC II, p. 73. Cf. also the 'Ophite' formula quoted by Origen, *Contra Celsum*, 6.27: 'I have been anointed with white chrism from the Tree of Life'. For the Syrian tradition, see R. Murray, *Symbols of Church and Kingdom*, 2nd edn, Edinburgh, 2006, pp. 115–16, 125, 322.

[37] See in particular E. Haenchen, 'Das Buch Baruch: Ein Beitrag zum Problem der christlichen Gnosis', *Zeitschrift für Theologie und Kirche*, 50, 1953, pp. 123–58, repr. in idem, *Gott und Mensch*, Tübingen, 1965, pp. 299–334; M. Marcovich, 'Justin's *Baruch*: A Showcase of Gnostic Syncretism', in idem, *Studies in Graeco-Roman Religions and Gnosticism*, Leiden, 1988, pp. 93–119; and, most recently, R. van den Broek, 'Gospel Tradition and Salvation in Justin the Gnostic', *Vigiliae Christianae*, 57, 2003, pp. 363–88 (with bibliography in n. 5).

[38] See R. van den Broek, 'The Shape of Edem According to Justin the Gnostic', *Vigiliae Christianae*, 27, 1973, pp. 35–45, repr. in idem, *Studies in Gnosticism and Alexandrian Christianity*, Nag Hammadi and Manichaean Studies, 39, Leiden, 1996, pp. 131–41.

god called 'the Good' and an intermediary, demiurgic Father who is also called Elohim. The latter desired Edem, and from their union twelve male and twelve female angels were born. These angels, Justin says, are what the book of Genesis refers to as the trees of paradise. For Justin paradise thus means the world in general; the paternal angels/trees represent the higher, i.e. rational, components of the cosmos, whereas the maternal ones are its lower, i.e. irrational parts. Humans were created with a similar set of components, taking spirit from Elohim and soul from Edem.

As Justin relates, Elohim surveyed his creation and, ascending to its uppermost boundary, became aware of the still higher realm of the Good and decided to stay there. He abandoned Edem, who was enraged and resolved to take her anger out on the humans. She let her twelve angels spread passions and vices among humankind, especially the angel Naas, who became a constant instigator of sin and evil. The world was ruled by the angels, who flowed throughout the cosmos[39] causing all kinds of sickness and natural disasters. Elohim responded by sending his own angel Baruch to advise and support the human spirit against the power of Naas. His mission was unsuccessful, however, since he was unable to prevent Adam and Eve, Moses, or the prophets from being seduced by Naas. Even Heracles, who laboured heroically against the twelve angels, eventually succumbed to erotic seduction. In the end, only Jesus was able to obey Baruch. For this reason Edem crucified him, but Jesus abandoned on the cross the body he had been given by Edem, and went up to the Good.

In Justin's interpretation, then, paradise is equivalent to the cosmos as a whole. Edem is something like Physis,[40] or a world soul deprived of rationality, a condition represented by her being abandoned by Elohim. Her trees represent the passions. This is in some respects similar to the *Apocryphon of John,* where, as we have seen, the trees of paradise also symbolize human passions.[41] On the other hand, Justin does not share the *Apocryphon's* exegetical reversal of the story of the Fall. On the contrary, Baruch advises Adam and Eve to eat from all the trees, because even if they represent passions, they are not sinful.[42] The Tree of Knowledge, however, must be avoided, because that tree is Naas and leads to sin. As it happened, Adam and Eve did eat from the Tree of Knowledge, which means that they were seduced by Naas to commit acts of fornication and homosexuality. From that moment on, humanity became victim to the vices of Naas. Thus Justin, unlike the *Apocryphon of John,* accepts in a broad sense the traditional values

[39] This is what Moses alluded to, Justin says, when he spoke about the rivers of paradise.

[40] She is not unlike the figure of Physis in the Hermetic *Poimandres*; cf. J. Büchli, *Der Poimandres: Ein paganisiertes Evangelium*, Tübingen, 1987, pp. 88–9. Van den Broek, 'The Shape of Edem' (n. 37 above), highlights affinities to the Hellenistic Isis-Thermouthis, whereas Marcovich, 'Justin's *Baruch*' (n. 37 above), p. 97, finds the closest model for Edem in the Hesiodean Echidna (who is in fact identified with Physis by Damascius, *In Phaedrum* 142, in L. G. Westerink, *The Greek Commentaries in Plato's Phaedo*, II, *Damascius II*, Amsterdam, 1977, p. 362).

[41] The doctrine also appears in Manichaeism according to the *Acta Archelai* XI (= Epiphanius, *Panarion,* 66.29.1).

[42] Hippolytus, *Refutatio*, 5.26.22.

of the Fall narrative: eating from the Tree of Knowledge was a fault and led to sin. Though he interprets paradise as the world and its trees as passions, he does not demonise the passions, as does the *Apocryphon,* but rather sees them as inevitable components of the human soul living in the cosmos.[43] On the other hand, the ultimate goal of humankind, as was shown by the example of Jesus, is to abandon the soul created by Edem, release the internal spirit and ascend like Elohim to the pure realm of the Good. It is a vision of redemption which involves not only the need for the human soul to overcome the physical world, but also the necessity for the Old Testament deity to be transferred to a higher sphere and for the mundane meaning of the Old Testament narrative to be revealed and transcended. In this sense, Justin's rewriting of the paradise story is gnostic too.

Anatomical Allegories

Two curious allegorical interpretations of paradise found in gnostic sources may also be briefly recorded here. The treatise known as *The Great Announcement* (*Apophasis Megale*), attributed by Hippolytus to Simon Magus,[44] explains paradise anatomically as the maternal womb in which the human being is fashioned, the rivers being the arteries through which the foetus is nourished, etc. This exegesis is not particularly gnostic, just allegorical, and we therefore need not dwell further on it in the present context, except to note that a psychologizing understanding of the allegory, as the expression of a nostalgic desire for paradise in the form of a *regressus ad uterum,* would be misplaced.[45]

A different anatomical allegory is applied in the so-called Naassene treatise,[46] where paradise is interpreted as the human head and Edem as the brain, from which a river flows out into the four organs of sense located in the head. This, again, is not a typically gnostic exegesis and need not further detain us here.

Paradises Beyond the Cosmos

Before ending this brief survey I wish to direct attention to certain texts where the concept of paradise does not refer to the imperfect creation of the demiurge but, instead, has very positive connotations. In the Valentinian *Tripartite Tractate,* a paradise is said to be located above the cosmos, in the intermediary region between the planetary hebdomad and the realm of the divine Fullness. This is the region of Sophia, or, to be precise, the Logos, which is the name used for the fallen aeon in this particular text. According to the Valentinian

[43] As van den Broek emphasises in his 'Gospel Tradition' (n. 37 above), pp. 367–9, creation and procreation are not seen as intrinsically bad by Justin. Lettieri's interpretation of paradise as a prison, along the lines of the *Apocryphon of John*, in 'The Ambiguity of Eden' (n. 2 above), pp. 41–3*,* is therefore somewhat misleading.

[44] Hippolytus, *Refutatio*, 6.14.7–9.

[45] It may, perhaps, be more appropriate to understand the womb metaphor in, say, Ephrem's hymns on paradise along these lines, as does J. Delumeau, *History of Paradise*, New York, 1995, pp. 32, 35. See Sergey Minov's essay in this volume.

[46] Hippolytus, *Refutatio*, 5.9.14–15.

myth, the Saviour was sent out to stabilize the fallen aeon: he cured its passions and gave it a vision of the transcendent world. The vision of the Saviour, who carried within him the entire divine Pleroma, inspired the Logos to perform acts of thanksgiving and praise. These acts became spiritual children, images of the Pleroma he had seen. The Logos then proceeded to create an abode for himself and his offspring:

> Then he arranged the abode of those he had brought forth by way of glorification, which is called 'paradise', 'enjoyment', 'delight full of nourishment', and 'delight <of> the preexistent ones', reproducing the image of all the good things which exist in the Fullness.[47]

We see that the full biblical vocabulary of paradise is put into use in order to describe this hypercosmic realm. It is not a garden, of course, and there are no trees – not even metaphorical ones. In the text which follows, the sphere of the Logos is further described as a 'kingdom', a 'city' and a 'church'. In other words, paradise here appears conflated with the ideas of the Kingdom of Heaven, the Heavenly Jerusalem and the heavenly Church. This paradise possesses, no doubt, all the soteriological connotations which the conflation implies; on the other hand, the term also functions as part of a cosmographic vocabulary. The *Tripartite Tractate* expounds a hierarchy of being where all entities are replicated at each successive level of reality: the highest level of the Fullness is copied as 'images' on the spiritual level of the Logos, which in turn is reflected as 'likenesses' in the psychic realm and as 'imitations' in the sensible world of matter. It follows, then, that paradises must exist on all these levels, and as a matter of fact the paradise of the psychic demiurge is a likeness of the one which exists in the region of the Logos: 'With him [i.e. the demiurge] is also a paradise and a kingdom and everything else which exists in the aeon before him'.[48] Moreover, all his subordinate powers replicate in their own realms the features of the demiurge's dominion: paradises, kingdoms, rests etc.[49]

All these lower paradises are thus increasingly faint copies of the real paradise which must exist in the Fullness itself. That paradise, which is obviously an entirely supra-sensual place, offering enjoyments of a purely spiritual and noetic kind, is not explicitly designated as such in the *Tripartite Tractate.* Nevertheless, the language of paradisiacal delights is unmistakably present in an earlier passage, which describes the Son's revelation of the Father to the aeons inside the Fullness: 'He was given them as delight and nourishment, joy and abundant illumination'.[50] The *gnosis* provided by the Son for the aeons is described in terms of the fruits growing on the trees of paradise. Even the Father's knowledge of himself may be portrayed in such terms: 'That which he conceives, sees, and speaks is nourishment and delight, truth, joy, and rest'.[51]

[47] NHC I, p. 96.
[48] Ibid., p. 101.
[49] Ibid., p. 102.
[50] Ibid., p. 65.
[51] Ibid., p. 55.

In consequence, it is not surprising that the Valentinian *Gospel of Truth* is able to use the metaphor of paradise explicitly to describe even the highest level of oneness which the spiritual person may hope to enjoy with the Father:

> He is good. He knows his plants because he planted them in his paradise. And his paradise is his place of rest. Paradise is the perfection within the Father's thought, and the plants are the thoughts of his meditation.[52]

The idea of paradise was thus used in a variety of ways by the gnostics. There is no single gnostic understanding of paradise; the notion of paradise as a trap and a prison for the human spirit is only one of many interpretations of the idea, both positive and negative. In this, as in most other respects, gnosticism was a far from monolithic phenomenon.

[52] NHC I, pp. 36–7.

Diabolizing the Garden of Eden: Re-Interpretations of Jewish Pseudepigraphy in Medieval Christian Dualism

Yuri Stoyanov

The cosmographic and theological provenance and contextualization of the notion of paradise (and the related revisions/re-writing of the canonical Eden narrative in Genesis 2–3) in medieval Christian dualism (specifically, Paulicianism, Bogomilism and related groups in the Eastern Orthodox world as well as Catharism in Western Christendom) have not so far been analyzed systematically as an area of study of their own. Due to the long-standing inattention to this problematic an essential and potentially very promising sphere in the study of medieval Christian dualism and the development and transmutations of medieval ideas about paradise has remained virtually unexplored.

Given the insufficient state of research on this particular subject matter, the present article does not intend to provide a detailed and in-depth discussion of the medieval Christian dualist re-interpretations of traditional inherited paradise themes against the background of the respective treatment of these themes in Gnosticism, Manichaeism and early and medieval Christianity. Its purpose rather will be to draw attention to some of the most characteristic and cosmographically relevant concepts of paradise in medieval Christian dualist traditions and address the crucial (but presently not always solvable) problem of their literary and theological provenance. It will also take into consideration the increasing amount of textual evidence of the role played by various early Jewish (extant in Christian redactions) and early Christian pseudepigrapha (preserved and circulated in various versions in medieval Eastern Christendom), in the formation and elaboration of medieval Christian dualist etiological, cosmological, satanological, Christological and biblical history traditions. Therefore one of the main research questions to be addressed will be whether at least some elements of the complex of notions and narratives regarding paradise in medieval Christian dualism are traceable to and had been mediated by such pseudepigraphic works.

It will be necessary to begin with a brief introduction to the manifestations and varieties of medieval Christian dualism and their religio-historical provenance. Following the dissemination of varied Gnostic dualist and related teachings during Late Antiquity, the residues and actual and potential carriers of such traditions in early medieval Christendom become increasingly problematic to detect and recognize with any certainty. However, proto-Gnostic, Gnostic-like and dualist-leaning notions and narratives were preserved in the medieval versions of the already mentioned early Jewish and Christian pseudepigraphic

The Cosmography of Paradise: The Other World from Ancient Mesopotamia to Medieval Europe, Warburg Institute Colloquia 27, 2016

works. Despite being prohibited and ostensibly suppressed, such pseudepigraphic texts survived and continued to be circulated (predominantly in Eastern Christendom) in heterodox, sectarian, monastic or learned milieux. In suitable circumstances such pseudepigraphic notions and narratives could prompt revitalizations of such doctrinal and theological attitudes and nuances through the literal borrowing of their themes and imagery. A resulting idiosyncratic or creative exegesis then could spread from these works to the canonical scriptures, a process carrying the potential for conceptualizing new heterodoxies and heresies.

The current state of knowledge and research indicate that the beginning of medieval Christian dualism can be traced to the rise of Paulicianism in early medieval Armenia, with its characteristic religious complexity, tensions and antagonisms. In the wake of its Christianization at the beginning of the fourth century Armenia had become an arena of intense political and religious contention in the centuries-long imperial rivalry between the Christianized East Roman (Byzantine) and Sasanian Persian Empires. The evidence that following the introduction of progressively severe anti-heretical (particularly anti-Manichaean) legislation in the Christianized East Roman realm, Christian dissident groups and communities, branded heterodox and heretical by its institutionalized clerical elites, sought and found refuge in Armenian lands, is still uncertain and awaits a fresh and up-to-date reappraisal. Hence the origins and early dissemination of Paulicianism (which need to be carefully explored against the backdrop of the distinctive currents and sub-currents in the early medieval religious development in Armenian-speaking areas in the Caucasus and eastern Asia Minor) are still very obscure and the reconstruction of its basic outlines still faces some insurmountable difficulties.

Although uneven, the evidence of Paulicianism's spread and fortunes in Byzantine-ruled Anatolian areas (which it entered reportedly early in its development) allows much better insights into its turbulent history (which included periods of toleration under the Byzantine iconoclastic emperors) and the kind of challenges which the movement posed for the Byzantine secular and ecclesiastical authorities. Byzantine polemicists principally regarded and condemned Paulicianism as a revival of Manichaeism: by that time accusations of Manichaeism in Byzantium were employed not only against genuine or supposed Christian heretics but also to denounce and criminalize religious and political adversaries. Consequently Byzantine polemics against Paulician doctrines as 'Manichaean' and dualist have triggered continuing debates concerning the doctrinal provenance of Paulician teachings and religious practices.

While stating that the Paulicians adhered to vigorous iconoclasm and refuted the sacraments and clerical hierarchy of the normative church, Byzantine polemical works also time and again portrayed the Paulicians as veritable 'Manichaeans' and ascribed to them the radical dualist doctrine of two gods or principles, the evil creator of the present material world and the good god of the future world. The Paulician Christology is reported as being of the Docetic kind: Christ's incarnation was seen as illusory and the Virgin Mary was not extolled as the mother of Christ but as the 'Heavenly Jerusalem'.[1]

The diverse references to the Paulicians in Armenian and Byzantine sources have attracted contradictory and conflicting readings as to whether they were originally dualist or adopted dualism later in their history.[2] Prioritizing the Armenian sources, Nina Garsoïan has contested the influential and most commonly held view that Paulicianism arose as a Christian dualist heresy. In a far-reaching reappraisal of Paulician history and teachings she has presented a series of arguments that both the dualist and Docetic doctrines reflect later innovations in the Paulician movement.[3] Such reconstruction of Paulician religious development, however, has not been able to present convincing evidence for any such hypothetical Paulician religious re-orientation, its timing and possible causes. Consequently, the theory of a secondary dualist reformation within Byzantine Paulicianism and the attempted harmonization of the primary sources to attain such reconstruction has met intense and sustained criticism. This thesis, therefore, has remained a theoretical construct which needs to be supported by much more solid data before it can be considered on a par with the other approaches to Paulician dualism.[4]

The Byzantine heresiological record of Paulician teachings continuously ascribe to them the type of dualism between the creator-god and ruler of this world and the concealed god of the world to come who could be also recognized as lord of heavens. Since the early stages of the critical study of Paulician dualism historiographical and theological disputes have persisted concerning the authenticity and credibility of these reports and whether such dualist teachings could have stemmed from the direct or indirect impact of Manichaean,

[1] The main Byzantine sources on Paulicianism have been edited and collected in 'Les Sources grecques pour l'histoire des Pauliciens d'Asie Mineure', eds C. Astruc, W. Conus-Wolska, J. Gouillard, P. Lemerle, D. Papachryssanthou and J. Paramelle, *Travaux et mémoires,* 4, 1970, pp. 1–227. A valuable selection of Byzantine sources for Paulician history and teachings have been translated, with commentaries, in *Christian Dualist Heresies in the Byzantine World c. 650–c. 1450*, eds J. Hamilton and B. Hamilton, Y. Stoyanov (assist. ed.), Manchester, 1998, pp. 57–114, 139–42, 166–75, 259–60. Generally, on the Byzantine sources for Paulicianism, cf. H. Grégoire, 'Les Sources de l'histoire des Pauliciens', *Bulletin de la Classe des Lettres et des Sciences Morales et Politiques,* 5e sér., 22, 1936, pp. 95–114; R. M. Bartikian, *Istochniki dlia izucheniia istorii pavlikianskogo dvizheniia*, Erevan, 1961, ch. 2, pp. 55–102; N. Garsoïan, *The Paulician Heresy, A Study of the Origin and Development of Paulicianism in Armenia and the Eastern Provinces of the Byzantine Empire*, The Hague and Paris, 1967, ch. 1; P. Lemerle, 'L'Histoire des Pauliciens d'Asie Mineure d'après les sources grecques', *Travaux et mémoires*, 5, 1973, pp. 1–137.

[2] For a discussion of the posited references to, and discussions of, Paulicianism in the Armenian sources, cf. Bartikian, *Istochniki* (n. 1 above), ch. 1; Garsoïan, *The Paulician Heresy* (n. 1 above), ch. 2; Hamilton, Hamilton and Stoyanov, *Christian Dualist Heresies* (n. 1 above), Appendix 2, 'Armenian Sources and the Paulicians', pp. 292–3.

[3] Garsoïan, *The Paulician Heresy* (n. 1 above), with a proposed critical reconstruction of Paulician history, ch. 3, pp. 112–51, and of Paulician doctrine, ch. 4, pp. 151–86; Garsoïan, 'Byzantine Heresy. A Reinterpretation', *Dumbarton Oak Papers*, 25, 1971, pp. 87–114; Garsoïan, 'L'abjuration du moine Nil de Calabre', *Byzantinoslavica: revue internationale des études byzantines*, 35, 1974, pp. 12–27.

[4] See Lemerle, 'L'Histoire' (n. 1 above), pp. 12ff. and passim; L. Barnard, 'The Paulicians and Iconoclasm', in *Iconoclasm: Papers Given at the Spring Symposium of Byzantine Studies, March, 1975*, eds A. Bryer and J. Herrin, Birmingham, 1977, pp. 75–83, especially 81; I. P. Coulianu, *The Tree of Gnosis: Gnostic Mythology from Early Christianity to Modern Nihilism*, San Francisco CA, 1992, pp. 192–4; Hamilton, Hamilton and Stoyanov, *Christian Dualist Heresies* (n. 1 above), Appendix 2, 'Armenian Sources and the Paulicians', pp. 292–3.

Marcionite or other dissenting Christian groups.[5] Lately, though, rather than trying to reconstruct sectarian historical continuities and genealogies on the basis of the often-used and abused contact-diffusion model, recent research has been also exploring the possibility that the Paulician dualist version of Christianity developed independently via a spiritualist and allegorical reading of the New Testament in sixth- and seventh-century Armenian heterodox sectarian milieux (while its dualist element also could have borne the impact of Zoroastrian, Christian heterodox and possible Gnostic-related survivals on the late antique and early medieval Armenian religious scene).[6]

Despite some substantial gaps and obscure periods the evidence of the subsequent history of the expansion of Paulicianism in Byzantine Anatolia and the re-settlement and migrations of Paulician communities in the Balkans and their later fortunes provides some important indications of the dynamics of the growth, spread and suppression of religious dualist heterodoxy and heresy along the Caucasus – Anatolia – Balkan axis. The development of medieval Christian dualism entered a new, major phase with the emergence of the Bogomil heresy in the first half of the tenth century in the newly Christianized Bulgarian kingdom. Characteristically, medieval Eastern Orthodox polemicists often routinely applied to Bogomilism the heresiological stereotype of 'Manichaean heresy' or tended to describe it as a mixture of earlier heresies; such perceived combinations could refer to combinations of Manichaeism and Paulicianism, Paulicianism and Massalianism or Manichaeism and Massalianism[7] (denounced as heretical, Massalianism, with its

[5] For earlier and more recent arguments that Manichaeism was in many respects a direct ancestor of Paulicianism, which is seen as also experiencing some Marcionite influences, see D. Obolensky, *The Bogomils. A Study in Balkan Neo-Manichaeism*, Cambridge, 1948, pp. 44–7, followed by D. Angelov, *Bogomilstvoto*, Sofia, 1993, pp. 83, 97, n. 57; for arguments that Paulician dualism may have been a development of Marcionite teachings, cf. A. von Harnack, *Marcion: Das Evangelium vom fremden Gott: eine Monographie zur Geschichte der Grundlegung der Katholischen Kirche,* 2nd edn, Leipzig, 1924, pp. 382–3 (with reservations); Grégoire, 'Les Sources' (n. 1 above); J. Anastasiu, *Oi paulikianoi*, Athens, 1959, pp. 153ff.; M. Loos, 'Le Mouvement paulicien à Byzance', *Byzantinoslavica*, 25, 1964, pp. 55–6; M. Loos, *Dualist Heresy in the Middle Ages*, Prague, 1974, pp. 34–5; for arguments for Gnostic influences on Paulician dualism, see J. I. von Döllinger, *Beiträge zur Sektengeschichte des Mittelalters*, I, *Geschichte der gnostisch-manichäischen Sekten im frühen Mittelater*, Munich, 1890, repr. Darmstadt, 1968, pp. 2–3; H. Söderberg, *La Religion des Cathares: études sur le gnosticisme de la basse antiquité et du moyen âge*, Uppsala, 1949, pp. 52ff.; cf. the cautious approach to the problem of possible Manichaean and/or Marcionite impact on Paulicianism in S. Runciman, *The Medieval Manichee. A Study of the Christian Dualist Heresy*, Cambridge, 1947, pp. 46–59; cf. the views of Coulianu, *The Tree of Gnosis* (n. 4 above), pp. 190–96, who, while treating Paulicianism as a 'popular Marcionism', argues that the Marcionite influence need not have been a direct historical one. See also Lemerle, 'L'Histoire des Pauliciens' (n. 1 above), pp. 132–5, for a discussion of the parallels and the important differences between Marcionism and Paulicianism, and an emphasis on Paulician reinstatement of evangelical Christianity and the Pauline tradition.

[6] Cf. Hamilton, 'Historical Introduction', in Hamilton, Hamilton and Stoyanov, *Christian Dualist Heresies* (n. 1 above), pp. 7–8; Y. Stoyanov, *The Other God. Dualist Religions from Antiquity to the Cathar Heresy*, London and New Haven CT, 2000, pp. 125–9; on 5th-century reports of trends and movements in Armenian Christianity perceived as Gnostic-related, see V. N. Nersessian, *The Tondrakian Movement: Religious Movements in the Armenian Church from the Fourth to the Tenth Centuries,* London, 1987, pp. 79ff.

[7] See, for example, Theophylact Lecapenus, *Epistula*, ed. I. Duichev, 'L'epistola sui Bogomili del patriarca constantinopolitano Teofilatto', in *Mélanges E. Tisserant,* Vatican City, 1964, pp. 89–91; Anna Comnena, *Alexiad* XV.10.3–4, ed. B. Leib, Paris, 1945, III, pp. 227ff.; Pseudo-Psellus, *De daemonibus*, ed. P. Gautier, 'Le *De Daemonibus* du Pseudo-Psellus', *Revue des études Byzantines,* 38, 1980, pp. 105–94.

distinctive anthropological dualism had spread along a comparable east-west axis of diffusion from north-east Mesopotamia to Syria, Armenia and Asia Minor). Such precise heresiological diagnoses of Bogomilism can be exceedingly deceptive but sometimes have been espoused in earlier scholarship on Balkan dualist heresies. As a result medieval heresiological definitions could be reiterated in modern scholarly versions and thus Manichaean, Marcionite or even separate Gnostic traditions had been variously assumed to have exercised impact on the formation of Bogomil dualist teachings.[8] At the same time, current and growing trends of research on the various processes accompanying the accelerated formation of Slavo-Byzantine Orthodox theology, culture and learning in the late ninth and tenth centuries have pioneered new, promising approaches to Bogomil origins. This new research has made available important new material and source analyses showing that the translation of the scriptures in a language sufficiently close to the vernacular in this period went together with the translation and diffusion of non-canonical pseudepigraphic texts which crucially contributed to the conceptualization and elaboration of Bogomil theological dualism.[9]

Extant sources indicate that by the mid-eleventh century Bogomilism had entered and spread eastwards into Byzantine Anatolia, expanding also into peninsular Greece, the central and western Balkans, especially in Dalmatia and Bosnia. In the crusading era when contacts and interchange between Eastern and Western Christendom greatly increased Bogomilism is reported to have made missionary inroads further to the west and to have exercised a formative impact in the spheres of theology and cultic observances on the first medieval Western Christian dualist groups which became identified as 'Cathars'. The problem of the doctrinal orientation of early Cathar groups and their links with East Christian dualism has been lately subjected to fresh reappraisal which on occasion has vigorously questioned the cumulative evidence of primary sources that by the last three decades of the twelfth century both Bogomil and the related Cathar dualist communities divided into two strands: a monarchian and a radical dualist trend. Earliest Bogomil dualism had a monarchian or moderate character in clear contrast with the mature Paulician radical dualist dogma of the two principles, the evil creator of this world and the good Lord of the world to come. The interaction between Bogomilism and Catharism, on one side, and popular forms of Christian belief (in Eastern Europe also

[8] See the summary of earlier and more recent approaches to the provenance of Bogomil dualism in Stoyanov, *The Other God* (n. 6 above), pp. 161–6.

[9] For earlier views about this process cf., for example, I. Ivanov, *Bogomilski knigi i legendi*, Sofia, 1925; E. Turdeanu, 'Apocryphes bogomiles et apocryphes pseudo-bogomiles', *Revue d'Histoire des Religions*, 138, 1950, 1, pp. 22–52 and 2, pp. 176–218; N. Minissi, 'La tradizione apocrifa e le origini del bogomilismo', *Ricerche slavistiche*, 3, 1954, pp. 97–113; see the critical survey of more recent explorations of this phenomenon in Stoyanov, *The Other God* (n. 6 above), pp. 260–87 passim; Y. Stoyanov, 'Apocryphal Themes and Apocalyptic Traditions in Bogomil Dualist Theology and their Implications for the Study of Catharism', PhD diss., University of London, 2000, pp. 67–73, 213–36; see also the bibliography on the subject of the interrelations between Bogomilism and apocryphal literature in A. Orlov, 'Selected Bibliography on the Transmission of the Jewish Pseudepigrapha in the Slavic Milieux', in A. Orlov, *Selected Studies in the Slavonic Pseudepigrapha*, Leiden, 2009, pp. 203–435 (418–23).

pre-Christian survivals) should also be taken into consideration as well as the likely existence of more monistically-inclined Bogomil groups and heresiarchs.

The records of all reported versions of Bogomil and Cathar dualism highlight their marked anti-cosmic and anti-somatic nature. One of the characteristic features of medieval Christian dualism is that both versions of its Bogomil and Cathar trends present notions and narratives, variously related, for instance, to extra-biblical embellishments of the Genesis creation and flood stories, apocryphal and heretical diabologies and Christologies, etc., that are not found in the teachings of anti-clerical, dissident and heretical groups and individual figures preceding the emergence of Bogomilism and Catharism. Some of the notions and narratives, however, can be found in the various pseudepigraphic works that came to be translated and circulated in various Slavo-Byzantine contexts and milieux prior to and during the formulation and embellishment of Bogomil dualist theology.[10] Indeed some of the most important notions in Bogomilism which also later became influential in Western Catharism (such as its multi-heaven cosmology as well as core elements of its satanology, cosmogony and eschatology) were appropriated from such pseudepigraphic works as *The Book of the Secrets of Enoch* (*II Enoch*),[11] *The Apocalypse of Abraham*[12] and *The Greek Apocalypse of Baruch* (*III Baruch*).[13]

[10] See the analysis of the evidence of primary sources, including critical surveys of existing studies touching on this problematic, in Stoyanov, 'Apocryphal Themes and Apocalyptic Traditions' (n. 9 above), chs 2–4.

[11] The first edition of the apocalypse as a whole was prepared by A. I. Popov (based on a late 17th-century Russian manuscript of the Long recension): A. I. Popov, *Bibliograficheskie materialy sobrannye A. N. Popovym, Chteniia v imperatorskom obshchestve istorii i drevnostei Rossiiskikh*, 3.9, Moscow, 1880, pp. 66–139, soon after which was published for the first time a manuscript of the short recension: S. Novaković, 'Apokrif o Enohu', *Starine*, 16, 1884, pp. 67–81. The subsequent discoveries of more manuscripts belonging to both recensions led to a continuous textual debate focused on the problem of which one of two is closest to the original Slavonic translation of the apocalypse and, occasionally, whether there exists a third, intermediate version. For a bibliography of the editions, translations and studies of *II Enoch*, see Orlov, 'Selected Bibliography' (n. 9 above), pp. 222–43.

[12] Like *II Enoch* and *The Ladder of Jacob, The Apocalypse of Abraham* is extant only in Slavonic manuscripts. The Slavonic version of *The Apocalypse of Abraham* has been preserved in a more or less full form in nine Russian manuscripts, the earliest of which date from the 14th century and was published separately by N. Tikhonravov, *Pamiatniki otrechennoi russkoi literatury*, Moscow, 1863, I, pp. 32–53, and by I. I. Sreznevskii, *Drevnie pamyatniki russkogo pis'ma i iazyka: obshchee povremennoe obozrenie*, St Petersburg, 1861–63, cols 648–65. Recent critical editions of the apocalypse were published separately by B. Philonenko-Sayar and M. Philonenko, *L'Apocalypse d'Abraham: Introduction, texte slave, traduction et notes*, Paris, 1981, and by R. Rubinkiewicz, *L'Apocalypse d'Abraham en vieux slave: Introduction, texte critique, traduction et commentaire*, Lublin, 1987. The recent important textual critical study of the apocalypse includes an English translation of the text: A. Kulik, *Retroverting Slavonic Pseudepigrapha: toward the Original of the Apocalypse of Abraham*, Atlanta GA, 2004, pp. 9–37. For a bibliography of the editions, translations and studies of *The Apocalypse of Abraham*, see Orlov, 'Selected Bibliography' (n. 9 above), pp. 246–56.

[13] *The Greek Apocalypse of Baruch (III Baruch)* has been a subject of academic study for more than a century. The text of a Slavonic version of the apocalypse was published (from a 15th-century Serbian manuscript) for the first time by S. Novaković, 'Otkrivenjie Varuhovo', *Starine,* 18, 1886, pp. 203–9, and an edition of the Greek text by M. R. James, 'The Apocalypse of Baruch' in *Apocrypha Anecdota II* (TS 5/1), ed. J. A. Robinson, Cambridge, 1897, pp. li–lxxi; 83–94. The subsequent discoveries and publications of more manuscripts of *III Baruch* led to a continuous textual debate focused on the problem of the relationship between the Greek and Slavonic versions and the Slavonic textual tradition. The study of *III Baruch* was greatly enhanced by the critical editions of the Greek version of the apocalypse by J.-C. Picard, *Apocalypsis Baruchi Graece*, Leiden, 1967, and its Slavonic version by H. Gaylord, 'The Slavonic Version of 3 Baruch', PhD diss., Hebrew University of Jerusalem, 1983. These were

The role of the Garden of Eden narratives, paradise and its cosmography in medieval Christian dualism thus cannot be appropriately assessed without a proper consideration and understanding of its main dualist currents and the pseudepigraphic impact on a number of their teachings. As a rule medieval Eastern and Western polemical literature against moderate Christian dualism dwells on its principal teaching of the Devil (while recognized as a secondary and lesser power to God the Father) as a maker or architect of the visible world[14] – the Devil thus can be envisaged also as a creator of paradise.[15] The creation of paradise and its place in the configuration of the material cosmos in the records of medieval moderate dualist teachings is following a tradition of biblical exegesis exemplified by earlier Gnostic traditions in which the Creator, the God of the Old Testament, is recognized as a secondary agency, an inferior demiurge who has raised a rebellion in heaven against the superior, invisible Father who exists above him.

The theme of the creation of paradise by a secondary 'satanic' demiurge, is emphasized and embellished in the Bogomil pseudepigraphon, *Interrogatio Iohannis* (the most comprehensive and important internal source of medieval moderate Christian dualism).[16] The fashioning of paradise in the pseudepigraphon is largely synchronized with the other acts of creation in Genesis 1–2 (which the text attributes to the secondary 'diabolical' power) accomplished in the aftermath of Satan's fall from heaven to the firmament as a result of his revolt against the Father. *Interrogatio Iohannis*'s adaptation and modification of the Genesis Garden of Eden narrative is intended to provide it with a new theodicy and as in its earlier Gnostic appropriations, the focus lies on the Edenic pre-history of the first

followed by the major studies of the apocalypse by D. C. Harlow, *The Greek Apocalypse of Baruch (3 Baruch) in Hellenistic Judaism and Early Christianity*, Leiden, 1996, and most recently, A. Kulik, *3 Baruch: Greek-Slavonic Apocalypse of Baruch*, Berlin and New York, 2010, which includes a very valuable new English translation of, and commentary on, the apocalypse, pp. 89–386. For a bibliography of the editions, translations and studies of *The Apocalypse of Abraham*, see Orlov, 'Selected Bibliography' (n. 9 above), pp. 278–84.

[14] See, for example, Presbyter Cosmas's 10th-century *Sermon Against the Heretics*, newly edited in Iu. Begunov, *Kozma prezviter v slavianskikh literaturakh*, Sofia, 1973, pp. 305, 327 and 331; Moneta of Cremona, *Adversus Catharos et Waldenses libri quinque*, ed. T. A. Ricchini, Rome, 1743, *Book 1*, *Preface*, pp. 2ff.

[15] See, for example, Euthymius of Peribleptos's *Epistola* (an important source on 11th-century Anatolian Bogomilism and related groups; its text is preserved in five manuscripts but only two contain the whole text. The letter is contained in *PG* CXXXI, cols 47–58, but is erroneously attributed to a later theologian, Euthymius Zigabenus (see note 25 below). Another edition is to be found in G. Ficker, *Die Phundagiagiten: Ein Beitrag zur Ketzergeschichte des byzantischen Mittelalters*, Leipzig, 1908, pp. 3–86 (the reference to the belief he attributes to the Bogomils/Phundagiagites, that the Devil is maker on paradise, is on p. 33: ...οὐ μὴν ἀλλά ὅτι καὶ τὸν παράδεισον αὐτός φασιν ἐποίησεν καὶ τὸν ἄνθρωπον ἔπλασην).

[16] The apocryphon is extant only in Latin and divides into two main versions; the first version derives from a manuscript once in the archives of the Inquisition at Carcassonne but subsequently destroyed: it survives in two late manuscripts and one printed text which was published in J. Benoist, *Histoire des Albigeois et des Vaudois ou Barbets*, Paris, 1691, I, pp. 283–96. Subsequent reprints included that of J. C. Thillo, *Codex apocryphus Novi Testamenti,* Leipzig, 1832, I, pp. 884–96. The second version is represented solely by a manuscript preserved in the National Library of Austria (Vienna), apparently dating from the 12th-13th centuries. The Carcassonne version was reproduced alongside the Vienna version by M. Sokolov, *Slavianskaia kniga Enokha pravednago*, Moscow, 1910, pp. 165–75; also by Ivanov, *Bogomilski knigi* (n. 9 above), pp. 73–87, R. Reitzenstein, *Die Vorgeschichte der christlichen Taufe*, Leipzig and Berlin, 1929, pp. 297–311, and in the most recent critical edition of text, E. Bozóky, *Le Livre secret des cathares,* Paris, 1980, pp. 41–94.

human couple, the identity of the Edenic serpent who tempted Eve to eat the prohibited fruit of the Tree of Knowledge of Good and Evil and the nature of the Tree itself.[17]

In its rendition of the Genesis paradise story, and the fall of Adam and Eve, *Interrogatio Iohannis* furnishes, moreover, a new version of the tradition developed in earlier Gnostic systems regarding the Gnostic demiurge's (or his demonic archons's) sexual encounter with, or assault on, Eve (or her 'likeness'): this finds parallels in Jewish rabbinic traditions concerning Eve's seduction by Samael/Satan (who could be identified or associated with the Edenic snake) and their progeny (*Pirke de-Rabbi Eliezer* 22; *Targum-Pseudo-Jonathan* 5:1–3).[18] In *Interrogatio Iohannis*'s Edenic narrative Satan made paradise with the unequivocal design of seducing the first human couple to commit sin[19]; he planted in the middle of it a bed of reeds and created the serpent;[20] he forbade Adam and Eve to eat the fruit of good and evil but then he entered the serpent and tempted Eve into sin[21] (as embellished in one of the glosses of the Vienna version of the pseudepigraphon, posing as a 'beautiful youth'[22]). Subsequently, Satan also infected Adam with a lust for intercourse

[17] On Gnostic attitudes to, and revisions of, the Genesis paradise narrative and related conceptualizations of the theme of paradise, see, for example, R. van den Broek, 'The Shape of Edem According to Justin the Gnostic', *Vigiliae Christianae*, 27, 1973, pp. 35–45; M. Tardieu, *Trois mythes gnostiques: Adam, Éros et les animaux d'Égypte dans un écrit de Nag Hammadi (II, 5)*, Paris, 1974, P. S. Alexander, 'The Fall into Knowledge: The Garden of Eden/paradise in Gnostic Literature', in *A Walk in the Garden*, eds P. Morris and D. Sawyer, Sheffield, 1992, pp. 91–104; J. Magne, *From Christianity to Gnosis and From Gnosis to Christianity: An Itinerary through the Texts to and from the Tree of Paradise*, Atlanta GA, 1993 (Magne's thesis sees the reciprocal exegesis of the Emmaus and paradise stories as the key to the origins of Gnosticism and Christianity, arguing for a new chronology of these interpretations in the context of Jewish-Gnostic-Christian interrelations; Magne asserts that the paradise narrative in the Gnostic writings 'is the fundamental, central core around which and for which everything is organized', p. 73, and the Gnostic movement 'would have sprang directly ... from the exegesis of the paradise narrative revealed by the Emmaus narrative ...', p. 59); T. Rasimus, *Paradise Reconsidered in Gnostic Mythmaking: Rethinking Sethianism in Light of the Ophite Evidence*, Leiden, 2009.

[18] For Gnostic elaborations of the theme of Eve's seduction or attempted seduction by the demiurge, his oppressive archons or the chief archon in the Nag Hammadi tracts, see *The Apocryphon of John* 24:8–28; *The Apocalypse of Adam* 66:25–66:10; *The Hypostasis of the Archons*, 89:17–28; *On the Origin of the World* 116:5–117:20; For a comparative survey of the theme of Eve's seduction in Gnostic (original and reported) and Jewish traditions, see G. A. G. Stroumsa, *Another Seed: Studies in Gnostic Mythology*, Leiden, 1984, pp. 35–53. For further evidence concerning Gnostic and Gnostic-related teachings about Eve's seduction by the demiurge or his archons (or authorities), which maintained their currency among the schismatic Audians in the Holy Land, see H.-C. Puech, 'Fragments retrouvés de l'Apocalypse d'Allogéne', repr. in H.-C. Puech, *En quête de la Gnose, I: La Gnose et le temps et autres essais*, Paris, 1978, pp. 271–300.

[19] *Interrogatio Iohannis* (Vienna version, 94–5), ed. Bozóky, *Le Livre secret* (n. 16 above), p. 58: 'Initiator autem peccati cum sua seductione ita fecit: plantavit paradisum et misit hominess intus et precepit eis ne comederent ex eo.'

[20] *Interrogatio Iohannis* (Vienna version, 96–9), ed. Bozóky, *Le Livre secret* (n. 16 above), p. 60: 'Diabolus intravit in paradisum et plantavit arundem in medio paradisi et de sputo suo fecit serpentem et precepit ei in arundine manere et sic diabolus ascondebat sapientiam sue fraudis ut non viderent deceptionem suam.'

[21] *Interrogatio Iohannis* (Carcassonne version, 93–9), ed. Bozóky, *Le Livre secret* (n. 16 above), p. 60: 'Et intrabat et loquebatur ad eos dicens: De omni fructu qui est in paradise, comedite, de fructu vero scientie boni et mali nolite comedere. Et iterum intravit dyabolus in serpentem nequam et seduxit angelum qui erat in forma mulieris et effidit super ea concupiscentiam peccatorum et fecit concupiscentiam suam cum Aeva in cauda Serpentis.'

[22] *Interrogatio Iohannis* (Vienna version, Gloss 7, 31–3), ed. Bozóky, *Le Livre secret* (n. 16 above), p. 90: 'Serpens non fuit ut serpens sed sicut homo. Fecit enim se sicut pulchrum adolescentem et preceptum diaboli introivit in paradisum per canam et decepit mulierem et fornicatus est cum ea cum cauda.'

with Eve and they begot together the 'children of the devil and of the serpent' and Satan's reign, being dependent on procreation, would last until the consummation of this age.[23]

Thus in *Interrogatio Iohannis* the Devil's fashioning of paradise, his seduction of Eve and transmission to Adam of a 'lust for debauchery' (resulting in physical procreation) marks a vital phase of his stratagems to confine humanity's divine substance in matter. In his earlier stratagems during the anthropogonic process he forced the angels of the first and second heaven to enter bodily captivity in mortal, respectively male and female, forms.[24] Triggering (and enacting with Eve) a 'longing for sin' and 'lust for debauchery' into paradise, Satan secures and reinforces humanity's imprisonment in matter. The 'satanic' Eden's position in the creation and structure of the material universe is thus determined by its crucial role in the process of the imprisoning the protoplasts in the flesh and subjecting humanity to the disastrous (for its spiritual substance) chain of physical procreation.

A shorter version of the paradise episode of Eve's seduction by Satan also appears in the very highly valuable early twelfth-century theologico-polemical source on contemporaneous moderate dualist Bogomilism, Euthymius Zigabenus's *Panoplia Dogmatica*.[25] In Zigabenus's version the creation of Adam and Eve was enacted in a kind of cooperation between Samale/Satanael and God – while Satanael moulded their bodies, God gave them the breath of life. However, Satanael (Samael) became jealous of them and began plotting against their descendants, which represented the first breach in his contract with God concerning the creation of man and the fate of his generations. As in *Interrogatio Iohannis*, he slipped into the serpent and through its medium he tricked and seduced Eve after which she begot from him twins, Cain and his sister Calomena, whereas Abel is born after her later human union with Adam.[26] Significantly, elements of this paradise narrative of Eve's seduction by Satan and (on certain occasions) the consequent Cainite lineage are related, or alluded to, in some of the principal Catholic polemical accounts of moderate Cathar dualism.[27]

[23] *Interrogatio Iohannis* (Carcassonne version, 101–4), ed. Bozóky, *Le Livre secret* (n. 16 above), p. 62: 'Et iterum effudit dyabolus in angelo qui erat in Adam, venenum suum et concupiscentiam que generat filios serpentis et filios dyaboli usque ad consummationem huius seculi.'

[24] *Interrogatio Iohannis* (Vienna version, 87–90), ed. Bozóky, *Le Livre secret* (n. 16 above), p. 58: 'Et precepit angelo secundi celi introire in corpus luti et tulit de eo et fecit alium corpus in forma mulieris precepitque angelo primi celi introire in illum. Angeli ploraverunt multum videntes super se formam mortalem esse in divisis formis.'

[25] Euthymius Zigabenus, *Panoplia Dogmatica, PG* CXXX; the Bogomil section comprises cols 1289–1331; another version of the Bogomil section is also edited by Ficker in *Die Phundagiagiten* (n. 15 above), pp. 89–111.

[26] ... καὶ εἰσρυῆναι τοῖς ἐγκάτοις τοῦ ὄφεως, καὶ ἐξαπατῆσαι τὴν Εὖαν καὶ συγγενέσθαι αὐτῇ, καὶ ποιῆσαι ἔγκυον, ἵνα τὸ σπέρμα τούτου προλαβὼν κατακυριεύῃ τοῦ Ἀδαμιαίου σπέρματος...Τὴν δέ ταχέως ὠδινήσασαν ἀπογεννῆσαι τὸν Καῒν εκ τῆς συνουσίας τοῦ Σαταναὴλ, καὶ ἀδελφὴν δίδυμον ὁμοιότροπον, ὄνομα ἀυτῇ Καλωμενά. Ζηλοτυπήσαντα δὲ τὸν Ἀδὰμ συνηλθεῖν καὶ αὐτὸν τῇ Εὔᾳ καὶ γεννῆσαι τὸν Ἀβὲλ, ὃν ἀνελὼν εὐθὺς ὁ Καῒν τὸν φόνον εἰς τὸν βίον εἰσήγαγε ... Zigabenus, *Panoplia Dogmatica* (n. 25 above), col. 1297.

[27] Bonacursus, 'Manifestatio haeresis catharorom quam fesit Bonacursus', *PL* CCIV, cols 775–92 (776); Moneta of Cremona, *Adversus Catharos*, ed. T. Ricchini (n. 14 above), p. 145; Georgius, *Disputatio inter Catholicum et Paterinum hereticum*, eds E. Martène and U. Durand, in *Thesaurus novus anecdotorum*, V, Paris, 1717, cc. 1705–53 (1710–11); Salvo Burci, *Liber supra Stella*, ed. I. da Milano, 'Il 'Liber supra Stella' del piacentino Salvo Burci contro i catari e altri correnti ereticali', *Aevum*, 19, 1945, pp. 218–341 (314). The interrelations between the

The theme of Eve's seduction by a secondary diabolical power, Satan (or Samael), in paradise is present not only in the earlier and already mentioned Jewish and Gnostic traditions but also in *II Enoch* 31:6. In this section of *II Enoch* Satanael (to be transformed into Satan) starts plotting against Adam, makes his entry into paradise and seduces Eve.[28] It is possible, but not certain, that this narrative in *II Enoch* may represents itself a Bogomil or Bogomil-influenced interpolation; otherwise, given the fact that *II Enoch* served as an important source for Bogomil teachings, this diabological section in *II Enoch* undoubtedly would have influenced or enhanced the Bogomil preoccupation with, and elaboration of, this paradise theme. Furthermore, the association between the Devil, a paradise tree (whether the Tree of the Knowledge of Good and Evil or some other tree) and the sin of Adam and Eve (understood as related to lust or sexual union) appears in the Slavonic version of *III Baruch* 4:8 where Satanael plants in paradise the vine which is defined as the tree via which the serpent seduced Adam and Eve, representing also the sinful desire that Satanael spread over them.[29]

The Apocalypse of Abraham 23:5–12 offers a similar association: after Abraham queries the identity of Eve's seducer, and what was the fruit of the tree, he is granted a vision of paradise in which he sees Adam entwined with Eve under a tree. The fruit of the tree has the appearance of a bunch of grapes, held and fed to the couple by a dragon with human hands and feet, recognized as Azazel. Thus elements of the paradise-related complex of notions, variously narrated and emphasized in the *Interrogatio Iohannis*, *II Enoch* and the Slavonic version of *III Baruch* – namely the tree that caused the Fall, the sexual union of Adam and Eve and the figure of God's adversary (in the case of the *The Apocalypse of Abraham*, Azazel) – are intertwined in Abraham's synthetic vision of the Garden of Eden.[30]

representation of the association between the sexual union of Eve with Satan and Adam with the forbidden fruit of the Tree of Knowledge of Good and Evil is observed by Bozóky, *Le Livre secret* (n. 16 above), p. 136.

[28] И разоумѣ дїаво҃, ꙗко дроугы миръ хоштѫ сьтворити, занеже повинѫло се е҃ Адамоу земли и обладати и цр҃твовати ими…Тацѣм҃ ѡбразом҃ вънидет҃ въ раи и прѣльсти Еввѡ, Адамоу же не прикоснѫ сѧ, *II Enoch* 31:6 (Long version), MS Belgrade, National Library of Serbia, No. 321, reproduced in *Slavianskaia kniga Enokha*, p. 31; Ivanov, *Bogomilski knigi* (n. 9 above), p. 178.

[29] …и пакы рѣхъ азъ вароухъ къ ангелоу покажи ми дрѣво юже прѣльсти змиꙗ ювъгѫ и адама… прѣвоѥ дрѣво юсть лоза, въторою юст дрѣво похоть грѣховьна, юже излиꙗ сатанаиль на ювъгѫ и на адама; и сего ради проклѧт богъ лозѫ зане бе ѭ сатанаиль въсади и тоѭ прѣльсти прьвозьданаго адама и ювъгѫ, *III Baruch* 4:8 (Slavonic version), MS Belgrade, National Library of Serbia, No. 651, published in Ivanov, *Bogomilski knigi* (n. 9 above), p. 196. For the variant readings of the passage in other Slavonic manuscripts of *III Baruch*, see Gaylord, 'The Slavonic Version of 3 Baruch' (n. 13 above), pp. 46–7 (the Greek texts of *III Baruch* also identify the tree that deceived Adam with the vine but lack the theme of three as a 'sinful desire').

[30] И смотрихъ въ образование и течесте очи мои на страну овощника Едемь и видѣхъ ту мужа велика зѣло въ висоту и страшна въ щириню, бесприкладна зракомъ, съплетшася съ женою яже и та сравняшеся мужесцѣм зрацѣ и възрастѣ. И бѣсте стояща подъ единемъ садомъ едемьскимъ. И бяше плодъ сада того яко видъ грозда виничьна. И за садомъ бяше стоя яко змии образомъ, руцѣ. же и нозѣ. имы подобно чл(о)в(ѣ)ку и крилѣ на плещю его…И держаше в руку гръздъ садовныи. Залагаше оба яже видѣхъ съплетшася. И рекохъ: «Кто еста съплетшася к себе или кто есть сыи меж нима или что есть плодъ егоже ѣста, Крѣпце Превѣчне? И реч(е): "Се есть свѣтъ чл(о)в(ѣ)чъ – се есть Адамъ. И се есть помышьление ихъ на земли - си есть Евьга. А еже межю има, то есть бесцестие начинания ихъ въ погыбель - самъ Азазилъ", *The Apocalypse of Abraham*, 23:5–9; text in Philonenko-Sayar and Philonenko, 'L'Apocalypse d'Abraham' (n. 12 above), p. 88.

The association between paradise and its primal tree(s) with the human carnality made explicitly in *Interrogatio Iohannis* (and the monarchian Christian dualism it exemplifies)[31] is thus largely shared in the Slavonic version of *III Baruch* and *The Apocalypse of Abraham* and very likely entered Bogomil sacred history and cosmography narratives under the influence of these pseudepigrapha. Earlier Gnostic traditions, as developed, for example, in the Nag Hammadi tract, *Apocryphon of John* and Justin the Gnostic's *Book of Baruch* also present parallels to this association.[32] Eve is also linked with the Tree of Knowledge in another Nag Hammadi tract, *The Origin of the World*, which narrates that while being sexually pursued by archons, Eve hid herself within the Tree of Knowledge, leaving behind for them only an earthly resemblance of herself. This link finds an intriguing analogy in one of the fifty articles of Patarene/'Manichaean' beliefs reportedly renounced by three 'heretical' Bosnian noblemen before Cardinal Torquemada in 1461.[33] Article 12 of the list concerns the belief that the Tree of Knowledge of Good and Evil was envisaged as a woman and after Adam sinned with her, was expelled from paradise. It is, of course, possible that this belief could have been attributed to the Bosnian noblemen on the basis of inquisitorial records of earlier medieval dualist teachings.[34] But approached in the context of the already mentioned pseudepigraphic narratives this material suggests that these teachings were elements of a more comprehensive and Gnostic-leaning reworking of the Genesis paradise story than the extant evidence allows one to reconstruct. This impression is also reinforced by Cathar teaching attributed to the Cathars in Georgius, *Disputatio inter Catholicum et Paterinum hereticum*, allegorizing the Tree of Knowledge of Good and Evil as a woman's womb.[35] That specific teaching becomes the target of a bitter polemical attack in the *Disputatio*, vigorously denying that woman can be associated with paradise, or the tree in its midst with her womb.

Another confirmation to this effect comes from a gloss to the *Interrogatio Iohannis*'s paradise section in its Vienna version which elaborates with slightly more detail the episode of the planting of paradise (regarding the type of trees planted) and adds the

[31] See, for example, the association between the forbidden tree and fornication in the moderate Cathar dualist teachings recounted in *De heresy catharorum*, ed. A. Dondaine, 'La Hiérarchie cathare en Italie, I: "*De heresy catharorum in Lombardia*"', *Archivum fratrum praedicatorum*, 19, 1949, ch. 1, pp. 306–12 (310): 'Et dicunt quod comestio ligni prohibiti fuit fornicatio ... '

[32] Justin the Gnostic's *Book of Baruch* is reproduced in Hippolytus, *Refutatio Omnium Haeresium*, 5.26–7 (it is uncertain whether his presented, reproduced text is full or partial).

[33] These Patarene articles are refuted in Juan Torquemada, *Symbolum pro informatione manichaeorum*, eds N. L. Martinez and V. Proano, Burgos, 1958. An English translation and a useful commentary are in J. V. A. Fine Jr, *The Bosnian Church. A New Interpretation*, New York and London, 1975, pp. 355–7, repr. London, 2007, pp. 280–86. Attention to this article in the context of earlier and similar Gnostic beliefs was drawn by S. Gero, 'The Seduction of Eve and the Trees of Paradise: A Note on a Gnostic Myth', *The Harvard Theological Review*, 71, 3/4, 1978, pp. 299–301.

[34] As argued by Fine, *The Bosnian Church* (n. 33 above), p. 282.

[35] Georgius, *Disputatio inter Catholicum et Paterinum hereticum*, eds Martène and Durand, *Thesaurus novus anecdotorum* (n. 27 above), V, 1710.

characteristic cosmographic notion that Satan surrounded it by a ring of fire and emphasizing its essentially negative and evil nature.[36] Some of the motifs of *Interrogatio Iohannis*'s episode of Satan's planting of paradise also seem certainly interrelated with similar themes which emerge in medieval Slavonic apocryphal texts of a compilatory nature such as the cycle of *The Legend of the Wood of the Cross*, attributed to Gregory of Nazianzus[37]; the erotapocritic work, *Razumnik*[38]; and in the sixteenth-century South Slavonic cosmogonic text *On the Beginning of the World*.[39]

The first two texts contain the theme of Satanael/Samael stealing seeds from paradise and planting them on his own; he was exposed and challenged by God to confront the tree he has stolen and planted and as a result of his misdeed he was blackened and expelled from Eden by the very tree itself.[40] Aside from the theme of Satan planting a vine and importing sinful passion in paradise to deceive Adam, the third apocryphal legend, *On the Beginning of the World*, provides more analogies to *Interrogatio Iohannis* in its narrative of Satanael's entering paradise and assuming the shape of a serpent with the head of a maiden to lead Adam and Eve astray.[41] Further study is clearly needed to explore in greater depth all these evident intertextual links between pseudepigrapha-derived notions and narratives in Bogomilism and these particular cycles of apocryphal works, with their focus on cosmogony, cosmology and biblical history. Such study would certainly shed new light on the provenance and main traits of the paradise material shared, rewritten and re-interpreted by these two currents of medieval appropriation and reworking of earlier pseudepigraphic texts.

Apocryphal accounts of the creation of paradise and revisions of the Genesis story also can be found in other pseudepigraphic texts circulated in the medieval Slavonic Orthodox world such as the Slavonic version of the *Life of Adam and Eve*, *The Sea of Tiberias* and

[36] *Interrogatio Iohannis* (Vienna version, Gloss 6, 23–30), ed. Bozóky, *Le Livre secret* (n. 16 above), p. 90.

[37] Recent edition in A. Miltenova, 'Tekstologicheski nabliudeniia vŭrkhu dva apokrifa (Apokrifen tsikŭl za krŭstnoto dŭrvo pripisvan na Grigorii Bogoslov i apokrifa za Adam i Eva)', *Starobŭlgarska literatura*, 11, 1982, pp. 35–55 (44–55).

[38] Editions of the text include Ivanov, *Bogomilski knigi* (n. 9 above), pp. 264–73; A. Miltenova, *Erotapokriseis. Sŭchineniia ot kratki vŭprosi i otgovori v starobŭlgarskata literature* (Sofia, 2004).

[39] Published in Ivanov, *Bogomilski knigi* (n. 9 above), pp. 322–3.

[40] ...егда насаждаше гь раи. да иже повелѣвааше садити. самаиль крадѣше вси ѿ вьсего. и изышедь просипа на инѡмь мѣстѣ таж ѿ га. гь ре ти крадеши ѿ мене да бѫдеть тебѣ на прогнанїе. и изыде самаиль. и ре. ги бл̃ви, елика садихѡмь. гь ре бл̃венно и тоу есмь азь посрѣдѣ его. самаиль иде да види дрѣво, еже оукрадь посади. и егда виде дрѣво, тогда самаиль почрьнѣ и изьгна его дрѣво из раıа. и тогда нарече гь имѧ емꙋ дїаволь, texts (the 16th-century Lovchanski miscellany) respectively in A. Miltenova, *Erotapokriseis*, p. 456; Ivanov, *Bogomilski knigi* (n. 9 above), pp. 260–261. On the place of this particular passage on Satanael and the Tree in paradise in the manuscript tradition of *Razumnik*, see A. Miltenova, *Erotapokriseis. Sŭchineniia ot kratki vŭprosi i otgovori v starobŭlgarskata literatura*, Sofia, 2004, pp. 251–3.

[41] И Сатанаиль има злобꙋ на Ꙗдама и на Євгꙋ и васади лозꙋ и похоть древнꙋ да преласти Ꙗдама. И такы сташе Ꙗдамь и Євга ва раи, и дођe дıаволь и сатвори се змиѡмь. Ꙗ тадай биеше змиıа ꙋ раи и хођаше на рапꙋ како и човекь на ногꙋ и збаше се красна девица, тере се сатвори девичномь змиѡмь и главомь девичномь, и дође напастовати Євгꙋ, *On the Beginning of the World*, ed. Ivanov, *Bogomilski knigi* (n. 9 above), p. 323; on these parallels between the paradise material in *Interrogatio Iohannis* and *On the Beginning of the World*, see ibid., *Bogomilski knigi*, p. 70.

the *Battle between Archangel Michael and Satanael*. It has been continuously argued that some versions of these works display traces of medieval dualist editorial interventions and interpolations.[42] Moreover, certain traits of the paradise material in these apocryphal works possess interesting cosmographic features but as they lack discernible dualist elements or tendencies, it would be premature and risky to discuss this material in the framework of the paradise tradition in medieval Christian dualism.[43] What is certain is that future study of the patterns of intertextual relations between these texts and Bogomil narratives' assimilation of the shared pseudepigraphic heritage in medieval Byzantine and Orthodox Slavonic contexts would open new vistas for the exploration of the evolution of ideas about the cosmography of paradise in normative, heterodox and heretical medieval Eastern Orthodox milieux.

It is worth noting in this context that recent studies of the paradise material in *III Baruch* 4 (particularly the specific narrative of individual angels planting trees in the Garden of Eden and its destruction by the waters of the Flood) have cast a fresh light on its dualist tendencies and potential which underlies and explains the medieval dualist scribes' predilection to adopt and revise some of its elements for their use. These recent analyses have shown that on the one hand, the list of angels planting trees and the type of trees they planted in the Slavonic version of *III Baruch* 4 exhibits continuity respectively with rabbinic traditions and Philo.[44] On the other hand, the notion of the five trees in paradise presents analogies not only in Philo but also in Gnostic and Manichaean traditions.[45] Significantly, the chronology and these two main strands of the paradise narrative in *III Baruch* 4 are also shared in the Qumran and Manichaean fragments of the *Book of Giants*[46] and a later and related text, the *Midrash of Shemhazai and Azael*.[47] As convincingly argued by Orlov,

[42] For a critical survey of the evidence for, and state of debate of, such dualist editorial interventions in these apocryphal works, see Stoyanov, 'Apocryphal Themes and Apocalyptic Traditions' (n. 9 above), pp. 114–191; in the case of the *Battle between Archangel Michael and Satanael*, the issue of dualist involvement in the text has been revisited by G. Minczew, 'Edin bogomilski tekst? Slovo na Sv. Ioan Zlatoust kak Mihail pobedi Satanail', *Palaeobulgarica*, 34/4, 2010, pp. 17–46; G. Minczew, 'John Chrysostom's Tale on How Michael Vanquished Satanael – a Bogomil text?', *Studia Ceranea*, 1, 2011, pp. 23–55.

[43] Significantly, although the *Battle between Archangel Michael and Satanael* recounts God's creation of paradise and attributes actual demiurgic acts to Satanael (he creates his own heavens, dark sun, moon and stars – a parallel satanic universe), these acts do not include the creation of an alternative paradise. On the dualist tendencies and type of cosmic dualism advanced in the apocryphon, see Stoyanov, 'Apocryphal Themes and Apocalyptic Traditions' (n. 9 above), pp. 120–22.

[44] Harlow, *The Greek Apocalypse of Baruch (3 Baruch)* (n. 13 above), pp. 125–6; A. Kulik, *3 Baruch. Greek-Slavonic Apocalypse of Baruch* (n. 13 above), pp. 198–208 passim; A. Kulik, 'Veritas Slavica: On the Value of Slavonic Evidence for the Early Apocalyptic Tradition', *Polata Knigopisnaia*, 38, 2010, pp. 1–65 (40–41).

[45] Kulik, *3 Baruch* (n. 13 above), pp. 204-6; Kulik, 'Veritas Slavica' (n. 44 above), pp. 40–41.

[46] Cf. the texts in W. B. Henning, 'The Book of the Giants', *Bulletin of the School of Oriental and African Studies*, 11, 1943, pp. 52–74; J. Milik and M. Black, *The Books of Enoch. Aramaic Fragments of Qumrân Cave 4*, Oxford, 1976, p. 317; L. Stuckenbruck, *The Book of Giants from Qumran: Texts, Translation, and Commentary*, Tübingen, 1997.

[47] Hebrew text and translation in Milik and Black, *The Books of Enoch* (n. 46 above), pp. 321–8; on the shared paradise material in *III Baruch* 4, the Qumran and Manichaean fragments of the *Book of Giants* and *Midrash of Shemhazai and Azael*, see A. Orlov, 'The Flooded Arboretums: The Garden Traditions in the Slavonic Version of 3 Baruch and in the Book of Giants', *Catholic Biblical Quarterly*, 65, 2003, pp. 184–201.

another vital theme is also shared by these two related strands of paradise material: the participation of a fallen angelic being (or beings) in the planting of the Garden of Eden – Samael (Satanael) in *III Baruch* – and the fallen angels of the Enochic tradition, the Watchers in the *Book of Giants*, respectively.[48] In the resultant paradise narrative paradise itself acquires an evident centrality in cosmic history and topography, being transformed into a primeval locus where a supernatural being (or beings) provoked the first revolt against the original divine order and authority.

As far as medieval Christian radical dualism is concerned, the extant evidence (internal and external) of its Eastern Christian variety is rather fragmentary, concerned predominantly with its theological, diabological and docetic tenets. Hence it does not allow even a partial reconstruction of its use and reappraisal of the paradise tradition comparable to the better evidenced moderate dualist paradise narrative. The corresponding records of Western Christian (Cathar) radical dualism are more informative, offering some interesting details and indications that its adoption and reworking of the paradise tradition was subject to a characteristic exegetical dynamics, with some clear implications for the cosmographic sphere. Drawing on the absolute dualist formula of two primal, causal and coevally opposed principles, medieval radical dualist narratives developed cosmogonies of two parallel universes. In the realm of evil the God of Darkness was the creator of the four elements, the visible heaven, the moon, the stars and everything on earth, while in the other realm the God of Light was the creator of everlasting, eternal things, four alternative elements of his own and another heaven, moon and stars.[49]

Cosmogonies of this kind in which the two spheres of light and darkness are coeternal and originally entirely separate possess an inbred tendency to attribute a transcosmic nature to a paradise located in the kingdom of Light, with all its 'heavenly' connotations. Such development may be discerned in the anonymous radical dualist 'Manichaean' treatise, segments of which are reproduced in Durand of Huesca's early thirteenth-century *Liber contra manicheos*.[50] In the sixth excerpt from the treatise the notions of 'new heaven and the new earth' of Isaiah 66:22, II Peter 3:13 and Revelation 21:1 are accordingly interpreted as the paradise of the celestial kingdom which serves as the location of the angelic court of the Father, the Tree of Life, river of life and holy Jerusalem. What is more, such celestial cosmographies in Cathar radical dualism also could include heavenly conflict narratives which have implications for the reappraisal of the paradise tradition in such medieval dualist

[48] Orlov, 'The Flooded Arboretums' (n. 47 above); cf. J. C. Reeves, *Jewish Lore in Manichaean Cosmogony*, Cincinnati OH, 1992, pp. 95–6; Stuckenbruck, *The Book of Giants* (n. 46 above), p. 114.

[49] See, for example, the system of Cathar absolute dualism recorded in Moneta of Cremona, *Adversus Catharos, Book 1* (n. 14 above), 'Preface', pp. 2–4.

[50] Published by C. Thouzellier, *Un Traité cathare inédit du début du XIIIe siècle d'après le* Liber contra manicheos *de Durand de Huesca,* Louvain, 1961, pp. 87–113.

settings. In such narratives Satan-Lucifer (seen on occasions as the son of an eternal evil god) is generally depicted as instigating an attack on the heaven of the good God, provoking the fall of the angels and the exile of angelic souls in human bodies. Significantly enough, in the teachings of the Cathar community of the Albanenses of North Italy, for example, he could also be described as sexually assaulting the celestial wife of the good God[51] which could parallel Eve's seduction by Samael/Satan in the paradise narratives of medieval moderate Christian dualism.[52]

A more detailed exploration of the Cathar radical dualist cosmography of the celestial paradise and the sexual aspect of its heavenly conflict narratives is needed before any conclusions could be reached about any possible correlation between the paradise narratives of the two principal currents of medieval dualism. Certain currents in medieval Christian dualism could also display a telling focus on the cosmography and soteriological role of terrestrial paradise. This is manifested, for instance, in the innovative teachings of the second person in the hierarchy of the moderate dualist Cathar Church of Concorezzo in North Italy, Desiderius,[53] whose embellishments in the monarchian Cathar paradise tradition as well as other doctrinal spheres eventually triggered a schism in this Cathar church. Desiderius ventured to advance and elaborate a more somatic variant of Bogomil-Cathar Christology by arguing that Jesus assumed a body of 'the stuff of Adam',[54] a teaching which contrasted sharply with traditional Bogomil-Cathar docetic doctrine that Christ never took a real physical body, but did so only in appearance. Desiderius, moreover, betrayed a markedly non-docetic concern with the location of the physical body of Jesus Christ, which, he asserted, was deposited in a terrestrial paradise. The Virgin Mary, John the Evangelist and the souls of the righteous dead also resided in the terrestrial paradise and were to stay there until the Day of Judgement when Christ will put his physical body again to rise and judge all good and evil.[55]

While Desiderius's teachings on terrestrial paradise are in obvious disagreement with the other extant medieval dualist reworkings of the paradise tradition, they may also reflect contemporaneous disputes and views on the nature, main features and cosmography of paradise in medieval Christendom. Medieval Christian dualism did not rise and evolve in isolation from contemporaneous intellectual currents and developing cultural trends, but variously interacted with them in diverse cultural and religious settings.

[51] The radical dualist teachings of the Albanenses are recorded in the treatise, 'Brevis summula contra herrores notatos hereticorum', ed. C. Douais, *La Somme des autorités à l'usage des prédicateurs méridionaux au XIIIe siècle*, Paris, 1896, pp. 114–43.

[52] For a brief discussion of such a possible interrelation between the Albanenses's and moderate dualist paradise narratives, see Stoyanov, *The Other God* (n. 6 above), pp. 276–7.

[53] Desiderius's teachings are recorded in Anselm of Alessandria, *Tractatus de hereticis*, ed. A. Dondaine, 'La Hiérarchie cathare en Italie, II: Le "Tractatus de hereticis d'Anselme d'Alexandrie"', *Archivum fratrum praedicatorum*, 19, 1949, pp. 308–25 (310–12).

[54] Anselm of Alessandria, *Tractatus de hereticis* (n. 53 above), p. 311.

[55] Ibid., pp. 311–12.

Recent studies have explored, for instance, the significance and evolution of high learning in Northern French,[56] Italian[57] and Languedoc Catharism,[58] highlighting the various patterns in which the Cathars played a role in the broader theological and educational trends of the late twelfth and early thirteenth centuries.[59] The study of whether, and to what degrees, reappraisals of the paradise tradition and its cosmography could have been a part of these insufficiently explored patterns of interaction between dualist heresy, literacy and intellectual pursuits of this period could indeed be one of the most promising areas for the continuing investigation of the transmutations of medieval perspectives on paradise.

The paradise narratives in medieval moderate and radical Christian dualism (as identified and explored so far) and their cosmographic features certainly lack the subtleties and intricacies of earlier Gnostic elaborations and re-conceptualizations of the paradise theme, with their cosmological, soteriological and eschatological dimensions and implications. But at least to some extent this may be due to the fact that not all relevant evidence (internal and external) of the role and principal re-contextualizations of the paradise narrative in medieval Christian dualism has been identified and explored. With the current level of research on this relevant evidence, it is difficult to determine as yet whether Eastern Christian dualism has assimilated in its theological and cosmological models more cosmographic material pertaining to paradise from the pseudepigraphic texts like *II Enoch* and *III Baruch* which partially had served as sources for these models.

The potential of these areas of study in a cosmographic context is indicated, for example, by the role of *II Enoch* in medieval Byzantine and Orthodox Slavonic speculations on the cosmography of paradise, as demonstrated by the close textual interrelations between its long version and the Greek and Slavonic manuscripts of the anti-Catholic satirical tract *The Disputation of Panagiot and Azimit*.[60] It is also significant that, even before its medieval circulation in Byzantine and Orthodox Slavonic literary (and at least in the Slavonic case, heretical sectarian) environments, the paradise material

[56] See, for example, P. Biller, 'Northern Cathars and Higher Learning', in *The Medieval Church: Universities, Heresy, and the Religious Life, Essays in Honour of G. Leff*, eds P. Biller and B. Dobson, Woodbridge, 1999, pp. 25–52.

[57] L. Paolini, 'Italian Catharism and Written Culture', in *Heresy and Literacy 1000–1530*, eds P. Biller and A. Hudson, Cambridge, 1994, pp. 83–103.

[58] P. Biller, 'The Cathars of Languedoc and Written Materials', in *Heresy and Literacy*, (n. 57 above), pp. 61–82, especially pp. 80–82.

[59] See the summary of the evidence of, and approaches to, this phenomenon in Stoyanov, *The Other God*, (n. 6 above), p. 261.

[60] For the Slavonic texts of *The Disputation of Panagiot and Azimit*, see A. N. Popov, *Istoriko-literaturnyi obzor drevnerusskikh polemicheskikh sochinenii protiv latinian*, Moscow, 1875, pp. 211–83; on the textual interrelations between the cosmographic paradise material in the long version of *II Enoch* and *The Disputation of Panagiot and Azimit*, cf. Sokolov, *Slavianskaia kniga Enokha*, II, pp. 136–44; A. Vaillant, *Le Livre des secrets d'Hènoch: Texte slave et traduction français*, Paris, 1952, pp. xvi–xvii; F. Andersen, 'The Sun in 2 Enoch', in *L'Église des deux Alliances: Mémorial Annie Jaubert (1912–1980)*, eds B. Lourié, A. Orlov, and M. Petit, Piscataway NJ, 2008, pp. 1–38 passim.

in the original version of *II Enoch* had been already used as a source for Manichaean dualist teachings about the heavenly paradise and the heavenly Tree of Life (as convincingly argued by J. C. Reeves[61]). Another related area that would profit from further exploration is whether the adoption of heavenly ascent notions from earlier pseudepigrapha and their recasting into new dualist versions of visionary mysticism (as attested among some Bogomil and Cathar circles[62]) included also re-interpretations of Paul's account of ascent to the third heaven and paradise (II Corinthians 12:1–4),[63] with the inevitable cosmographic implications of such dualist re-contextualizations.

A more detailed and methodical analysis of the paradise problematic in the sources for Eastern and Western Christian medieval dualism will thus prove fruitful in all these spheres of research. One of the main research questions of such future analysis will be to establish a more coherent picture of the nature of the cosmographic concerns accompanying the allegorical and inverted readings of the Genesis Eden story in medieval Christian dualist approaches to, and revisions of, the paradise tradition.

[61] J. C. Reeves, 'Jewish Pseudepigrapha in Manichaean Literature: The Influence of the Enochic Library' in *Tracing the Threads: Studies in the Vitality of Jewish Pseudepigrapha*, ed. J. C. Reeves, Atlanta GA, 1994, pp. 184–91.

[62] Stoyanov, *The Other God* (n. 6 above), 261–2; Stoyanov, 'Apocryphal Themes and Apocalyptic Traditions' (n. 9 above), p. 235.

[63] For a recent treatment of the paradise theme in II Corinthians 12:1–4 wider New Testament, pseudepigraphic and rabbinic frameworks, see Grant Macaskill, 'Paradise in the New Testament', in *Paradise in Antiquity: Jewish and Christian Views*, eds M. Bockmuehl and G. A. G. Stroumsa, Cambridge, 2010, pp. 64–81 (67–71).

Seeking paradise in the Egyptian desert

Dimitris J. Kyrtatas

Christian concerns regarding the nature and location of paradise began at a very early stage and had a long history.[1] To a certain extent, they were triggered by the absence of an authoritative account. For, strange as it may seem, although the story of Adam and Eve in Eden played an essential role in Christian salvation history, precious little was said about it in the New Testament, where the word παράδεισος, adopted by the Greek translators of the Hebrew Bible to describe the Garden of Eden, appears only three times. The expressions 'Kingdom of God' and 'Kingdom of Heaven' (referring to a wider notion than the earthly garden of the first human pair) were more frequently used but never fully explained.[2]

Responding to speculations, Paul had implied that it was best to avoid the topic altogether – although personally he had been granted (as he claimed) the privilege of an overview, or rather an overhearing (II Corinthians 12:2–4). The 'mystery of the Kingdom of God', Jesus was reported to have said, should not be explained to the multitude, but only to a select few (Mark 4:11).[3] At most, the disciples were told that the 'Kingdom of God' was like

> a mustard seed which, at the time of its sowing, is the smallest of all the seeds on earth. Yet once it is sown it grows into the biggest shrub of them all and puts out its big branches so that the birds of the air can shelter in its shade (Mark 4:30–32).

Other metaphors comparing the 'Kingdom of Heaven' to earthly kingdoms were hardly more informative.[4] Christians, therefore, could rely almost exclusively on the first and last books of the Christian Bible in its developed form, i.e. Genesis and Revelation. The problem was that, in both accounts, the information provided was rather vague and, to a certain extent, contradictory. Genesis actually provides two distinct versions, while the book of Revelation refers to an earthly as well as to a heavenly divine kingdom.

Just why the New Testament authors avoided the topic, while arousing the interest of their readers is not exactly clear. A precise description of paradise and its location, they might have thought, was not necessary. It is likely that in their view it was not the external

[1] See A. Scafi, *Mapping Paradise: A History of Heaven on Earth*, London and Chicago IL, 2006.

[2] See J. N. Bremmer, *The Rise and Fall of the Afterlife*, London and New York, 2002, pp. 109–27; D. J. Kyrtatas, 'Early Christian visions of paradise: Consideration on their Jewish and Greek background', in *Hellenic and Jewish Arts: Interaction, Tradition and Renewal*, ed. A. Ovadiah, Jerusalem, 1998, pp. 337–50; G. Macaskill, 'Paradise in the New Testament', in *Paradise in Antiquity: Jewish and Christian Views*, eds M. Bockmuehl and G. G. Stroumsa, Cambridge, 2010, pp. 64–81.

[3] For biblical quotations I use the *New Jerusalem Bible* translations.

[4] Cf. Matthew 18:23, 22:2.

The Cosmography of Paradise: The Other World from Ancient Mesopotamia to Medieval Europe, Warburg Institute Colloquia 27, 2016

characteristics of paradise that mattered so much as the sensation anticipated, something that by its nature evades definition. In any event, as it soon became obvious, the expectations of eternal bliss depended upon the ways the early Christians understood the resurrection: although most of them agreed that the righteous would enter the 'Kingdom of God' in their bodies, they were not all of the same mind on just how corporeal the eternal body would be - and hence advanced different views about the nature of paradise. Disagreements regarding the physical nature of paradise, in their turn, reflected the different interpretations of Jesus's message about life upon earth.

Early Christian speculations about paradise seem to have been vivid in Asia Minor and North Africa, but it was probably in Egypt that the controversy was most intense.[5] In the third century its Christian community was almost led to a schism, the leaders of the Church in Alexandria finding themselves in sharp conflict with many of their brethren living in the villages.[6] A century later the 'desert fathers' pondered deeply over the problem, preoccupied as they were to liberate themselves from carnal desire.[7] Some few (probably educated in Alexandria) appear to have thought that in the world to come, there would be no bodily needs; most, however, were convinced that the deprivations they voluntarily experienced upon earth would be rewarded by a life of eternal abundance in material terms. It was precisely such expectations, they thought, that gave meaning to their ascetic life. A story circulating in the desert around the end of the fourth century illustrates the character of such considerations.

* * *

Of the ascetic Macarius the Egyptian it was said that after much fasting he had once prayed to God

> to show him the paradise which Jannes and Jambres [also known as Mambres] had planted in the desert in their desire to make a copy of the true paradise. When he had wandered through the desert for three weeks, and not having eaten during this time was already fainting, an angel set him near the place. There were demons everywhere guarding the entrance of paradise and not allowing him to enter. The garden was very large, covering an enormous area. After he had prayed he made a bold effort and succeeded in entering. Inside the garden he found two holy men. They had entered by the same means themselves, and had already spent a considerable time there. When they had said a prayer, they embraced each other, overjoyed at the meeting. Then they washed his feet and set before him some of the fruit of paradise. He partook and gave thanks to God, marvelling at the size of the fruit and its varied colours ... In the middle of paradise ... there were three large springs which welled up from the depths and watered the garden and its huge trees, which were very productive and bore every kind of fruit that exists under the heavens.

[5] As the fragments of Papias make clear, speculations about paradise in Asia Minor were well on their way by the early 2nd century. On Africa, see J. N. Bremmer, 'Contextualizing heaven in third-century North Africa', in *Heavenly Realms and Earthly Realities in Late Antique Religions*, eds R. S. Boustan and A.Y. Reed, Cambridge, 2004, pp. 159–73.

[6] See D. J. Kyrtatas, *The Social Structure of the Early Christian Communities*, London, 1987.

[7] See Aline Rousselle, *Porneia: On Desire and the Body in Antiquity*, London, 1988, pp. 140–78.

Having spent seven days with the holy men, Macarius decided to return and bring the other monks with him, so that they might also enjoy the delight. As proof of his discovery he carried along with him some of the fruit. Back at the residence of his fellow monks, however, many fathers persuaded him to remain permanently in the desert. 'Could it not be that this paradise has come into being for the destruction of our souls?', they asked. 'For if we were to enjoy it in this life, we should have received our portion of good things while still on earth. What reward would we have afterwards when we come into the presence of God?'[8]

There is much in this story that calls for comment. First of all, since the paradise mentioned was a replica (ἀντίτυπον) all its details should more or less match those of the original. The only noticeable difference is that the creation of the magicians was available in this present life, while God's creation, offered as a reward after the final judgement, would last for eternity. The fathers did not reject the former as deficient, but as temporal. Of interest are also the implications regarding the creation and survival of this paradise, as well as the information provided about its entrance and its inhabitants. Significantly, no reproach was made against the holy men who had decided to spend their lives among trees and fountains. Choosing between the alternatives seems to have been the privilege of (virtuous) monks.

There is a clear moral in this simple story. As successors of the martyrs, the hermits were the most likely to inherit the 'Kingdom of God', but unlike them they kept reminding themselves that their own sufferings were voluntary. Having opted for a life of deprivation as the most secure path to everlasting delights, they should resist any temporary alleviation of their misery – let alone any temporal delight. Creating a paradise was not something totally beyond human power. Avoiding it was a far greater achievement. There is a further implicit message in the story. In God's kingdom the righteous would be rewarded according to their merits: those who had simply not sinned should not expect the same privileges as those who had sacrificed their lives for their salvation.

* * *

In its earliest recorded form the story was composed by an anonymous author, who paid the Egyptian monks a visit sometime from September AD 394 until January 395 – about five years after the death of Macarius – and published his account, best known by its Latin title *Historia Monachorum*, around 400.[9] We are fortunate, however, to possess a different version in which the story of the desert paradise was being told. Palladius, bishop first of

[8] *Historia Monachorum in Aegypto*, ed. A.-J. Festugière, Brussels, 1961, 21.5–12, pp. 125–6; English translation by N. Russell in *The Lives of the Desert Fathers*, London and Oxford, 1981, pp. 108–9. I have briefly dealt with the passage in my 'Early Christian visions of paradise' (n. 2 above). Cf. D. J. Kyrtatas, 'Paradise in Heaven and on Earth: Western Ideas of Perfect (Non) Organization', in *Myths, Stories and Organizations: Premodern Narratives for Our Times*, ed. Y. Gabriel, Oxford, 2004, pp. 66–79.

[9] Macarius the Egyptian lived from c. AD 300 to 390. See W. Harmless, *Desert Fathers: An Introduction to the Literature of Early Monasticism*, Oxford, 2004, pp. 194–6 and 275–9.

Helenopolis and later Aspuna, had heard of a variant account. His *Historia Lausiaca*, to use again the best known Latin title of a Greek text, was published some twenty years after the *Historia Monachorum*, but was based on notes taken down during his dwelling with the monks for more than nine years. Actually, Palladius (who in later years became a fervent supporter of John Chrysostom) had settled in Egypt about six years before the arrival of the anonymous author of the *Historia Monachorum*.[10]

Contradicting the anonymous author, Palladius does not ascribe the story to Macarius the Egyptian, but to his slightly younger namesake, Macarius the Alexandrian, an altogether different personality with whom he is, however, often confused.[11] (Incidentally, the two Macarii came from different cultural backgrounds, the former being a villager, and hence an Egyptian native speaker, the latter a citizen of Alexandria, and therefore known as πολιτικός, obviously with a Greek education.) Since Palladius came to know Macarius the younger well and claims to have personally heard the story from his mouth, his version should be considered as closer to the story as it was originally told.[12] The differences in the substance of the narrative are far more revealing.

Of Macarius the Alexandrian it was said that he had once desired to enter the garden tomb (κενοτάφιον) created by Jannes and Jambres, who had once great power with the Pharaoh. Having authority for long periods the two magicians had

> built their work with stones faced foursquare, and made their tomb (μνῆμα) there, and stored away much gold. They also planted trees, for the place [was] rather damp, and they dug a well besides. Since therefore the saint did not know the way, he followed the stars by a kind of guesswork, crossing the desert, as one does at sea.

After travelling for nearly nine days he approached the place and fell asleep. Waking up, he realized that seventy demons came out from the garden tomb to meet him, shouting and fluttering like ravens against his face and saying: 'What do you want, Macarius? What do you want monk? Why have you come to our place? You can't stay here.' The monk replied that he only wished to look around. After that he would walk away. The demons consented and allowed him to enter. Inside the garden tomb, Macarius

> found a little brazen jar suspended and an iron chain against the well, rusted already by time, and some pomegranates with nothing inside because they had been dried up by the sun. So ... he turned back and went on his way for twenty days.[13]

[10] Palladius was born around AD 363 and died between 420 and 430. He was ordained bishop of Helenopolis in Bithynia by John Chrysostom in 400 and moved to Aspuna in Galatia in 417.

[11] Macarius the younger died in AD 395 or later. See A. Guillaumont, 'Le Problème des deux Macaire dans les «Apophthegmata Patrum»', in *Irénikon*, 48, 1975, pp. 41–59.

[12] Significantly, in his Latin translation of the *Historia Monachorum*, Rufinus corrects the Greek original by attributing the story to Macarius the Alexandrian: see *PL* XXI, col. 453 (critical edition by Eva Schulz-Flügel, Berlin and New York, 1990).

[13] Palladius, *Lausiac History*, ed. C. Butler, Hildesheim, 1967, 18.5; English translation by W. K. Lowther Clarke, London, 1918, pp. 78–80.

The extremely rare word 'garden tomb' (otherwise only known from inscriptions) used by the bishop in place for paradise, must have been rather obvious, since the burial places of eastern monarchs were occasionally located in fenced gardens, called paradises.[14] In this account the creation of the magicians, when discovered by the monk, was found (surprisingly) already dry, rusted and abandoned – clearly no place for a carefree life with delights and food in abundance. It certainly did not cross (this) Macarius's mind to invite others as well. Having achieved his goal, he honoured his commitment to the demons and left. He clearly had no reason to remain any longer in a deserted spot.

This version has no obvious moral. Apart from a tortuous march in the desert to satisfy his curiosity, the monk did not face any real challenge, coming to an understanding with the demons. It is not even clear why Palladius included the story in his narration. Unless he wished to make it clear that the earthly paradise, about which much was being rumoured, was already dead and useless. The reader, however, is left perplexed: Why was an unpleasant and inhospitable area, such as that, guarded by so many demons? And why was Macarius initially prevented from entering?

The answers to such questions are partly supplied by a third version in which the story was told. The *Life of Macarius of Alexandria*, written somewhat later by another anonymous author, clearly depends upon the *Historia Lausiaca* but adds further interesting information. Regarding Macarius's motives, it implies that the monk had in mind to meet the demons of the place rather than examine the garden they were guarding. Actually, as we are told, the demons he encountered had been placed there by the magicians of old. Since they were hoping to use their creation after their death for their own bliss, as they understood it after an Egyptian fashion, Jannes and Jambres had stored treasures of gold as well as other abundant provisions. Hence they wished to safeguard their possessions.

This version makes some sense but leaves other problems unsolved. According to the *Vita* the demons were terrified when they met the daring visitor. They cried out to Macarius that they did not harm him. They questioned him: he had his desert, from which he had chased other demons, their kin. Why then, they insisted, the hermit should disturb them? As a hermit, he should be content with the desert.[15] But if the garden was already dead and the treasures long forgotten, why were they guarding it and why were they terrified?

* * *

Although Jannes and Jambres hardly survived as significant figures in Christian lore, their identity was perfectly clear in the Roman age. It was believed that they were the unnamed Egyptian magicians mentioned in the book of Exodus, where they opposed Moses and

[14] Cf. Strabo 15.3 on the tomb of Cyrus.

[15] *Vita Sancti Macarii Alexandrini, PG* XXXIV, col. 188.

Aaron by performing similar, though less potent, miracles (7:11, 7:22, 8:14–15, 9:11). Around the second century BCE, they acquired names and a full-scale biography, which was eventually written down in book form. Their story probably originated among some Jewish circles of Palestine, but spread in Egypt and other parts of the Roman Empire.

Fragments of their long lost *Apocryphon*, preserved on papyrus, have been recently found and published along with an extensive commentary.[16] As they stand, these fragments make little sense, but interpreted in the light of latter traditions they give a general idea of the story once told. One of the details that has come to light is that the garden, as a burial place, was used for necromantic practices. This may explain why the area was guarded and why some visitors, especially Christian monks, were prevented from entering. Their presence would pollute the region and cancel its magic quality.

As the three distinct versions demonstrate, Christians did not make a consistent use of the story and nor were they interested in its original meaning or moral. They probably knew the saga well already from an early age since the author of II Timothy refers to the magicians by name, implying that his readers would readily understand his allusion (3:8–9). It is likely (though far from certain) that in this late New Testament letter, the ideas attributed to Jannes and Jambres were thought to have inspired some contemporary Christians, namely Hymenaeus, Philetus and, possibly, Alexander, who believed that the resurrection of the dead had already taken place (II Timothy 2:18, 4:14; cf. I Timothy 1:20). If that were so, paradise would have already been functional and the righteous would be enjoying their rewards in it. Such ideas were confronted by others who drew a clear line between the present world and the world to come. The reference to the old Egyptian magicians served to illustrate the difference between an authentic garden of bliss and its similar but deceptive replication.

The *Apocryphon* was certainly known around the middle of the third century. It is mentioned by Origen, among others, who had found out that the Platonist philosopher Numenius had read it as well.[17] Since Origen has precious little to say about the content of the book it is not exactly clear what he made of it. But his views regarding the nature of paradise are well known. Above all he rejected the idea of carnal pleasures after the resurrection.[18] Certain persons, he argued,

> refusing the labour of thinking, and adopting a superficial view of the letter of the law, and yielding rather in some measure to the indulgence of their own desires and lusts, being disciples of the letter alone, are of opinion that the fulfilment of the promises of the future are to be looked

[16] See A. Pietersma, *The Apocryphon of Jannes and Jambres the Magicians. P. Chester Beatty XVI (with New Editions of Papyrus Vindobonensis Greek inv. 29456+29828 verso and British Library Cotton Tiberius B.v f 87)*, edited with introduction, translation and commentary, with full facsimile of all three texts, Leiden, New York and Cologne, 1994. Also G. Schmelz, 'Zwei neue Fragmente des Apokryphons über die Zauberer Jannes und Jambres', in *Atti del XXII Congresso Internazionale di Papirologia*, Florence, 2001, pp. 199–212.

[17] Origen, *Contra Celsum* 4.51.

[18] See H. Crouzel, *Origen*, Edinburgh, 1989, pp. 248–57.

> for in bodily pleasure and luxury; and therefore they especially desire to have again, after the resurrection, such bodily structures [*carnes*] as may never be without the power of eating, and drinking, and performing all the functions of flesh and blood ... And consequently, they say, that after the resurrection there will be marriages, and the begetting of children, imagining to themselves that the earthly city of Jerusalem is to be rebuilt, its foundations laid in precious stones, and its walls constructed of jasper, and its battlements of crystal; that it is to have a wall composed of many precious stones.[19]

Such views as expressed by Origen and the Alexandrian Christian school were vigorously attacked by Christians living in the Egyptian villages of Arsinoe (present Fayum), who insisted that the promises of the sacred texts should indeed be interpreted literally. Their arguments were written down by one of their bishops, called Nepos, who composed a book with the characteristic title *Refutation of the Allegorists*. This book, it was thought, contained hidden mysteries. Since it has completely perished we know next to nothing about its content and its possible use of old legends. We may surmise, however, that the most important mystery revealed was the idea it advanced about the nature of paradise: it would in many ways resemble what was already in existence upon earth.

The Alexandrian bishop Dionysius (a disciple of Origen) wrote extensively against these ideas in a pamphlet called *On the promises* – some sections of which have been preserved. He also met the Egyptian Christians in their villages to confront their leaders, but achieved only partial success.[20] A century later, at the time of the Macarii, the topic was being still hotly debated among Christians, often disguised in the form of personal experiences. The different versions preserved may serve to illustrate the different values attributed to an earthly paradise.

* * *

Interestingly, the widely read *Life of Pachomius* provides yet a fourth and much more important version of the story once told about a paradise created in the Egyptian desert.[21] Although lacking the visionary and imaginative character of the other reports, Pachomius's accomplishment, as recorded, exercised a far greater and lasting influence throughout the Middle Ages. It is actually this version that probably clarifies the underlying meaning of the debates and controversies over the nature of paradise that remains rather implicit in the Jannes and Jambres myths.

As the *Life of Pachomius* claims, the ascetic movement was greatly inspired by the courage and endurance of the martyrs of former days, especially those under the last and

[19] Origen, *De Principiis* 2.11.2, transl. F. Crombie, Edinburgh, 1869, pp. 145–6, clearly referring to the book of Revelation and its literal interpretation.

[20] Dionysius was bishop from AD 248 to 265. Eusebius, *Ecclesiastical History* 7.24. See Kyrtatas, *The Social Structure* (n. 6 above), pp. 175–8.

[21] Pachomius lived from c. AD 292 to c. 348. Other versions of the story are also found in the lives of various monks.

probably most harsh persecutions.[22] Since, as it was generally believed, martyrs had secured their entrance to paradise, and were indeed expected to ascend to heaven immediately upon their death, not having to await the general resurrection, numerous Christians were hoping to obtain a similar privilege by mortifying their bodies themselves.[23] The first and most decisive step to be taken by all potential ascetics was retreat from the world and avoidance of the pleasures of life. The next: adoption of a most austere and, as far as possible, deprived way of living. In a typical reaction, having been invited to participate in a luxurious family feast, Theodore, Pachomius's most faithful disciple, absolutely refused. He kept reminding himself that, should he enjoy that food, he would never receive the everlasting goods of the true life. He was repeating the arguments of the fathers who discouraged Macarius from returning to the earthly paradise he had discovered.[24]

Pachomius did not invent the idea of Christian asceticism. Others before him, and most notably Anthony, had done much to promote the ideals of retreat from the world and of self-deprivation. His own invention was something far more complex and at once accessible to large numbers of dedicated Christians. Without abandoning the spirit of the first hermits, Pachomius dedicated his life to the establishment of a community, the so-called κοινωνία.[25] An organized community living within a clearly defined area, with social relations, a hierarchical structure, productive and consuming activities, as well as rules, that should be faithfully observed by all the members of a monastery.[26]

Because of his great personal powers, Pachomius was regarded as a continuator of the former prophets. While proceeding to report his first initiatives in establishing a monastery, his *Life* presents him as a miracle maker. Indeed, by referring to the well-known controversy of Moses with the Egyptian magicians, the *Life* clearly alludes to the story of Jannes and Jambres.[27] Pachomius, who had encountered Macarius the Alexandrian and conversed with him, invited his fellow monks to create with him a community of their own that would resemble paradise in many ways.[28] This was a message that appears to have been readily understood. In joining him, new arrivals are reported

[22] *First Greek Life of Pachomius* 1, 13 (critical edition by Fr. Halkin, *Sancti Pachomii Vitae Graeca*, Bruxelles, 1932).

[23] For an early declaration of the privileged position of martyrs in the afterlife, see *The Shepherd of Hermas*, vision 3.1.9, text and English translation included in *The Apostolic Fathers*, transl. K. Lake, Cambridge MA and London, 1913, II, p. 29. The idea was further developed by Tertullian and others.

[24] *Third Greek Life of Pachomius* 45 (critical edition by Halkin, *Sancti Pachomii Vitae Graeca*, n. 22 above). Cf. note 8 above.

[25] See Philip Rousseau, *Pachomius: The Making of a Community in Fourth-Century Egypt*, Berkeley CA, Los Angeles CA and London, 1985.

[26] *First Greek Life of Pachomius* 25, 95 (n. 22 above).

[27] Ibid., 17, 21.

[28] On the encounter, see Palladius, *Historia Lausiaca* 18.12 (n. 13 above). Cf. K. B. Copeland, 'The earthly monastery and the transformation of the Heavenly city in Late Antique Egypt', in *Heavenly Realms and Earthly Realities* (n. 5 above), pp. 142–8.

to have said to themselves: 'Let us die and live again with this man', as if resurrected into a new existence.[29] That the monastery conceived in such a way had strict rules came as no surprise. Moses was best known as a law-giver and, besides, rules and regulations were already given to Adam and Eve while they were still dwelling in paradise.

* * *

The common aim of all ascetics in Christian Egypt as well as elsewhere was eternal life. All ascetics furthermore agreed that this could be achieved by abandoning the pleasures of this present, temporal existence and, thus, by avoiding the temptations of the Devil. But they were not all of the same mind regarding the true nature of paradise. Some thought that it will be only spiritual, with no marriages or feasting, interpreting the biblical accounts in a spiritual or allegorical sense. Others read the descriptions of paradise in a rather literal manner, expecting in heaven a recompense of all the delights that they had by necessity and, mostly, by free choice, forsaken during their mortal existence.

Those who retold and modified the Jannes and Jambres stories, as they have survived in the extant literature, were in various degrees opposing the conception of a carnal paradise. To discredit the views of their opponents they advanced a reasonable argument. If the expectation of devoted Christians was a life of abundance and bodily pleasure, why await it in the distant future? What was wrong with a sinless and yet leisurely life in the present world? They knew, of course, that sinless pleasures could not be achieved upon earth, but they mostly appealed to the common understanding that an ascetic life with no hardship and deprivation was meaningless.

Pachomius came up with an even better idea. The monastery he conceived and organized could become both: an instrument of salvation, as well as a foreshadowing of future existence. It was not only a means to an end, but also an end in itself. Just like the paradise of Adam and Eve, it had clear boundaries and appropriate regulations. Until his intervention, most ascetics thought that they should spend their time in agony and internal turmoil – although some few may have found peace of mind in the desert. The Pachomian monastery presupposed instead an intricate rhythm of communal work and prayer that could secure sentiments of peace and joy to most, if not all, dedicated monks.[30] In due course, seen through a Pachomian perspective, the lives of all early ascetics and monks came to be thought of as the 'Paradise of the Holy Fathers', a paradise far greater and more profound by far than that of the Egyptian magicians Jannes and Jambres.

[29] *First Greek Life of Pachomius* 25 (n. 22 above).
[30] See Harmless, *Desert Fathers* (n. 9 above), p. 122.

Gazing at the Holy Mountain: Images of Paradise in Syriac Christian Tradition*

Sergey Minov

Pour qu'une montagne puisse jouer le rôle de Mont Analogue, … il faut que son sommet soit inaccessible, mais sa base accessible aux êtres humains tels que la nature les a faits. Elle doit être unique et elle doit exister géographiquement. La porte de l'invisible doit être visible.

René Daumal, *Le Mont Analogue*

The notion of paradise was deeply ingrained in the Christian imagination, beginning from the earliest stages of its development. Rooted in the biblical account of Genesis 2–3, the image of the Garden of Eden became a key element of Christian thought and allowed its thinkers to address a wide range of issues and themes.[1] In the course of history, the Christian views on paradise underwent profound changes, as they were transformed and adapted to the needs of the moment in different Christian cultures during different epochs.

This essay represents a modest attempt to bring together several traditions on the Garden of Eden that were current among the Syriac-speaking Christians during the period of Late Antiquity.[2] Due to the limitations of space, no attempt will be made to offer even a cursory overview of the wide gamut of views on paradise that developed among the Christians of late antique Syria-Mesopotamia.[3] Instead, I shall focus only on

* Author's Note: This article is based on the presentation given at the conference *Mondes clos: Cultures et jardins;* Université de Genève, Genève, April 20–21, 2011. I am grateful to the organizers of this conference, Prof. Philippe Borgeaud and Daniel Barbu, for their hospitality. I would like also to express my gratitude to Prof. Sebastian Brock for his comments on an earlier draft of this paper.

Abbreviations used in this essay:
CSCO = *Corpus Scriptorum Christianorum Orientalium.*
MT = Masoretic.
Syr. = *Scriptores Syri.*
Subs. = *Subsidia.*

[1] Literature on this subject is immense. See, for example, J. Delumeau, *History of Paradise: The Garden of Eden in Myth and Tradition*, transl. M. O'Connell, New York, 1995; J. Daniélou, 'Terre et Paradis chez les Pères de l'Église', *Eranos-Jahrbuch*, 22, 1953, pp. 433–72; G. G. Stroumsa, '*In Illo Loco*: Paradise Lost in Early Christian Mythology', in *Genesis and Regeneration: Essays on Conceptions of Origins*, ed. Sh. Shaked, Jerusalem, 2005, pp. 110–26.

[2] For a general introduction into Syriac Christianity during this epoch, see L. van Rompay, 'The East (3): Syria and Mesopotamia', in *The Oxford Handbook of Early Christian Studies*, eds S. A. Harvey and D. G. Hunter, Oxford, 2008, pp. 365–86; W. S. McCullough, *A Short History of Syriac Christianity to the Rise of Islam*, Chico CA, 1982; R. Murray, *Symbols of Church and Kingdom: A Study in Early Syriac Tradition*, Cambridge, 1975.

[3] On various aspects of paradise in Syriac Christian tradition, see G. A. Anderson, *The Genesis of Perfection: Adam and Eve in Jewish and Christian Imagination*, Louisville KY, 2001; P. Bettiolo, 'Adamo in Eden e la liturgia celeste: temi della meditazione cristiana nella Siria del IV secolo, tra Afraate e il Liber graduum', *Rivista di storia e letteratura religiosa*, 37/1, 2001, pp. 3–27; R. A. Kitchen, 'Syriac Additions to Anderson: The Garden of Eden in the *Book of Steps* and Philoxenus of Mabbug', *Hugoye*, 6/1, 2003, §§1–47; T. Kronholm, *Motifs from Genesis 1 – 11 in the Genuine Hymns of Ephrem the Syrian*, Uppsala, 1978; S. Minov, 'The Question of Sexuality in Paradise

those aspects of paradise imagery that seem to be specific to Syriac Christianity and mark it as different from other ancient Christian traditions of interpretation of the Garden of Eden.

With this in mind, I am going to analyze three particular aspects of paradise found in Syriac Christian texts from Late Antiquity that provide us with an opportunity to bring into focus different cultural forces that were at work during this formative period in the history of Syriac Christianity. The first aspect to be discussed is the idea of paradise as a sacred, temple-like, place. The second is the image of paradise as a cosmic mountain. And, in conclusion, I will examine the use of feminine imagery with regard to the Garden of Eden.

1. Paradise as a Temple

The earliest and most elaborated example of the application of Temple imagery to paradise comes from the writings of Ephrem the Syrian, one of the most famous Syriac authors, who was active during the fourth century. In Ephrem's *Hymns on Paradise* we find several references to paradise that evoke the Jerusalem Temple. On several occasions Ephrem refers to the bipartite division of the Temple space into the 'sanctuary' (ܡܩܕܫܐ) and the 'Holy of Holies' (ܩܘܕܫ ܩܘܕ̈ܫܐ) as reflected symbolically in the inner structure of paradise.[4] Further on Ephrem calls Jerusalem 'sanctuary' (ܒܝܬ ܡܩܕܫܐ) to be a 'type for paradise' (ܛܘܦܣܐ ܕܦܪܕܝܣܐ).[5]

In his use of cultic imagery in relation to paradise Ephrem does not limit himself to the idea of typological correspondence between the two. At some point he interprets the whole drama of Adam's transgression and fall through the lens of the Temple and priestly ministry. Thus, in order to justify the divine prohibition to Adam against eating from the Tree of Knowledge, Ephrem explains it as a pedagogical stratagem, the purpose of which was that Adam had first to prove his suitability as priest by performing service in the outer part of the sanctuary of paradise. Only after that would he be permitted to enter and serve in the sanctuary's inner part, where God himself dwells:

> God did not permit Adam to enter that innermost Tabernacle (ܡܫܟܢܐ ܕܠܓܘ); this was withheld, so that first he might prove pleasing in his service of that outer Tabernacle

as Reflected in Late Antiquity Biblical Exegesis: Between the Sages and the Fathers of the Church', in *A Garden Eastward in Eden--Traditions of Paradise: Changing Jewish Perspectives and Comparative Dimensions of Culture*, ed. R. Elior, Jerusalem, 2010, pp. 158–72 (in Hebrew); R. Murray, 'The Lance which Re-opened Paradise: A Mysterious Reading in the Early Syriac Fathers', *Orientalia Christiana Periodica*, 39/1, 1973, pp. 224–34; A. Palmer, 'Paradise Restored', *Oriens Christianus*, 87, 2003, pp. 1–46; E. ten Napel, '"Third Heaven" and "Paradise": Some Remarks on the Exegesis of 2 Cor. 12, 2–4 in Syriac', in *V Symposium Syriacum, 1988*, ed. R. Lavenant, Rome, 1990, pp. 53–65; L. van Rompay, 'Memories of Paradise: The Greek "Life of Adam and Eve" and Early Syriac Tradition', *Aram*, 5, 1993, pp. 555–70.

[4] Ephrem the Syrian, *De paradiso*, III.17; IV.2; *Des heiligen Ephraem des Syrers Hymnen de Paradiso und contra Julianum*, ed. E. Beck, *CSCO* 174, *Syr.* 78, Louvain, 1957, pp. 12–13; *Saint Ephrem. Hymns on Paradise*, transl. S. P. Brock, Crestwood, NY, 1990, pp. 96–8. See discussion in Kronholm, *Motifs from Genesis* (n. 3 above), p. 72.

[5] Ephrem the Syrian, *De paradiso*, XV.8; ed. Beck (n. 4 above), p. 64; transl. Brock (n. 4 above), p. 184.

(ܡܫܟܢܐ ܕܠܓܘ); like a priest (ܐܝܟ ܟܗܢܐ) with fragrant incense (ܒܥܛܪܐ ܡܒܣܡܐ), Adam's keeping of the commandment (ܢܛܪܗ ܕܦܘܩܕܢܐ) was to be his censer (ܦܝܪܡܐ); then he might enter before the Hidden One into that hidden Tabernacle (ܡܫܟܢܐ ܟܣܝܐ).[6]

Ephrem compares Adam's access to this innermost part of paradise to the cultic duties of the Jewish High Priest, who was permitted to enter the Holy of Holies only once a year, during Yom Kippur:

> The symbol of paradise (ܐܪܙܗ ܕܦܪܕܝܣܐ) was depicted by Moses who made the two sanctuaries, the sanctuary (ܡܩܕܫܐ) and the Holy of Holies (ܩܘܕܫ ܩܘܕܫܐ); into the outer one (ܒܪܝܐ) entrance was permitted, but into the inner (ܓܘܝܐ), only once a year. So too with paradise, God closed off the inner part, but He opened up the outer, wherein Adam might graze.[7]

In another place, Ephrem draws a parallel between the transgression of Adam and the sin of Uzziah, the king of Judah, who tried to usurp the priestly prerogatives of the descendants of Aaron and offer an incense-sacrifice in the sanctuary, but was punished with leprosy (see II Chronicles 26:16–21):

> In the midst of paradise God had planted the Tree of Knowledge to separate off, above and below (ܕܢܦܪܘܫ ܠܥܠ ܘܠܬܚܬ), sanctuary (ܡܩܕܫܐ) from Holy of Holies (ܩܘܕܫ ܩܘܕܫ̈ܐ). Adam made bold to touch and was smitten like Uzziah: the king became leprous, Adam was stripped. Being struck like Uzziah, he hastened to leave: both kings fled and hid, in shame of their bodies.[8]

According to Ephrem's symbolic interpretation of Genesis's account of the Fall, the sin of the eating of the forbidden fruit by Adam was in its essence an attempt to get unauthorized access to the restricted part of God's sanctuary.[9]

Ephrem was not the only Syriac author who ascribed priestly status to Adam or presented him performing priestly-like ministry in paradise. In the *Book of Steps*, an anonymous Syriac work, which is also considered to have been produced during the fourth century, it is related that as a consequence of their fall Adam and Eve lost their 'heavenly ministry' (ܬܫܡܫܬܐ ܫܡܝܢܝܬܐ), which they 'used to fulfill like angels and together with the angels' (ܕܐܝܟ ܡܠܐܟܐ ܥܡ ܡܠܐܟܐ ܡܫܡܫܝܢ ܗܘܘ).[10] It is remarkable that in this passage both the Syriac noun and the verb used by the author to describe

[6] Ephrem the Syrian, *De paradiso*, III.16; ed. Beck (n. 4 above), p. 12; transl. Brock (n. 4 above), p. 96.

[7] Ephrem the Syrian, *De paradiso*, III.17; ed. Beck (n. 4 above), p. 12; transl. Brock (n. 4 above), p. 96.

[8] Ephrem the Syrian, *De paradiso*, III.14; ed. Beck (n. 4 above), p. 11; transl. Brock (n. 4 above), p. 95. Cf. also *De paradiso*, XV.9–10; ed. Beck (n. 4 above), p. 64; transl. Brock (n. 4 above), p. 185.

[9] This idea is expressed also in *De paradiso*, IV.2, where Adam's transgression is presented as an attempt to enter without permission the 'Holy of Holies' (ܩܘܕܫ ܩܘܕܫܐ), which is also described as the 'inner tabernacle' (ܡܫܟܢܐ ܕܠܓܘ); ed. Beck (n. 4 above), p. 13; transl. Brock (n. 4 above), pp. 97–8.

[10] *Liber Graduum*, XXVIII.9; *Liber Graduum*, ed. M. Kmosko, Paris, 1926, col. 800; *The Book of Steps: The Syriac Liber Graduum*, transl. R. A. Kitchen and M. F. G. Parmentier, Kalamazoo MI, 2004, p. 318.

Adam's occupation are derived from the root *šmš* that was used to convey a wide range of meanings related to liturgical or priestly service.[11]

Later on, we find the imagery of paradise as the place of cultic ministry having been embraced by the author of the *Cave of Treasures*, an elaborated paraphrase of the biblical *Heilsgeschichte*, written in Syriac during the sixth or early seventh century.[12] The *Cave*'s author places special emphasis on the sanctity of paradise, referring to it routinely as the 'holy mountain' (ܛܘܪܐ ܩܕܝܫܐ).[13] Moreover, for him, 'Eden is the Holy Church' (ܥܕܝܢ ܐܝܬܘܗ̇ ܥܕܬܐ ܩܕܝܫܬܐ).[14] Adam on several occasions is called 'priest' (ܟܗܢܐ).[15] It is the sacred precincts of paradise, understood under Ephrem's influence as a mountain, where the newly-created Adam had to realize his priestly vocation:

> And because Adam was priest, and king, and prophet (ܡܠܟܐ ܘܟܗܢܐ ܘܢܒܝܐ), God brought him into paradise that he might minister in Eden, the Holy Church (ܕܢܫܡܫ ܒܗ ܒܥܕܝܢ ܥܕܬܐ ܩܕܝܫܬܐ).[16]

In order to buttress this claim with a scriptural proof-text, the author of the *Cave* resorts to an interpretation, rare in Christian exegetical tradition, of God's commandment of tilling and keeping the Garden of Eden given to Adam in Genesis 2:15. According to him, the verb 'to till' (*plaḥ* in Syriac, which is equivalent of the Hebrew *ʿāḇaḏ*) in this verse expresses the idea of quasi-priestly liturgical ministry that Adam was supposed to perform:

> even as the blessed man Moses testifies concerning him (i.e. Adam), saying, 'That he might serve God by means of priestly ministration with praise (ܕܢܦܠܚܝܘܗܝ، ܒܝܕ ܬܫܡܫܬܐ ܟܗܢܝܬܐ ܒܬܫܒܘܚܬܐ), and that he might keep that commandment which had been entrusted to him by the compassion of God'.[17]

[11] Thus, in the Peshitta version of the Old Testament this root is used not only to translate the general Hebrew verb *šārēt* 'to minister, serve', but also for such a specific verb referring to priestly service as *kahēn* 'to minister as a priest'; cf. Leviticus 7.35, 16.32.

[12] For a discussion of the date and provenance of this work, see C. Leonhard, 'Observations on the Date of the Syriac *Cave of Treasures*', in *The World of the Aramaeans III: Studies in Language and Literature in Honour of Paul-Eugène Dion*, eds P. M. M. Daviau et al, Sheffield 2001, pp. 255–93; S. Minov, *Syriac Christian Identity in Late Sasanian Mesopotamia: The Cave of Treasures in Context*, Ph. D. dissertation, The Hebrew University of Jerusalem, 2013, pp. 31–86.

[13] *Cave*, V.18, 25; VII.12, 18; IX.5; X.6; XI.11; XII.8, 11, 13; XIII.3; XVI.13; XVII.7, 18; *La Caverne des Trésors: les deux recensions syriaques*, ed. S.-M. Ri, CSCO 486, Syr. 207, Louvain, 1987, pp. 42, 44, 60, 72, 76, 86, 90, 92, 98, 120, 130, 134. If not marked otherwise, Syriac text of the *Cave* is quoted according to the work's best textual witness Or^{A}, i.e. British Library MS Add. 25875, as it is reflected in the apparatus of Ri's edition. For explanation of this choice as well as general discussion of the state of the *Cave*'s text, see Minov, *Syriac Christian Identity* (n. 12 above), pp. 21–31.

[14] *Cave* III.21; ed. Ri, 1987 (n. 13 above), p. 26; *The Book of the Cave of Treasures*, transl. E. A. W. Budge, London, 1927, p. 62. On the paronomasia between *ʿēden*, 'Eden' and *ʿēdᵉtā*, 'church' that underlies this assertion, see G. A. Anderson, 'The Cosmic Mountain: Eden and Its Early Interpreters in Syriac Christianity', in *Genesis 1–3 in the History of Exegesis: Intrigue in the Garden*, ed. G. A. Robbins, Lewiston NY, 1988, pp. 187–224 (208–10).

[15] *Cave* II.18 (Oc.); IV.1; ed. Ri (n. 13 above), pp. 19, 28–9. Cf. also *Cave* V.27, where Adam is called the 'first priest' (ܟܗܢܐ ܩܕܡܝܐ); ed. Ri (n. 13 above), p. 44; transl. Budge, 1927 (n. 14 above), p. 70.

[16] *Cave* IV.1; ed. Ri (n. 13 above), p. 28; transl. Budge (n. 14 above), pp. 62–3.

[17] Ibid., p. 63.

There is reason to think that this heavenly priestly ministry was envisaged by the author of the *Cave* in a manner very similar to that of the *Book of Steps*, namely, as a liturgy of praise performed together by angels and men. Testimony to that is his description of the course of life pursued by Adam's son Seth and his descendants. He relates that, while they were still dwelling on the lower slopes of the mountain of paradise, their only occupation was 'to praise and glorify God, with the angels,' as they could hear the 'voices of the angels who were singing praises in paradise'.[18] In that function the Sethites, living in purity and holiness, served as a replacement for those angels that fell together with Satan and had been expelled from heaven.

* * *

We must turn our attention next to the question of the origin of these Christian traditions about paradise as a temple or sanctuary and Adam as a priest. It was quite early, already in the process of development of biblical tradition, that the Garden of Eden of the Genesis narrative (Chapters 2–3) began to be understood and described in terms of sacred or cultic space. This phenomenon should be considered in a broader context of the close connection between Temple and creation that is found in the Bible itself and was further developed in the post-biblical Jewish literature.[19] One finds parallels drawn between the Temple of Jerusalem and the Garden of Eden already in some biblical books.[20] Among the most explicit cases of this tendency should be counted Chapter 28 of Ezekiel, where the cultic dimension of Eden is brought to the fore.[21]

Later on, during the Second Temple period, the notion of paradise as a sanctuary was elaborated even more. As far as ancient Jewish exegesis is concerned the most important witness on how this issue was treated is the *Book of Jubilees*, where Eden is described as the holiest place on the earth, to which the Temple laws of purity are applied.[22] For

[18] *Cave* VII.4–6; ed. Ri (n. 13 above), pp. 56–8; transl. Budge (n. 14 above), p. 74.

[19] See M. Weinfeld, 'Sabbath, Temple, and the Enthronement of the Lord: The Problem of the Sitz im Leben of Genesis 1:1–2:3', in *Mélanges bibliques et orientaux en l'honneur de M. Henri Cazelles*, eds A. Caquot and M. Delcor, Neukirchen and Vluyn, 1981, pp. 501–12; G. J. Wenham, 'Sanctuary Symbolism in the Garden of Eden Story', in *Proceedings of the Ninth World Congress of Jewish Studies. Division A: The Period of the Bible*, Jerusalem, 1986, pp. 19–25; J. D. Levenson, 'The Jerusalem Temple in Devotional and Visionary Experience', in *Jewish Spirituality: From the Bible through the Middle Ages*, ed. A. Green, New York, 1986, pp. 32–61 (51–3); M. Himmelfarb, 'The Temple and the Garden of Eden in Ezekiel, the Book of the Watchers, and the Wisdom of ben Sira', in *Sacred Places and Profane Spaces: Essays in the Geographics of Judaism, Christianity, and Islam*, eds J. Scott and P. Simpson-Housley, New York, 1991, pp. 63–78.

[20] See T. Stordalen, *Echoes of Eden: Genesis 2–3 and Symbolism of the Eden Garden in Biblical Hebrew Literature*, Leuven, 2000, pp. 457–9.

[21] See Himmelfarb, 'The Temple and the Garden of Eden' (n. 19 above), pp. 64–6; Stordalen, *Echoes of Eden* (n. 20 above), pp. 350–56.

[22] See J. R. Levison, *Portraits of Adam in Early Judaism: From Sirach to 2 Baruch*, Sheffield, 1988, pp. 93–5; G. A. Anderson, 'Celibacy or Consummation in the Garden? Reflections on Early Jewish and Christian Interpretations of the Garden of Eden', *Harvard Theological Review*, 82/2, 1989, pp. 121–48 (129–31); J. T. van Ruiten, 'Eden and the Temple: The Rewriting of Genesis 2:4–3:24 in *The Book of Jubilees*', in *Paradise Interpreted: Representations of Biblical Paradise in Judaism and Christianity*, ed. G. P. Luttikhuizen, Leiden, 1999, pp. 63–94.

example, the author of *Jubilees* establishes a connection between the halakhic laws of impurity after childbirth and the story of creation of Adam and Eve, while explicitly characterizing the Garden of Eden as the place which is 'the holiest in the entire earth, and every tree which is planted in it is holy'.[23] On another occasion it is related about Noah that 'he knew that the Garden of Eden was the holy of holies and the dwelling of the Lord'.[24]

The notion of paradise as a sacred space is found also in a number of sectarian works from the Qumran library. In a most pronounced manner it is expressed in the *Damascus Document*, whose author depends on *Jubilees* and quotes it almost verbatim, saying that 'holy is the garden of Eden, and every shoot which is in its middle, is holy'.[25] A similar understanding of the Garden of Eden is found in *4QFlorilegium*, where it is called the 'temple of man' (מקדש אדם),[26] and in some passages from the *Hodayot* Hymns.[27]

In addition to these traditions there are various ancient Jewish sources where Adam is represented as a priest. As in the case of paradise, the reworking of Adam's image along priestly lines began as part of the process of inner-biblical exegesis. Of especial importance in that context are the two oracles on the king of Tyre in Ezekiel 28, a passage already mentioned, where several parallels between the depiction of this figure and Adam can be recognized.[28] This process was continued by the post-biblical Jewish writers. Thus, it is noteworthy that in the Septuagint version of Ezekiel the list of the precious stones in Ezekiel 28:13 was changed from the original nine to twelve. It seems that the Greek translator of Ezekiel corrected the number of stones on the basis of the description of the high priest's breastplate in Exodus 28:17–20, thus bringing the priestly dimension to the image of the primal human even more to the fore.[29] An example of the

[23] *Jubilees*, 3.8–14; *The Book of Jubilees*, *CSCO* 511, *Scriptores Aethiopici* 88; transl. J. C. VanderKam, Louvain, 1989, pp. 17–18. For an analysis of this tradition, see J. M. Baumgarten, 'Purification after Childbirth and the Sacred Garden in 4Q265 and Jubilees', in *New Qumran Texts and Studies: Proceedings of the first Meeting of the International Organization for Qumran Studies, Paris 1992*, ed. G. J. Brooke, Leiden, 1994, pp. 3–10; M. Segal, 'Law and Narrative in *Jubilees*: The Story of Entrance to the Garden of Eden Revisited', in *Meghillot: Studies in the Dead Sea Scrolls I*, eds M. Bar-Asher and D. Dimant, Jerusalem, 2003, pp. 111–25 (in Hebrew).

[24] *Jubilees*, 8.19; transl. VanderKam (n. 23 above), p. 53. Cf. also *Jubilees*, 4.26.

[25] 4Q265, line 14: קדוש גן עדן וכל האב אשר בתוכו קודש; *The Dead Sea Scrolls Study Edition*, eds F. García Martínez and E. J. C. Tigchelaar, Leiden, 1997–8, p. 550.

[26] 4Q174 1 I, 21, 2 line 6; eds García Martínez and Tigchelaar (n. 25 above), p. 352. On this expression, see M. O. Wise, '*4QFlorilegium* and the Temple of Adam', *Revue de Qumran*, 15/1–2, 1991, pp. 103–32 (123–32); Baumgarten, 'Purification' (n. 23 above), pp. 8–9.

[27] Cf. 1QH[a] XVI lines 10–13; eds García Martínez and Tigchelaar (n. 25 above), p. 180.

[28] See D. E. Callender, *Adam in Myth and History: Ancient Israelite Perspectives on the Primal Human*, Winona Lake IN, 2000, ch. 3, where he concludes that 'the tradition upon which Ezekiel bases his lamentation in 28.11–19 understands the primal human in priestly terms, or, perhaps better put, in "intermediary' terms"' (p. 132). S. D. Ricks, 'The Garment of Adam in Jewish, Muslim, and Christian Tradition', in *Judaism and Islam: Boundaries, Communication and Interaction: Essays in Honor of William M. Brinner*, eds B. H. Hary et al., Leiden, 2000, pp. 203–25 (212), characterizes this passage 'as an early attempt to connect Adam's clothing with that of the high priest'.

[29] See on this Callender, *Adam in Myth and History* (n. 28 above), pp. 102–3.

juxtaposition of Adam's figure and high-priest could be found in the *Book of Ben Sira*.[30] Adam is represented as performing priestly duties in such apocryphal works as *Jubilees* (3.26–27) and the Greek *Life of Adam and Eve* (29.3–6), where he offers an incense-sacrifice after expulsion from paradise.[31] Another cluster of traditions, centered on sacrificial activity of Adam after the Fall, appears later on in rabbinic writings, such as midrash *Genesis Rabbah* (34.9) and others, where the first man is reported to perform animal sacrifice.[32] The idea of human participation in the angelic heavenly liturgy is well attested in such writings from Qumran as the *Hodayot* or the *Songs of Sabbath Sacrifice*, and is deeply rooted in the self-image of this sectarian community as angels.[33]

Even such a specific exegetical motif as the cultic interpretation of the verb 'to till' of Genesis 2:15 in the *Cave of Treasures* has a precedent in rabbinic literature. In *Genesis Rabbah*, this verb is also interpreted as referring to sacrificial activity: '*to till it and to keep it* – these are sacrifices; "you shall serve God" (Exodus 3:12); "you shall observe to offer unto me" (Numbers 28:2).'[34]

All this makes it clear that the imagery of the Garden of Eden as temple, and Adam as priest, existed in Jewish tradition long before it appears in Syriac sources. Given the fact, noted by several scholars, that during the early stages of its development Syriac Christianity experienced considerable influence from Judaism,[35] it seems justified to regard these traditions in the works of Ephrem and other Syriac-speaking authors as a part of the Jewish matrix of Syriac Christianity. Unfortunately, our knowledge of the mechanisms of cultural interaction between different religious groups of the late antique Near East is rather limited and does not always allow us to determine the exact channels of transmission for the literary traditions under discussion. As one ponders over the possible ways of transmission of such extra-biblical material from Jews to Christians in

[30] Cf. Sirach 49.16–50.1. See also J. K. Aitken, 'The Semantics of "Glory" in Ben Sira: Traces of a Development in Post-Biblical Hebrew', in *Sirach, Scrolls, and Sages*, eds T. Muraoka and J. F. Elwolde, Leiden, 1999, pp. 1–24 (5–10).

[31] On this tradition, see J. C. VanderKam, 'Adam's Incense Offering (Jubilees 3:27)', in *Meghillot: Studies in the Dead Sea Scrolls V–VI*, eds M. Bar-Asher and E. Tov, Jerusalem, 2008, pp. 141*-56*; P.-B. Smit, 'Incense Revisited: Reviewing the Evidence for Incense as a Clue to the Christian Provenance of the *Greek Life of Adam and Eve*', *Novum Testamentum*, 46/4, 2004, pp. 369–75.

[32] Cf. *Babylonian Talmud*, treatises *Avoda Zara* 8a and *Shabbath* 28b; *Targum Pseudo-Jonathan* on Genesis 8.20; midrash *Avoth de-Rabbi Nathan* A 1.

[33] See D. Dimant, 'Men as Angels: The Self-Image of the Qumran Community', in *Religion and Politics in the Ancient Near East*, ed. A. Berlin, Bethesda MD, 1986, pp. 93–103; E. G. Chazon, 'Liturgical Communion with the Angels at Qumran', in *Sapiential, Liturgical and Poetical Texts from Qumran*, eds D. K. Falk et al., Leiden, 2000, pp. 95–105.

[34] *Genesis Rabbah* 16.5: ד"א לעובדה ולשמרה אילו הקורבנות תעבדון את האלהים תשמרו להקריב לי; *Midrash Bereshit Rabba*, eds J. Theodor and Ch. Albeck, 2nd rev. edn, Jerusalem, 1965, p. 149.

[35] See S. P. Brock, 'Jewish Traditions in Syriac Sources', *Journal of Jewish Studies*, 30/2, 1979, pp. 212–32; H. J. W. Drijvers, 'Syrian Christianity and Judaism', in *The Jews among Pagans and Christians in the Roman Empire*, eds J. Lieu et al., London, 1992, pp. 124–46; R. B. ter Haar Romeny, 'Hypotheses on the Development of Judaism and Christianity in Syria in the Period after 70 C.E.', in *Matthew and the Didache: Two Documents from the Same Jewish-Christian Milieu?*, ed. H. van de Sandt, Assen, 2005, pp. 13–33.

Syria-Mesopotamia, two considerations should be taken into account. first, one should be aware of the centuries-long history of uninterrupted Jewish presence in this region, including such centres of Syriac Christian culture as Nisibis and Edessa.[36] Second, we do know that some of the post-biblical works, such as *Jubilees*, where these traditions figure prominently, were translated into Syriac and exerted considerable influence upon the Syriac tradition of historiography and biblical exegesis.[37]

To conclude this section I would like to point out that adoption of the temple imagery, in relation to paradise, by Syriac Christian writers might be facilitated by the influence of other, inner-Christian factors: such, for instance, as the notion of the eschatological temple in heaven. This image appears in the New Testament, where the author of the Epistle to the Hebrews presents Jesus's redemptive mission as a priestly act of entering the heavenly temple, through which he makes this 'sanctuary' accessible to the believers.[38] While relevance of this imagery for the idea of paradise located in heaven is immediately obvious, further investigation is needed in order to unravel in detail the ways in which these two traditions interacted with each other in Syriac sources, as well as their relation to the notion of the perpetual liturgy performed by the angels in heaven.

2. Paradise as a Mountain

Another aspect of paradisiacal imagery in Syriac tradition that deserves our attention is the notion of the Garden of Eden as a cosmic mountain. As far as cosmological traditions of Late Antiquity are concerned, such an approach to representing paradise seems to be attested only in the works of Syriac-speaking authors, thus setting Syriac Christian tradition apart from the rest of Christian and Jewish cultures.

The earliest examples of this cosmological idea come again from the writings of Ephrem. It is his *Hymns on Paradise* that provide the most explicit references to paradise as a mountain of cosmic dimensions. For example, already in the first strophe of the *Hymns*, Ephrem represents paradise as an elevated mountain whose summit's height surpasses that of the other mountains. Accordingly, when the waters of the Flood covered the whole earth, this mountain alone remained unaffected by the catastrophe:

[36] See J. B. Segal, 'The Jews of North Mesopotamia before the Rise of Islam', in *Studies in the Bible Presented to Professor M.H. Segal*, eds J. M. Grintz and J. Liver, Jerusalem, 1964, pp. 32*-63*; D. Noy and H. Bloedhorn, *Inscriptiones Judaicae Orientis*, III, *Syria and Cyprus*, Tübingen, 2004.

[37] On *Jubilees* in Syriac, see E. Tisserant, 'Fragments syriaques du Livre des Jubilés', *Revue Biblique*, 30, 1921, pp. 55–86, 206–32. For a useful discussion of transmission of apocryphal works in Syriac-speaking milieu, see J. C. Reeves, 'Exploring the Afterlife of Jewish Pseudepigrapha in Medieval Near Eastern Religious Traditions: Some Initial Soundings', *Journal for the Study of Judaism*, 30/2, 1999, pp. 148–77.

[38] Cf. Hebrews 9:24, 10:19–21. See G. W. MacRae, 'Heavenly Temple and Eschatology in the Letter to the Hebrews', *Semeia*, 12, 1978, pp. 179–99; K. L. Schenck, *Cosmology and Eschatology in Hebrews: The Settings of the Sacrifice*, Cambridge, 2007, pp. 144–81. On early Jewish traditions of this kind, see D. E. Aune and E. Stewart, 'From the Idealized Past to the Imaginary Future: Eschatological Restoration in Jewish Apocalyptic Literature', in *Restoration: Old Testament, Jewish, and Christian Perspectives*, ed. J. M. Scott, Leiden, 2001, pp. 147–77 (163–9).

> With the eye of my mind I gazed upon paradise; the summit of every mountain (ܪܘܡܐ ܕܟܠ ܛܘܪ̈ܝܢ) is lower than its summit (ܪܘܡܗ), the crest of the Flood reached only its foothills (ܥܩܒ̈ܘܗܝ); these it kissed with reverence before turning back to rise above and subdue the peak of every hill and mountain. The foothills of paradise it kisses, while every summit it buffets.[39]

The imagery of paradise as a mountain is not limited to the *Hymns*, but can be found also through Ephrem's prose *Commentary on Genesis*. For example, while describing how the four rivers of Genesis 2:10–14 make their way from the place of their origin in Eden to the surface of the inhabited earth through the subterranean channels, he relates that 'paradise is set on a great height' (ܒܪܘܡܐ ܪܒܐ ܣܝܡ ܦܪܕܝܣܐ).[40] Further on, in his explanation of Genesis 4:8, Ephrem refers to the 'mountain of the outskirts of paradise' (ܛܘܪܐ ܕܫ̈ܘܦܠܝ ܦܪܕܝܣܐ) as the possible place where Cain and Abel dwelt.[41]

A particularly remarkable aspect of Ephrem's representation of paradise is that he conceives this cosmic mountain as encircling the whole universe. Thus, in the verses 8 and 9 of the same first strophe of the *Hymns*, Ephrem brings evidence both from nature and from Scripture in order to support this unusual cosmological idea that does not appear in the Bible. first of all, he points to the halo that surrounds the moon as a visual icon of paradise:

> Resembling that halo which surrounds the moon (ܫܘ ܠܗܬܐ ܕܣܗܪܐ) we should look upon paradise as being circular (ܟܪܝܟ) too, having both sea and dry land encompassed within it (ܚܒܝܫܝܢ ܠܓܘ ܡܢܗ).[42]

In addition to that Ephrem interprets symbolically the 'crown of gold' (MT זֵר זָהָב; Peshitta ܟܠܝܠܐ ܕܕܗܒܐ) that Moses commanded to make for the altar, according to Exodus 30:3, as referring to paradise that surrounds the created world:

> Moses made a crown (ܟܠܝܠܐ) for that resplendent altar; with a wreath entirely of gold did he crown (ܡܟܠܠ) the altar in its beauty. Thus gloriously entwined is the wreath of paradise (ܟܠܝܠܐ ܕܦܪܕܝܣܐ) that encircles the whole creation (ܟܪܝܟ ܠܒܪ̈ܝܬܐ).[43]

The idea of a circular spatial arrangement of paradise is expressed again in the second strophe of the *Hymns*, where it is likened to a girdle that encircles the earth:

> Paradise girds the loins of the world (ܐܣܪ ܚܨ̈ܝ ܗܘ̇ ܕܥܠܡܐ), encircling the great sea (ܠܝܡܐ ܣܚܐ ܪܒܐ); neighbor to the beings on high, friendly to those within it, hostile to those without.[44]

[39] Ephrem the Syrian, *De paradiso*, I.4; ed. Beck (n. 4 above), p. 2; transl. Brock (n. 4 above), pp. 78–9.

[40] Ephrem the Syrian, *In Genesim*, II.6, *Sancti Ephraem Syri in Genesim et in Exodum commentarii*, ed. R. M. Tonneau, *CSCO* 152, *Syr.* 71, Louvain, 1955, p. 29; *St. Ephrem the Syrian. Selected Prose Works*, transl. E. G. Mathews and J. P. Amar, Washington DC, 1994, p. 101.

[41] *In Genesim*, III.5; ed. Tonneau (n. 40 above), p. 49; transl. (modified) Mathews and Amar (n. 40 above), p. 126.

[42] Ephrem the Syrian, *De paradiso*, I.8; ed. Beck (n. 4 above), pp. 2–3; transl. Brock (n. 4 above), p. 80.

[43] Ephrem the Syrian, *De paradiso*, I.9; ed. Beck (n. 4 above), p. 3; transl. Brock (n. 4 above), p. 81.

[44] Ephrem the Syrian, *De paradiso*, II.6; ed. Beck (n. 4 above), p. 6; transl. Brock (n. 4 above), p. 87.

A word should be said about theological significance of the mountain imagery of paradise in Ephrem's works. A great advantage in using this image is that it makes paradise both terrestrial and celestial, and, thus, provides Ephrem with a convenient way of reconciling two different conceptual pictures of paradise that were current among Christians during Late Antiquity. One, 'literal', approach, based on the first chapters of Genesis, envisaged paradise as an earthly garden from which the four famous rivers issue and from which Adam was expelled. This approach was hardly compatible with another popular notion: the spiritualized idea of paradise as an eschatological place of rest, where the disembodied souls of the deceased saints and righteous await the final resurrection as if in a state of sleep.[45] The theological genius of Ephrem was that by presenting paradise as a mountain, that stands at the intersection of the two spheres and whose different levels correspond to different stages of spiritual progress, he offered a persuasive and visualizable resolution of this contradiction, without resorting to the theory of two distinct paradises.[46]

The influence of Ephrem's mythopoetic representation of paradise can be recognized in a number of later Syriac compositions, mostly of an apocryphal nature, as their authors seem to feel less pressure to conform to more rationalistic and dogmatically restricting theological patterns that were developed in the Greek Christian world and began to dominate the mainstream Syriac theological discourse from the fifth century onwards. There are several post-Ephremian descriptions of paradise produced in a Syriac-speaking milieu that feature the two distinguishing aspects of Ephrem's vision of paradise: namely, an understanding of it as the highest mountain and the circular arrangement of this mountain in space.

The image of paradise as a mountain located on the perimeter of the earth appears in the fifth-century Syriac version of the *Transitus Mariae*, an apocryphal work that describes the last days and post-mortem fate of the Virgin Mary. There, at the beginning of the fifth book, where Mary is brought to heaven and placed in paradise, a developed description of the 'paradise of Eden' (ܦܪܕܝܣܐ ܕܥܕܢ) is offered.[47] Paradise is represented as a mountain surrounded by other mountains (ܛܘܪܐ ܕܚܕܪܝܢ ܠܗ). It is described as the 'great mountain' (ܛܘܪܗ ܪܒܐ), situated 'at the outer edge' (ܒܣܘܦܐ ܕܒܪܝܬܐ) of the earth, whose 'foundation' (ܫܬܐܣܬܗ) is said to be placed upon the earth. In a way similar to Ephrem the Syriac editor of the *Transitus* relates that the Flood did not cover this 'holy mountain' (ܛܘܪܐ ܩܕܝܫܐ) and reached only its 'lower grounds' (ܫܦܘܠܘܗܝ). The reason

[45] On this notion, characteristic for the Syriac Christian tradition, see M. J. Dal Santo, *Debating the Saints' Cult in the Age of Gregory the Great*, Oxford Studies in Byzantium, Oxford, 2012, pp. 237–320.

[46] On this aspect of Ephrem's image of paradise, see Brock, *Saint Ephrem. Hymns on Paradise* (n. 4 above), pp. 50–53; I. Ortiz de Urbina, 'Le Paradis eschatologique d'après S. Ephrem', *Orientalia Christiana Periodica*, 21, 1955, pp. 467–72; S. J. Shoemaker, *Ancient Traditions of the Mary's Dormition and Assumption*, Oxford, 2002, pp. 184–8.

[47] *Apocrypha Syriaca: The Protevangelium Jacobi and Transitus Mariae*, ed. A. S. Lewis, London, 1902, pp. 154–5 (Syriac).

for that was that God himself was standing in his glory, 'fixed' on the paradise of Eden, and observing the course of the catastrophe. It is noteworthy that no other early textual witness of this apocryphal work features this motif, which leads us to regard it as a unique inner-Syriac development within the *Transitus* tradition.[48]

Later on we find the idea of mountainous paradise in the *Cave of Treasures*. While describing the state of Adam and Eve in the Garden of Eden before their fall, the author of the *Cave* relates that paradise is situated 'above' (ܠܥܠ), that it is thirty spans 'higher than all the high mountains' (ܡܥܠܝ ܡܢ ܟܠܗܘܢ ܛܘܪ̈ܐ ܪ̈ܡܐ), and that it 'surrounds the whole earth' (ܚܕܪ ܠܟܠܗ ܐܪܥܐ).[49] Furthermore, similarly to Ephrem and the *Transitus*, the author of the *Cave* endorses the opinion that the Flood did not inundate the Garden of Eden. According to him, when the waters of the deluge brought Noah's ark to the 'lower grounds of paradise' (ܫ̈ܦܘܠܝ ܦܪܕܝܣܐ), they received a blessing from the latter and piously retreated, once they had kissed the 'skirts of paradise' (ܚܕ̈ܢܝ ܦܪܕܝܣܐ).[50]

It seems that the notion of the mountainous paradise has exerted some influence also on the author of the Syriac cosmological tractate, ascribed to Pseudo-Dionysius the Areopagite.[51] This composition, the main thrust of which is anti-astrological and anti-magical polemic, was written originally in Syriac, most likely during the sixth or early seventh century.[52] At a certain point, while describing the season of winter, the author of this work refers to the mountain range, called 'the Breasts of the North', which apparently is located at the northern extremity of the inhabited earth. Beyond these mountains there is no human presence, as they are bordered by the 'fiery river' (ܢܗܪܐ ܕܢܘܪܐ) and the ocean (ܐܘܩܝܢܘܣ). The ocean encircles the whole earth like a wall (ܫܘܪܐ) and is devoid of any animal life. It is beyond this ocean that the author locates the 'paradise of gods' (ܦܪܕܝܣܐ ܕܐܠܗ̈ܐ), the place to which the angels bring the souls of deceased saints.[53] While there are no explicit references to mountain imagery in this depiction of paradise, one might still recognize there some residual traces of this idea. Thus, when the tractate's author describes the process of post-mortem transfer of the souls of saints to this place, he refers to the 'mansions that surround the paradise'

[48] For an analysis of paradise material in the different versions of *Transitus*, see Shoemaker, *Ancient Traditions* (n. 46 above), pp. 179–203.

[49] *Cave* III.15; ed. Ri (n. 13 above), p. 24; transl. Budge (n. 14 above), p. 60.

[50] *Cave* XIX.3–4; ed. Ri (n. 13 above), p. 146; transl. Budge (n. 14 above), p. 113.

[51] This work was first published by M.-A. Kugener, 'Un Traité astronomique et météorologique syriaque attribué à Denis l'Aréopagite', in *Actes du XIVᵉ Congrès international des orientalistes, Alger 1905, IIᵉ partie: Section II (Langues sémitiques)*, Paris, 1907, pp. 137–98. Slightly later it was reedited and translated into English by G. Furlani, 'A Cosmological Tract by Pseudo-Dionysius in the Syriac Language', *Journal of the Royal Asiatic Society*, 49/2, 1917, pp. 245–72.

[52] It should not be confused with the writings from the original Pseudo-Dionysian corpus that was produced in Greek somewhere towards the end of the 5th century and translated into Syriac already during the 6th century by Sergius of Reshaina. On this, see P. Sherwood, 'Sergius of Reshaina and the Syriac Versions of the Pseudo-Denis', *Sacris Erudiri*, 4, 1952, pp. 174–84; I. Perczel, 'The Earliest Syriac Reception of Dionysius', *Modern Theology*, 24/4, 2008, pp. 557–71.

[53] Pseudo-Dionysius, ed. Kugener (n. 51 above), pp. 154–5.

(ܐܘܢ̈ܘܬܐ ܕܣܚܪܝܢ ܠܦܪܕܝܣܐ).[54] The circular arrangement of these 'mansions' in space is most likely derived from the idea of paradise as a mountain on the slopes of which they are located, as we find for example in the *Cave of Treasures* (XIII.19), where God brings the biblical patriarch Enoch to the 'delectable mansions that surround paradise' (ܐܘܢ̈ܘܬܐ ܪ̈ܓܝܓܬܐ ܕܚܕܪ̈ܝ ܦܪܕܝܣܐ),[55] or in the sixth-century Syriac version of the *Transitus Mariae*, where the 'chambers of the just' (ܓܢ̈ܘܢܐ ܕܙܕܝ̈ܩܐ) are said to be situated around the paradisiacal mountain.[56]

Finally, it might be added that diffusion of the notion of paradise as a mountain in Syriac Christianity was not limited exclusively to the literary sources. Influence of this idea can be recognized in the fifth-century floor mosaic from the Church of the Holy Martyrs at the village of Tayibat al-Imam in Central Syria. There, in the third register of the eastern panel, the four rivers of paradise are depicted that flow from a mountain upon which a spread-winged eagle is depicted.[57]

* * *

Having observed dissemination of the image of paradise as a circular cosmic mountain among Syriac-speaking writers, let us address the question of the origins of this peculiar cosmological idea which, as has already been emphasized, seems to be absent from any other Christian culture of Late Antiquity.

The understanding of paradise as a mountain is attested in pre-Christian Jewish sources. In fact it appears already in the Old Testament where, in the famous lamentation over the prince of Tyre from the Chapter 28 of Ezekiel, we find in the verses 13–14 'Eden, the garden of God' (עֵדֶן גַּן־אֱלֹהִים) being equated with 'the holy mountain of God' (הַר קֹדֶשׁ אֱלֹהִים).[58] This lead was followed and further developed by several Jewish writers from the Second Temple period. Thus, the biblical characteristic of paradise as a mountain was adopted by the author of *Jubilees*. In his treatment of the Noah narrative he relates that the Flood's 'water did not come on any of the land of Eden' and lists the 'Garden of Eden' together with three mountains as the 'four places on earth that belong to the Lord'.[59] Another

[54] Ibid., p. 155, line 3.

[55] *Cave*, ed. Ri (n. 13 above), p. 104; transl. Budge (n. 14 above), pp. 95–6.

[56] 'The Departure of my Lady Mary from this World', ed. W. Wright, *Journal of Sacred Literature and Biblical Record*, NS VI, 12, 1865, pp. 417–48; VII, 13, 1865, pp. 110–60 (44–5 [Syriac]).

[57] See A.-R. Zaqzuq and M. Piccirillo, 'The Mosaic Floor of the Church of the Holy Martyrs at Tayyibat al-Imam – Hamah in Central Syria', *Liber Annuus*, 49, 1999, pp. 443–64; R. Wisskirchen, 'Der Adler auf dem Paradiesesberg. Zum Bodenmosaik im Ostteil der Kirche der 'Heiligen Märtyrer' in Tayibat Al-Imam/Hama (Syrien)', *Jahrbuch für Antike und Christentum*, 48–49, 2005–2006, pp. 154–63.

[58] One can recognize in this passage convergence of two distinctive traditions that were current in the ancient Near East: the mythological idea of mountains as the dwelling place of deities and the function of gardens as cultic sites. On the former, see Sh. Talmon, 'הַר,' in *Theological Dictionary of the Old Testament*, eds G. J. Botterweck et al., Grand Rapids MI, 1978, III, pp. 427–47 (442–4); R. J. Clifford, *The Cosmic Mountain in Canaan and the Old Testament*, Cambridge MA, 1972; on the latter, see Stordalen, *Echoes of Eden* (n. 20 above), pp. 111–38.

[59] *Jubilees*, 4.24, 26, transl. VanderKam (n. 23 above), pp. 28–9. It is noteworthy that in the surviving Syriac fragments of *Jubilees*, the Garden of Eden in 4.26 is explicitly called the 'mountain of Paradise' (ܛܘܪܐ ܕܦܪܕܝܣܐ); Tisserant, 'Fragments syriaques' (n. 37 above), p. 73.

example of the mountainous paradise imagery from this period is found in the mythical geography of *I Enoch*.[60] There, during his second heavenly journey to the north-western extremity of earth, Enoch observes a mountain range that consists of the 'seven magnificent mountains' that are made of precious stones. On the peak of the highest of these mountains, which resembles the 'throne of the Lord,' a beautiful fragrant tree grows, from whose fruits life will be given to the righteous on the day of eschatological consummation. Several scholars regarded this tree as derived from the Tree of Life of Genesis 2–3 and the whole setting as being modeled after that of paradise.[61]

Basing themselves on this evidence some scholars have treated the idea of paradise as a mountain in Syriac sources either as a direct development of the biblical imagery, or as an instance of Jewish influence upon Syriac Christianity.[62] However, while it is possible that these biblical and post-biblical traditions could contribute to the development of the tradition of the mountainous paradise among Syriac Christians,[63] referring to them alone can hardly help us to explain the genesis of this motif as a whole. There are several considerations that prevent us from being satisfied with one of these explanations. First of all, it has to be stressed that while such texts as Ezekiel and *Jubilees* were read in other Christian cultures as well, the fact remains unexplained why, during the period under consideration, the concept of paradise as a mountain gains popularity nowhere else, save in the Syriac-speaking milieu. The second, and more important, objection is that none of the biblical or post-biblical Jewish authors describes the mountain of paradise as *surrounding* the earth. In that connection it has also to be underscored that none among the Syriac authors, who use the image of paradise as a mountain, considers it to be the *axis mundi*. The author of the *Cave*, for example, adheres to the widespread Christian idea of Golgotha in Jerusalem as the centre of the world.[64]

Looking for a possible background for the notion of the mountainous paradise in Syriac Christianity some scholars have pointed to the cosmological and cultic traditions of the Ancient Near East in general, and the mythology of the world-mountain

[60] *I Enoch* 24.2–25.7, *The Ethiopic Book of Enoch*, ed. M. A. Knibb, Oxford, 1978, II, pp. 112–14.

[61] See P. Grelot, 'La Géographie mythique d'Hénoch et ses sources orientales', *Revue biblique*, 65, 1958, pp. 33–69 (42); G. W. E. Nickelsburg, *1 Enoch 1: A Commentary on the Book of 1 Enoch, Chapters 1–36; 81–108*, Minneapolis MN, 2001, pp. 313–15; J. J. Collins, 'Before the Fall: The Earliest Interpretations of Adam and Eve', in *The Idea of Biblical Interpretation: Essays in Honor of James L. Kugel*, eds H. Najman and J. H. Newman, Leiden, 2004, pp. 293–308 (302–4).

[62] For the former approach, see Anderson, 'The Cosmic Mountain' (n. 14 above). The latter position has been taken by Brock, *Saint Ephrem. Hymns on Paradise* (n. 4 above), pp. 50–51.

[63] Especially the idea, which appears for the first time in *Jubilees*, employed at times by Ephrem that the waters of the Flood did not cover paradise.

[64] Cf. *Cave* IV.3; V.11. Gary Anderson misinterprets the *Cave* IV.2–3, when he claims that it describes the Tree of Life as located in the 'center of the earth' (Anderson, 'The Cosmic Mountain' [n. 14 above], p. 199). It is not the Tree of Life that is located in the middle of the earth, according to the plain meaning of this passage, but the Tree of the Cross, of which the former is only a prefiguration.

symbolized by the Sumero-Akkadian *ziqqurat* in particular.[65] This theory, however, faces the same difficulty as the previous, since it is far from certain that the idea of the cosmic mountain that *surrounds* the earth was actually present in the ancient Near Eastern cosmological tradition. Although one finds in ancient Mesopotamian sources several passages that might lend themselves to such an interpretation, it is not expressed explicitly and unambiguously.[66]

The internal inconsistency of the image of the mountainous paradise in the writings of Ephrem, where seemingly mutually exclusive characteristics of being circular and having one summit are applied to it, has puzzled scholars. Sebastian Brock, who made an attempt to reconstruct how exactly Ephrem visualized paradise, described it as 'a conically shaped mountain whose base circumference encircles the "Great Sea"' so that 'our sky will thus be formed by the inside of the conical mountain of paradise'.[67] The contradictory character of this spatial construction made Brock to suggest that Ephrem deliberately presented his audience with an 'illogical' model of paradise that defies the ordinary human experience of time and space in order to challenge the Antiochene exegetical school literalist views on the biblical Garden of Eden that might be popular among Syriac Christians at that time.[68] While not denying such a possibility altogether I would like to point out that what seems illogical to a modern reader of Ephrem might not have seemed so to his intended audience, who could visualize the poet's references to paradise differently from us, in accordance with their own assumptions about mythical geography.

In order to shed light on these assumptions one should lend an ear to the opinion of those scholars who have pointed out the resemblance between the mountainous paradise of Ephrem and the *Cave of Treasures*, and Iranian mythological tradition where the idea of the circular cosmic mountain Alburz played an important role.[69] It is this aspect of the mythical geography of Iranians that allows us to make sense of the fact that the image of mountainous paradise is attested only among the Christians of Syria-Mesopotamia and not in any other Christian tradition.

[65] Murray, 'The Lance which Re-opened Paradise' (n. 2 above), p. 309. Cf. also G. Widengren, '"Synkretismus" in der syrischen Christenheit', in *Synkretismus im syrisch-persischen Kulturgebiet*, ed. A. Dietrich, Göttingen, 1975, pp. 38–64 (40).

[66] See W. Horowitz, *Mesopotamian Cosmic Geography*, Winona Lake IN, 1998, pp. 331–3.

[67] Brock, *Saint Ephrem. Hymns on Paradise* (n. 4 above), p. 54. A similar spatial interpretation of Ephrem's Paradise is offered by Palmer, 'Paradise Restored' (n. 3 above), pp. 2–3.

[68] Brock, *Saint Ephrem. Hymns on Paradise* (n. 4 above), pp. 54–5.

[69] Cf. A. S.-M. Ri, *Commentaire de la Caverne des Trésors: étude sur l'histoire du texte et de ses sources, CSCO* 581, *Subs.* 103, Louvain, 2000, pp. 40–41, 208. Ri follows the lead of G. Widengren, who discerns Iranian mythological background behind the image of the 'victorious mountain' (*mons victorialis*) in *Opus imperfectum in Matthaeum* II.2, *PG* LVI, cols 637–8 which, in its turn, is said to be related to the 'mountain of victory' (ܛܘܪ ܒܝ ܣܟܐ), another name of the paradise mountain in the *Cave* (XIV.1); see G. Widengren, *Iranisch-semitische Kulturbegegnung in parthischer Zeit*, Cologne, 1960, pp. 66–7, 80. Cf. also A. Toepel, *Die Adam- und Seth-Legenden im syrischen Buch der Schatzhöhle. Eine quellenkritische Untersuchung*, *CSCO* 618, *Subs.* 119, Louvain, 2006, p. 114, n. 32.

The notion of a cosmic mountain range that surrounds the earth was common to both Iranian and Indian cosmological traditions.[70] According to the Iranian views, mountains grew up from the surface of earth which was conceived as a round and flat plane. This process was imagined as similar to the growth of plants and the mountains were thought to have 'roots'. Zoroastrian cosmology, as it is reflected in the *Avesta* and the later Pahlavi writings, knows different mountains. The most prominent among them was the mount Alburz (Avestan *harā bərəzaitī*, Pahlavi *harburz*), the 'lofty watchpost,' as it grew out of the earth first.[71] This mythical mountain is mentioned already in the *Avesta*. It figures also in later Zoroastrian writings such as the Iranian *Bundahišn*.

In the most ancient layer of belief about the mount Alburz, it is represented standing in the middle of the world as a gigantic mountain, around whose peak revolve the stars, moon, and sun. However, already in ancient times this concept of Alburz was developed further and this name was given also to a great range of mountains that was thought to encircle the rim of the flat earth.[72] This cosmological idea is expressed already by the Young Avestan *Zamyād-Yašt* (19.1), according to which the mount Alburz, described as the first mountain that stood upon this earth, 'encircles entirely the western and eastern lands'.[73] Later on we find this notion in various Pahlavi sources as well. Thus, the author of the Iranian *Bundahišn* (5B.1) explicitly distinguishes between the mount Tēra that occupies the middle of the earth, and the mount Alburz situated 'around the world' (*pērāmōn ī gēhān*).[74] As the following narrative, i.e. 9.1–2, makes clear, he understood the former mountain to be an outgrowth of the latter. A similar cosmological scheme underlies the eschatological scenario proposed in the erotapocritic composition *Dādestān ī Dēnīg* (20.2), according to which the souls of the deceased shall pass over the bridge Činwad, which extends from the mount Dāitī, located in the 'middle of the earth' (*mayān ī gēhān*), in the region of Ērān-wēz, to the 'enclosure of Alburz' (*war ī alburz*).[75]

The circular arrangement in space is the crucial, but not the only, common feature that the mount Alburz of Iranian mythology shares with the paradisiacal mountain of the Syriac sources. Thus, similarly to the mountain of paradise, Alburz was thought to be higher than the rest of mountains and characterized accordingly as the 'chief of the summits' (*bālistān rad*).[76] In addition to that, both the Garden of Eden and the mount

[70] See M. Boyce, *A History of Zoroastrianism*, I, *The Early Period*, Leiden, 1975, pp. 132–7.

[71] See M. Boyce, 'Alborz. ii. Alborz in Myth and Legend', in *Encyclopaedia Iranica*, ed. E. Yarshater, London, 1985, I, pp. 811–13.

[72] This Iranian idea has a parallel in the ancient Indian cosmology, with its notion of *Lokāloka*, a mountain belt that surrounds all the continents of the earth. See W. Kirfel, *Die Kosmographie der Inder, nach Quellen dargestellt*, Bonn, 1920, pp. 121–2.

[73] *hā hama pairi.saēte frāpiiā̊ daŋ́hūš ā upaošaŋᵛhā̊sca*; *Der Zamyād-Yašt: Edition, Übersetzung, Kommentar*, ed. A. Hintze, Wiesbaden, 1994, p. 66.

[74] *Zand-Ākāsīh: Iranian or Greater Bundahišn*, ed. and transl. B. T. Anklesaria, Bombay [Mumbai], 1956, pp. 64–5.

[75] *Dādestān ī Dēnīg, Part I: Transcription, Translation and Commentary*, ed. M. Jaafari-Dehaghi, Paris, 1998, pp. 76–7. Cf. also Iranian *Bundahišn*, 30.1; Pahlavi *Vidēvdād*, 19.30.

[76] Iranian *Bundahišn*, 17.18; ed. Anklesaria (n. 74 above), pp. 154–5.

Alburz were invested with the cosmological function of being the sources of earthly rivers. While in the Bible it is the four rivers that issue forth from paradise according to Genesis 2:10–14, in the Iranian tradition it is the mythical river Ardvīsūr that pours down from the mountain Alburz into the ocean Frāxkard.[77] Moreover, while paradise in the book of Genesis functions as the place where the sacred Trees of Life and Knowledge grow, the mountain Alburz is said to be the place where the deity Baga put the sacred plant *haoma* to grow.[78] finally, while the Garden of Eden was understood as a sacred Temple-like enclosure, where the first humans offered incense sacrifice to God or angelic liturgy was performed, the mountain Alburz was likewise functioned as a cultic place, where such figures of the ancient Iranian mythology as Hōšang and Jamšēd offered sacrifices to the deities Arədvī Sūrā, Druvāspā, and Vayu.[79]

In the light of all these parallels as well as of the prominent place held by the idea of the cosmic mountain Alburz surrounding the inhabited earth in the Iranian mythical cosmology, it seems very likely that Syriac Christian interpreters of the Bible, beginning with Ephrem, developed their peculiar image of the biblical paradise as a circular cosmic mountain under the direct influence of this Iranian motif. If correct, this example provides a valuable testimony on the relatively little studied phenomenon of Syriac Christian acculturation to the Iranian culture.

* * *

One aspect of the notion of the mountainous paradise in Syriac sources still requires further elucidation: the close connection between the image of the circular cosmic mountain and the cosmological conception, so popular in the ancient world, of the cosmic Ocean surrounding the inhabited earth.[80]

The earliest attestation of influence exerted by this cosmological notion on the concept of paradise in Syriac literature appears in Ephrem's *Hymns on Paradise* where he states, concerning paradise, that it 'girds the loins of the world,' while at the same time 'encircling the great sea' (ܠܝܡܐ ܚܙܝܩ ܪܒܐ).[81] This laconic description seems to presume that, according to Ephrem, the earth is surrounded not only by the great circle of the paradisiacal mountain range, but also by the lesser circle, consisting of a body of water.[82]

[77] *Yašt*, 5.3; Iranian *Bundahišn*, 10.5–6; cf. also Iranian *Bundahišn*, 11.6; *Vidēvdād*, 21.

[78] *Yasna*, 10.4–10.

[79] *Yašt*, 5.21, 25; 9.3, 8; 15.7, 15. Cf. also traditions about the *yazata* Haoma as a priest of the gods, who offers on this mountain sacrifices to Mithra (*Yašt*, 10.88–89) and to Sraoša (*Yasna*, 57.19). See M. Boyce, 'Haoma, Priest of the Sacrifice', in *W.B. Henning Memorial Volume*, eds M. Boyce and I. Gershevitch, London, 1970, pp. 62–80.

[80] On the ancient Near Eastern and Greek views of Ocean, see A. Lesky, *Thalatta: Der Weg der Griechen zum Meer*, Wien, 1947; O. Kaiser, *Die mythische Bedeutung des Meeres in Ägypten, Ugarit und Israel*, Berlin, 1962; A. J. Wensinck, *The Ocean in the Literature of the Western Semites*, Amsterdam, 1918.

[81] Ephrem the Syrian, *De paradiso*, II.6; ed. Beck (n. 4 above), p. 6; transl. Brock (n. 4 above), p. 87.

[82] Andrew Palmer has recently tried to challenge the widely accepted interpretation of the passive participle *ḥziq* in this passage as conveying active meaning. He has suggested that it should be understood as passive and that the paradise mountain was envisaged by Ephrem not as 'encircling' Ocean, but as 'encircled' by it (Palmer, 'Paradise Restored' [n. 3 above], p. 2, n. 1). This attempt, however, is hardly convincing in light of the fact, noted by Syrian

Additional support for the assumption that Ephrem understood paradise to be located beyond Ocean comes from his *Commentary on Genesis*. There he deals with the question how the four great rivers, mentioned in Genesis 2:10–14, whose source is said to be in the Garden of Eden, find their way to our world. As Ephrem explains, after these rivers issue forth from their source on the mountain of paradise they enter some underground channels that go under the sea bed and, through them, arrive to the confines of the inhabited world.[83] This 'sea' (ܝܡܐ), under which the rivers of paradise are flowing, is nothing else but the 'great sea' (ܝܡܐ ܪܒܐ) surrounding the earth according to the quoted passage of the *Hymns*.

Later on we find more explicit references to the cosmic Ocean that surrounds the inhabited earth. One such reference comes from the *Cave of Treasures*. In the section dealing with Noah's Flood its author interprets the waters coming from 'the fountains of the great deep,' according to Genesis 7:11, as the waters of 'Ocean, that great sea which surrounds the whole realm' (ܐܘܩܝܢܘܣ ܝܡܐ ܪܒܐ ܗܘ ܕܚܕܪ ܠܟܠܗ ܐܘܚܕܢܐ).[84]

Two other works illustrate the wide popularity of this cosmologic belief among the Christians of Syria-Mesopotamia. One is the Syriac cosmological composition ascribed to the Babylonian priest Berossus, where we are told about the 'great Ocean that surrounds the whole earth'.[85] Another is the Syriac *Legend of Alexander*, a work written in northern Mesopotamia around AD 629–630. There, in the part dealing with Alexander's endeavour to explore the world, it is related how, during his journey to the northern boundaries of the inhabited world, the king met the local sages who inform him that paradise is securely separated from human access by the 'Ocean, the foetid sea' (ܐܘܩܝܢܘܣ ܝܡܐ ܣܪܝܐ), which is the 'great sea that surrounds all creation' (ܝܡܐ ܪܒܐ ܕܟܪܝܟ ܠܟܠܗ ܒܪܝܬܐ).[86] It should be noted that none of these two works contains the notion of paradise as a mountain.

The question of the origins of the idea of the circular cosmic Ocean in the Syriac Christian tradition is rather complicated. On the one hand one might regard this belief to be an example of Hellenistic influence upon Syriac culture, given the fact that the idea of the cosmic Ocean was extremely popular in the Greek cosmological tradition. There, it is attested as early as the first half of the first millennium BCE. Thus, one finds it in the works of Homer, where Ocean forms the edge of the circular disc of the earth, as illustrated by

grammarians, that this passive form is able to, and does actually, express active meaning. See Th. Nöldeke, *Compendious Syriac Grammar*, transl. J. A. Crichton, London, 1904, p. 220. In addition to that, Ephrem explicitly states in *De paradiso*, I.8 that 'sea and dry land encompassed *within* it (i.e. paradise)' (ܘܝܡܐ ܐܦ ܝܒܫܐ ܚܒܝܫܝܢ ܠܓܘ ܡܢܗ); ed. Beck (n. 4 above), pp. 2–3; transl. Brock (n. 4 above), p. 80.

[83] Ephrem the Syrian, *In Genesim*, II.6; ed. Tonneau (n. 40 above), p. 29; transl. Mathews and Amar (n. 40 above), p. 101.

[84] *Cave*, XVIII.10; ed. Ri (n. 13 above), p. 142; transl. Budge (n. 14 above), p. 112.

[85] ܐܘܩܝܢܘܣ ܪܒܐ ܗܘ ܕܚܕܪ ܠܗ ܠܟܠܗ ܐܪܥܐ; M. G. Levi Della Vida, 'Pseudo-Beroso siriaco', *Rivista degli Studi Orientali*, 3, 1910, pp. 7–43 (15).

[86] *The History of Alexander the Great, being the Syriac Version of the Pseudo-Callisthenes*, ed. and transl. E. A. W. Budge, Cambridge, 1889, pp. 266–7 (Syriac), 152–3 (English).

the shield of Achilles.[87] Later on, Herodotus ridicules some fellow Greek geographers, who draw the world 'encircled by the river of Ocean'.[88] Although no examples of the Greek maps criticized by Herodotus have survived, there is a literary description of one such work, with which the historiographer might have been acquainted. Agathemenus, a fourth-century BCE geographer, describes the first Greek map that was drawn by Anaximander (610–540), one of Thales's disciples, where 'swift flowing Oceanus completed a circle around the earth'.[89] Later on, this belief becomes a commonplace in the Greek cosmological tradition and could be found in works of various authors.

There is ample evidence that this cosmological notion entered the Syriac intellectual milieu through the translation of several Greek compositions that contained it. One such case is the Pseudo-Aristotelian cosmological tractate *De mundo* that features the Ocean, which surrounds the islands of the inhabited world.[90] This work was translated into Syriac in the sixth century by Sergius of Reshaina.[91] Another Greek composition that was translated into Syriac during Late Antiquity, containing the image of the circular cosmic Ocean, is the *History of Alexander* by Pseudo-Callisthenes.[92] While both these works had been translated into Syriac long after the time of Ephrem, even in his case the possibility of Hellenistic background for the image of the circular Ocean cannot be ruled out completely.[93]

At the same time one should also bear in mind that the notion of Ocean encircling the earth appears also in the Ancient Near Eastern cosmological tradition. For example, we find it in the Late Babylonian tablet British Museum 92687, dated to the late eighth or seventh century BCE. This tablet features a sketch map of the world, where the inhabited continent is depicted surrounded by the circular band indicated as 'salt-sea' (Akkad. *marratu*).[94] A parallel tradition is found in another Late Babylonian text, the so-called *Bilingual Creation of the World by Marduk*, according to which the deity Marduk creates dry land on a raft floating upon, and surrounded by, waters.[95] It is noteworthy that in distinction from its earlier synonym *tâmtu*, 'sea', the first-millennium

[87] *Iliad*, 18.607–8; 21.195–6; *Odyssey*, 11.13; 24.11–12. Cf. also Hesiod, *Theogony*, 789–91; *Shield of Heracles*, 314.

[88] Herodotus, *Histories*, IV.36; *Herodotus*, ed. and transl. A. D. Godley, Loeb Classical Library, London, 1926–30, II, pp. 234–5.

[89] *Geographi Graeci minores*, ed. K. O. Müller, Paris, 1882, II, pp. 471–2.

[90] Pseudo-Aristotle, *De mundo* 3 (393a16); *Il trattato sul cosmo per Alessandro attribuito ad Aristotele*, eds G. Reale and A. P. Bos, Milan, 1995, p. 186.

[91] For the relevant passage, see *Analecta syriaca*, ed. P. A. de Lagarde, Leipzig, 1858, p. 139.

[92] For this image in the Greek version of Pseudo-Callisthenes (II.41), see *Arriani Anabasis et Indica. Reliquia Arriani et scriptorum de rebus Alexandri M. fragmenta collegit, Pseudo-Callisthenis Historiam fabulosam*, eds F. Dübner and K. O. Müller, Paris, 1846, p. 91. For its counterpart in the Syriac version, see *The History of Alexander the Great*, ed. Budge (n. 86 above), p. 100.

[93] On Ephrem's acquaintance with Greek culture, see U. Possekel, *Evidence of Greek Philosophical Concepts in the Writings of Ephrem the Syrian*, *CSCO* 580, *Subs.* 102, Louvain, 1999.

[94] For the text and translation, see Horowitz, *Mesopotamian Cosmic Geography* (n. 66 above), pp. 20–25.

[95] See ibid., pp. 129–32.

Akkadian noun *marratu*, 'ocean,' was also used to refer to rivers and canals.[96] This suggests that the author of the Babylonian map understood the ocean surrounding the world not as a boundless sea, but as a relatively narrow body of water. As some scholars recognize vestiges of the Ancient Near Eastern beliefs in Ephrem's works[97] one might wonder whether the image of the circular ocean should be counted among them as well.

Moreover, it seems that the notion of the ocean surrounding the inhabited earth was known also in the Iranian mythological tradition. One thinks of the cosmic ocean Frāxkard, mentioned in Zoroastrian sources, starting with the *Avesta*.[98] According to the Pahlavi composition *Wizīdagīhā ī Zādsparam* (III.35) this body of water stretches out round Xwanirah, the central and largest among the seven kešvars, i.e. climes, into which the earth's surface is divided.[99] Again, in light of the influence of Iranian cosmological ideas on Ephrem's concept of paradise as the cosmic mountain, we cannot rule out the possibility of Iranian origins also for this belief.

In the light of this diverse evidence it seems impossible to establish with certainty which of the ancient cosmological systems – Greek, Near Eastern or Iranian – should be preferred in our search for the origins of the image of the circular cosmic Ocean in Syriac Christian sources. Whereas there are many examples of pre-Christian Greek ideas and concepts that found their way into Syriac Christian culture, there is also growing evidence for the indigenous ancient Near Eastern and Iranian traditions having been preserved by Syriac-speaking Christians. This ambiguity finds its best illustration in the figure of Ephrem, in whose works one finds vestiges of the all three cultures.

3. Paradise as a Womb and Breast

Let us consider the last distinctive aspect of the representation of paradise in Syriac Christian tradition: the use of feminine imagery in relation to the Garden of Eden. Practically all examples of this kind are provided by Ephrem who, in his writings, employs a broad array of feminine imagery.[100] Some of these images, such as those of womb and breast, are used by the poet in order to describe the Garden of Eden. All cases of Ephrem's use of feminine images in relation to paradise could be categorized into the two groups based on what connotations they bear: maternal or erotic.[101]

[96] See ibid., pp. 29–30.

[97] See A. Annus, 'The Survivals of the Ancient Syrian and Mesopotamian Intellectual Traditions in the Writings of Ephrem Syrus', *Ugarit-Forschungen*, 38, 2006, pp. 1–25.

[98] See A. Tafazzoli, 'Frāxkard', in *Encyclopaedia Iranica*, ed. E. Yarshater, New York, 2001, X, p. 201.

[99] *bahr-ē az zrēh ī Frāxkard pērāmōn ān gašt mayān*; *Anthologie de Zādspram*, eds P. Gignoux and A. Tafazzoli, Paris, 1993, p. 46.

[100] See K. E. McVey, 'Ephrem the Syrian's Use of Female Metaphors to Describe the Deity', *Zeitschrift für Antikes Christentum*, 5, 2001, pp. 261–88; S. Jones, 'The Womb and the Spirit in the Baptismal Writings of Ephrem the Syrian', *Studia Liturgica*, 33/2, 2003, pp. 175–93; C. Hélou, 'Le Symbolisme féminin dans la poésie de saint Éphrem', in *Saint Éphrem, un poète pour notre temps*, Antélias, Lebanon, 2007, I, pp. 143–54.

[101] For a thorough analysis of this material, see K. E. McVey, 'Images of Joy in Ephrem's Hymns on Paradise: Returning to the Womb and the Breast', *Journal of the Canadian Society for Syriac Studies*, 3, 2003, pp. 1–19.

As to the former, Ephrem evokes the images of womb and breast, when he wants to express the loving maternal care with which God surrounded the first humans in the Garden of Eden. For instance, in the *Letter to Publius*, where Ephrem exhorts his audience to follow the examples of the righteous men of past generations such as Adam and others, he relates how they 'luxuriated in the womb of the paradise of Eden' (ܒܥܘܒܗ ܕܦܪܕܝܣܐ ܕܥܕܢ).[102]

In a more developed form this maternal imagery is applied to paradise in the eleventh strophe of the *Hymns on Paradise*. There Ephrem compares the nourishment that Adam used to receive from the Garden of Eden before the Fall and expulsion to that of a baby attached to mother's breast:

> The air (ܐܐܪ) of paradise is a fountain of delight (ܡܒܘܥܐ ܕܒܘܣܡܐ) from which Adam sucked (ܝܢܩ) when he was young; its very breath, like a mother's breast (ܬܕܐ), gave him nourishment in his childhood. He was young, fair, and full of joy, but when he spurned the injunction he grew old, sad and decrepit; he bore old age as a burden of woes.[103]

Ephrem himself yearns for turning back to that carefree and protected state, when he metaphorically depicts his experience of reading the Genesis account of creation as an act of transfer from the 'womb of that book' (ܒܥܘܒܗ ܕܗܘ ܣܦܪܐ) to the 'womb of paradise' (ܒܥܘܒܗ ܕܦܪܕܝܣܐ).[104]

At times, however, enthusiastic in his desire to put into words the loving disposition of paradise towards humans, Ephrem shifts from maternal to erotic images. These come to the fore most prominently in the second strophe of the *Hymns*, where a picture of paradise that is unusual for Christian tradition is drawn. In this passage, Ephrem represents a personified paradise, who eagerly anticipates the arrival of the souls of righteous men after their death, by impersonating it as a woman that yearns for her lover:

> Blessed is he, who is desired (ܪܓܝܓܐ) by Paradise. Yes, Paradise yearns (ܪܐܓ) for the man whose goodness makes him beautiful; it engulfs him at its gateway, it embraces him in its bosom (ܒܥܘܒܗ ܡܚܒܒ ܠܗ), it caresses him in its very womb (ܒܟܪܣܗ ܡܚܝܪ ܠܗ); for it splits open and receives him into its inmost parts (ܒܓܘ ܓܘܝܘܗܝ). But if there is someone it abhors, it removes him and casts him out.[105]

What is striking in this description is that, in order to express the anticipating joy experienced by the hypostatized figure of Paradise, Ephrem uses ambivalent sexually-charged language. For example, the verb *rgg*, ܪܓܓ, 'to yearn,' used by the poet to

[102] Syriac text and transl. (modified) from S. P. Brock, 'Ephrem's Letter to Publius', *Le Muséon*, 89/3–4, 1976, pp. 261–305 (288).

[103] Ephrem the Syrian, *De paradiso*, X.1; ed. Beck (n. 4 above), p. 46; transl. Brock (n. 4 above), p. 154.

[104] Ephrem the Syrian, *De paradiso*, V.3; ed. Beck (n. 4 above), p. 16; transl. (modified) Brock (n. 4 above), p. 103.

[105] Ephrem the Syrian, *De paradiso*, II.1; ed. Beck (n. 4 above), p. 5; transl. Brock (n. 4 above), p. 85.

describe the eager anticipation of the saints by Paradise, is often used by Syriac authors to refer to sexual desire, as well as the noun 'desire' (ܪ ܓܬܐ), derived from this root. In addition to that, Ephrem refers to the 'bosom' (ܥܘܒܐ), 'womb' (ܟܪܣܐ) and 'intestines' (ܡܥܝ̈ܐ) of Paradise, in which it will caress the saints after their death. The erotic overtones of this imagery are hard to miss.

Another instance where Ephrem infuses the image of paradise with erotic connotations in order to describe the eschatological reward of the saints, is found in the seventh strophe of the *Hymns*. There, the righteous man who has abstained from wine during his life is promised that he will be greeted by the 'vines of paradise' (ܠܓܘ̈ܦܢܘܗܝ, ܕܦܪܕܝܣܐ) and enjoy their clusters, while the man who led a life of sexual abstinence and has 'never lain in any womb' (ܠܐ ܢܦܠ ܒܟܠ ܥܘܒܐ) will be welcomed 'into their pure womb' (ܠܓܘ ܥܘܒܗܝܢ ܕܟܝܐ).[106]

As it becomes clear from the quoted passages, Ephrem endows paradise with feminine traits mostly to express the idea of God's loving care for humans, whether in a protological context, as in the case of maternal images, or in an eschatological one, as in the case of erotic imagery. Looking at these images one must be aware, however, of their metaphorical character, as the poet himself warned his audience not to take these vivid descriptions of paradise literally, while acknowledging the limitations of human language in speaking of divine realities.[107]

As has been mentioned, it seems that no other Syriac-speaking authors from Late Antiquity save Ephrem dared to look at the Garden of Eden through the lenses of these female images. Neither do we find such language applied to paradise in the post-biblical Jewish or the mainstream Christian traditions of antiquity. All this presents a challenge for those who endeavour to explain the genesis of this vivid imagery.

Thinking about possible external agents of influence behind the feminine representations of paradise by Ephrem, it should be mentioned that there is at least one precedent for the Garden of Eden having been interpreted along the lines that bring to mind his approach. It comes from the description of the Gnostic mythological system of Simon Magus by the famous third-century heresiographer, Hippolytus of Rome. In the sixth book of his *Refutation of All Heresies*, Hippolytus exposes Simon's views on paradise, basing himself on a book that circulated under the heresiarch's name and was entitled *Apophasis Megale*, which means probably 'Great Revelation.' This exposition contains a detailed allegorical juxtaposition of the various parts of the female reproductive system with the description of the Garden of Eden in Genesis 2. Simon identifies paradise with

106 Ephrem the Syrian, *De paradiso*,VII.18; ed. Beck (n. 4 above), p. 29; transl. (modified) Brock (n. 4 above), p. 125. For an analysis of this passage, see McVey, 'Images of Joy' (n. 101 above), pp. 62–4.

107 Cf. Ephrem the Syrian, *De paradiso*, I.8; IV.7–9; XV.1. On this aspect of Ephrem's thought, see D. Bundy, 'Language and the Knowledge of God in Ephrem Syrus', *Patristic and Byzantine Review*, 5, 1986, pp. 91–103; P. S. Russell, 'Ephraem the Syrian on the Utility of Language and the Place of Silence', *Journal of Early Christian Studies*, 8/1, 2000, pp. 21–37; K. den Biesen, *Simple and Bold: Ephrem's Art of Symbolic Thought*, Piscataway NJ, 2006.

the 'womb' (μήτρα), Eden – with the 'after-birth' (χόριον), the source from which the four rivers flow – with the 'navel' (ὀμφαλός), while the rivers themselves – with the 'arteries' (ἀρτηρίαι) and 'veins' (φλέβες).[108] The main point of this allegory, rooted in the microcosm/macrocosm analogy, is to demonstrate that there is a close correspondence between the process of formation of a human embryo and the biblical account of the creation of Adam. Interesting as it might be, this allegorical scheme differs both structurally and conceptually from the Ephremian metaphors in order to regard this early Gnostic parallel as a possible source of the feminine imagery in Ephrem's picture of paradise.

It appears that in the case of Ephrem it would be more productive to focus on internal factors behind the feminine imagery of paradise and to look at it as an outcome of inner-Syriac theological development. As has been recognized by scholars, one of the distinctive features of early Syriac Christian theological discourse was its strong proclivity for applying feminine images to God.[109] While it stands in genetic relation to the Old Testament tradition of applying feminine-related imagery to the deity, Syriac Christian tradition shows a considerable degree of originality in coining new metaphors and images of this kind.[110] One of the earliest examples of this creativity comes from the *Odes of Solomon*, a collection of Christian hymns that was written in Syria during the second or third century. The unknown author of these hymns on several occasions represents God's loving care for those who believe in him using the imagery of breast-feeding.[111] Ephrem was an heir to this archaic theological language and did not hesitate to use and further develop feminine metaphors in relation to God. For example, we find maternal imagery

[108] Hippolytus of Rome, *Refutation of All Heresies*, VI.14.7–11; *Refutatio omnium haeresium*, ed. M. Marcovich, Berlin, 1986, pp. 219–20.

[109] See H. J. W. Drijvers, 'The 19th Ode of Solomon: Its Interpretation and Place in Syrian Christianity', *Journal of Theological Studies*, 31/2, 1980, pp. 337–55; J. Martikainen, 'Die Muttergestalt des Heiligen Geistes in der frühen syrischen Theologie', in *Makarios-Symposium über den Heiligen Geist: Vorträge der zweiten finnisch-deutschen Theologentagung in Karis 1984*, eds F. Cleve and E. Ryökäs, Åbo, 1989, pp. 114–27; S. P. Brock, 'The Holy Spirit as Feminine in Early Syriac Literature', in *After Eve: Women, Theology and the Christian Tradition*, ed. J. M. Soskice, London, 1990, pp. 73–88; G. Winkler, 'Überlegungen zum Gottesgeist als mütterlichem Prinzip und zur Bedeutung der Androgynie in einigen frühchristlichen Quellen', in *Liturgie und Frauenfrage. Ein Beitrag zur Frauenforschung aus liturgiewissenschaftlicher Sicht*, eds T. Berger and A. Gerhards, St Ottilien, 1990, pp. 7–29; S. A. Harvey, 'Feminine Imagery for the Divine: The Holy Spirit, the Odes of Solomon, and Early Syriac Tradition', *St Vladimir's Theological Quarterly*, 37, 1993, pp. 111–39; C. Hélou, 'Dieu au féminin dans la tradition syriaque', in *Le Visage de Dieu dans le patrimoine oriental*, Antélias, Lebanon, 2001, I, pp. 159–84; S. E. Myers, 'The Spirit as Mother in Early Syriac-Speaking Christianity', in *Women and Gender in Ancient Religions: Interdisciplinary Approaches*, eds S. P. Ahearne-Kroll et al., Tübingen, 2010, pp. 427–62.

[110] On the use of feminine imagery in relation to God in the Old Testament and later Jewish tradition, see E. Jacob, 'Traits féminins dans la figure du dieu d'Israël', in *Mélanges bibliques et orientaux en l'honneur de M. Mathias Delcor*, eds A. Caquot et al., Kevelaer, 1985, pp. 221–30; S. J. Dille, *Mixing Metaphors: God as Mother and Father in Deutero-Isaiah*, London, 2004; P. Schäfer, *Mirror of His Beauty: Feminine Images of God from the Bible to the Early Kabbala*, Princeton NJ, 2002; E. D. Haskell, *Suckling at My Mother's Breasts: The Image of a Nursing God in Jewish Mysticism*, Albany NY, 2012.

[111] Cf. *Odes of Solomon*, 8.14; 19.1–4; *The Odes of Solomon*, ed. and transl. J. H. Charlesworth, Oxford, 1973, pp. 40–42, 81–2.

applied to God in the fourth strophe of his *Hymns on Nativity*, where Ephrem refers to Jesus as to 'the Living Breast (ܬܕܐ ܚܝܐ) of living breath'[112] or says about Jesus's place in creation that 'in His womb dwell all creatures' (ܒܟܪܣܗ ܥܡܪ̈ܝܢ ܗ̈ܘܝ ܟܠ ܒܪ̈ܝܬܐ).[113] In the thirteenth strophe of these *Hymns*, where Ephrem rhetorically addresses those who seek to comprehend the true nature of Jesus, he reports the latter to be 'in heaven, that is, in the great womb of Divinity' (ܒܫܡܝܐ ܗܘ ܒܟܪܣܐ ܪܒܐ ܕܐܠܗܘܬܐ).[114]

Returning to the problem of origins of the feminine images of paradise in Ephrem's works, they emerge most likely as a result of the extension of these archaic feminine-gendered metaphors for God's loving attitude towards humanity or his creation in general. Ephrem had creatively transferred the inherited feminine images of God to the Garden of Eden, which was conceived by him to be *par excellence* an expression of the divine love for humankind.

There might be some additional factors, internal or external, that facilitated this transfer. In what concerns the former, the maternal metaphors of paradise converge well with the image, employed at times by Ephrem, of Adam and Eve as children.[115] As to the possible external factors I would like to draw attention once more to the Iranian mythological tradition, where one comes across the comparable use of maternal images to describe the deity's relation to the world. In the *Bundahišn*, for example, it is related about the foremost Zoroastrian deity, Ohrmazd, that in the two-fold act of creation he acted both as mother and as father, 'for when He nourished the creatures in the spirit (*mēnōg*) that was motherhood (*mādarīh*), when He created them in the material world (*gētīg*) that was fatherhood (*pidarīh*)'.[116] It is striking that both in the Iranian and Syriac Christian sources it is the spiritual aspect of deity that becomes associated with feminine. I believe that this expression of harmonious union of feminine and masculine aspects in the figure of the highest Zoroastrian deity opens a new possibility for looking at the popularity of the maternal metaphors in relation to God in Syriac Christian tradition.

112 Ephrem the Syrian, *De Nativitate*, IV.150; *Des heiligen Ephraem des Syrers Hymnen de Nativitate (Epiphania)*, ed. E. Beck, *CSCO* 186, *Syr.* 82, Louvain, 1959, p. 39; *Ephrem the Syrian. Hymns on the Nativity, Hymns Against Julian, Hymns on Virginity and on the Symbols of the Lord*, transl. K. E. McVey, New York, 1989, p. 100.

113 Ephrem the Syrian, *De Nativitate*, IV.154; ed. Beck (n. 112 above), p. 39; transl. (modified) McVey (n. 112 above), p. 100.

114 Ephrem the Syrian, *De Nativitate*, XIII.7; ed. Beck (n. 112 above), p. 74; transl. (modified) McVey (n. 112 above), p. 138.

115 Cf. Ephrem the Syrian, *De paradiso*, XV.12, 14. On this motif, see Murray, *Symbols of Church and Kingdom* (n. 2 above), pp. 304–6; Y. de Andia, 'Adam as a Child by Irenaeus of Lyons and in the St Ephrem's *Hymns of Paradise*', *The Harp*, 22, 2007, pp. 257–68; A. Salvesen, 'Infants or Fools in the Garden of Eden? An Ambiguity in Early Syriac Tradition', in *Hamlet on a Hill: Semitic and Greek Studies Presented to Professor T. Muraoka on the Occasion of His Sixty-fifth Birthday*, eds M. F. J. Baasten and W. T. van Peursen, Leuven, 2003, pp. 432–40.

116 Iranian *Bundahišn*, 1.59; ed. Anklesaria (n. 74 above), pp. 20–21; transl. from C. G. Cereti and D. N. MacKenzie, 'Except by Battle: Zoroastrian Cosmogony in the 1st Chapter of the Greater *Bundahišn*', in *Religious Themes and Texts of Pre-Islamic Iran and Central Asia: Studies in Honour of Professor Gherardo Gnoli on the Occasion of his 65th Birthday on 6th December 2002*, eds C. G. Cereti et al., Wiesbaden, 2003, pp. 31–59 (42).

Whereas the inner-Syriac roots of the maternal imagery of Ephrem's paradise are manifest, the erotic aspect of his images of the Garden of Eden presents more problems since no such metaphor in relation to God seems to be found in his compositions or in the works of his predecessors. Trying to explain these images one might recognize in them a distant echo of the popular eschatological notion of the heavenly 'bride-chamber' (ܓܢܘܢܐ), a metaphor used for the Kingdom of Heaven, which features prominently in the works of many Syriac Christian authors, including Ephrem.[117] In the *Hymns on Paradise* Ephrem on several occasions applies this eschatological image to the primordial paradise, describing it as the 'chaste bridal chamber' (ܓܢܘܢܐ ܕܢܟܦܘܬܐ) or the 'bridal chamber of glory' (ܓܢܘܢ ܫܘܒܚܐ).[118]

Another, more promising, avenue for understanding the genesis of Ephrem's erotic imagery of paradise opens up if we turn to the Iranian eschatological beliefs. One of the key roles in Zoroastrian eschatology is played by the mythological figure of *daēnā*, the female counterpart of man, who personifies a person's deeds, his inner self.[119] A detailed account of this figure is found, for example, in the Pahlavi *Vidēvdād* that presents a scene of the postmortem judgment, where the *daēnā* meets the souls of the deceased at the Činwad bridge. [120] She is described as a woman who is 'beautiful to look at' (*nēk pad didan*), 'well formed' (*hu-kard*), 'well shaped' (*hu-rust*), and 'full of desire' (*kāmagōmand*). While she 'drags the vile souls of the evil ones into the darkness', the souls of the just ones are taken by her 'across the Harborz (*tarist harborz*), through the Bridge of Compiler' to the deity Wahman, who introduces them to the Best Existence and to Ohrmazd.

Several aspects of this account bring to mind the descriptions of the Garden of Eden found in Ephrem's *Hymns on Paradise*. Thus, the erotic elements in the representation of paradise in the *Hymns* are paralleled by the erotic overtones in the description of the *daēnā*'s figure, noted by some scholars.[121] In addition to that, the *daēnā* is connected with the mythological mount Alburz, which, as we have seen above, inspired Ephrem's vision of paradise as a cosmic mountain. But the most striking similarity between the *daēnā* and Ephrem's personified Paradise is that both these female figures perform the same eschatological function of separation between the righteous and sinners during the post-mortem judgment.[122]

[117] On this motif, see Murray, *Symbols of Church and Kingdom* (n. 2 above), pp. 254–7; S. P. Brock, 'The Bridal Chamber of Light: A Distinctive Feature of the Syriac Liturgical Tradition', *The Harp*, 18, 2005, pp. 179–91.

[118] Ephrem the Syrian, *De paradiso*, XIII.3, 10; ed. Beck (n. 4 above), pp. 55, 57; transl. Brock (n. 4 above), pp. 169, 172.

[119] See M. Shaki, 'Dēn', in *Encyclopaedia Iranica*, ed. E. Yarshater, Costa Mesa CA, 1996, VII, pp. 279–81; W. Sundermann, 'Die Jungfrau der guten Taten', in *Recurrent Patterns in Iranian Religions: from Mazdaism to Sufism*, ed. P. Gignoux, Paris, 1992, pp. 159–73.

[120] *Vidēvdād*, 19.29–32; M. Moazami, 'The Confrontation of Zarathustra with the Evil Spirit: Chapter 19 of the *Pahlavi Vidēvdād*', *East and West*, 52, 2002, pp. 151–71 (158–9 [text], 166–7 [transl.]).

[121] A. Piras, 'Mesopotamian Sacred Marriage and Pre-Islamic Iran', in *Schools of Oriental Studies and the Development of Modern Historiography*, eds A. C. D. Panaino and A. Piras, Milan, 2004, pp. 249–59 (256).

[122] In Ephrem's works this idea is expressed explicitly in *De paradiso*, II.1, quoted above (n. 105), where the eschatological paradise is said not only to welcome the righteous ones, but also to reject sinners.

These parallels between the eschatological maiden of Zoroastrian mythology and the feminine Paradise of Ephrem's *Hymns* deserve, I believe, a serious attention. However, we must exercise certain caution before drawing any far-reaching conclusions about Zoroastrian influence on Ephrem in this case. As no one has yet addressed the question of Ephrem's acquaintance with Iranian culture in a systematic and methodologically rigorous manner, it might be wiser for the moment to treat these images as a result of the poet's own power of creative imagination.

Conclusion

The three cases of paradise imagery discussed shed light on the complicated cultural dynamics that were at play in the process of formation of the Syriac Christian tradition, where Jewish, Greek and Iranian elements were selectively appropriated and creatively adapted. In what concerns our examples, the motif of paradise as a temple was most likely inherited from the Jewish matrix of early Syriac Christianity. The image of paradise as a cosmic mountain that encircles the earth provides a good example of Iranian influence upon Syriac Christianity. The case of the feminine metaphors in relation to paradise is more complicated, as it seems to originate at the intersection of internal and external factors.

At this point it has to be emphasized that my (mostly 'genealogical') approach by no means exhausts the subject of the rich paradisiacal imagery that was current among Syriac Christians during Late Antiquity. One should take into account that the traditions considered above did not circulate in isolation, but formed a part of the much wider nexus of theologically charged images and metaphors, interdependent and mutually enriching. Some of these images are of particular importance for a deeper understanding of how the biblical notion of paradise evolved on Syrian soil. One such instance, i.e. the temple-related imagery, has been pointed out already. Another set of metaphors that might exert influence on the paradisiacal imagery comes from the area of Mariology. It seems that some sort of typological connection between Mary and paradise was known to Syriac Christian authors, as one comes across a number of similar images being applied to both of them. Of particular interest in that context is the characteristic of Mary as the 'second heaven' (ܫܡܝܐ ܕܬܪܝܢ), found in the works of some Syriac writers.[123] Furthermore, it is noteworthy that Ephrem likens the Virgin to the 'Holy of Holies' (ܩܘܕܫ ܩܘܕܫܐ) and Jesus, indwelling in her, to the 'High Priest' (ܪܒ ܟܘܡܪܐ), in a way very similar to how he handles paradise and Adam in it.[124] Also relevant are cases when

[123] Cf. Jacob of Serugh, *Homily on the Annunciation*; A. Shemunkasho, 'A Verse Homily of Jacob of Serugh on the Annunciation to the Mother of God: Manuscript Vatican Syr. 117, Nr. 12. (fol. 31a-32a)', in *Malphono w-Rabo d-Malphone: Studies in Honor of Sebastian P. Brock*, ed. G. A. Kiraz, Piscataway NJ, 2008, 575–602 (594). On this Mariological metaphor, see J. Puthuparampil, *Mariological Thought of Mar Jacob of Serugh (451–521)*, Kottayam, 2005, pp. 105–16.

[124] Ephrem the Syrian, *De Nativitate*, XVII.5; ed. Beck (n. 112 above), p. 88; transl. McVey (n. 112 above), pp. 154–5.

the image of 'bride-chamber' (ܓܢܘܢܐ) is used in connection with Mary.[125] As in the case of the temple imagery this rich and variegated material is in need of further investigation, both on its own terms and in its relation to the representation of paradise.

In focusing attention on the three particular images of paradise one of my goals was to offer a glimpse of the poetic and symbolic theological method that was characteristic for what one might call the 'archaic' current within Syriac Christianity during the period of Late Antiquity. This tradition, expressed prominently in the works of Ephrem, was dominant before it had been radically transformed as a result of the profound influence of Greek Christian culture during the fifth and sixth centuries. During this period many ideas and practices current in early Syriac Christianity were discarded in order to correspond to the standards set by the 'Western' Christian orthodoxy.[126]

The fate of Ephrem's heritage illustrates very well this process of Westernization of Syriac Christianity. As we have seen, some of the Ephremian metaphors of paradise, such as the sanctuary and mountain symbolism, appear in several later fifth- and sixth-century Syriac works, mostly of an apocryphal kind. At the same time it is striking that none of these peculiar images had left visible traces in the writings of such mainstream fifth-century Syriac authors as Narsai and Jacob of Serugh, who both wrote on paradise and were acquainted with the works of Ephrem. In my opinion this is nothing but a result of the deliberate ignoring of Ephrem as a theologian, whose daring language fell out of favour with the new Western-oriented tastes. Another expression of this tendency has been discussed recently by Lucas van Rompay, who pointed out the noticeable shift in attitude to the Christological language of Ephrem from a respectful to an indifferent one in the works of Philoxenus of Mabbugh, a sixth-century West-Syrian theologian.[127] This change reflects a general tendency among the West-Syrian theologians of this period towards the marginalization of Ephrem as an authority in Christological discussions. As a result of this 'westward turn' of Syriac Christianity a significant part of the rich legacy of Ephrem's symbolic theology, including his captivating imagery of paradise, became marginalized and survived only on the peripheries of what became the mainstream theology whether among the West or East Syrians.

[125] For examples, see S. P. Brock, 'Passover, Annunciation and Epiclesis: Some Remarks on the Term *aggen* in the Syriac Versions of Lk. 1:35', *Novum Testamentum*, 24/3, 1982, pp. 222–33 (228).

[126] On various aspects of this process, see A. Vööbus, *Celibacy: A Requirement for Admission to Baptism in the Early Syrian Church*, Stockholm, 1951, pp. 38–44; Ph. Wood, 'Syriac and the 'Syrians'', in *The Oxford Handbook of Late Antiquity*, ed. S. F. Johnson, Oxford, 2012, pp. 170–194 (182–6); S. Minov, 'Marriage and Sexuality in the *Book of Steps*: From Encratism to Orthodoxy', in *Breaking the Mind: New Studies in the Syriac Book of Steps* eds K. S. Heal and R. A. Kitchen, Washington DC, 2013, pp. 221–61 (259–60).

[127] L. van Rompay, '*Mallpânâ dilan Suryâyâ*. Ephrem in the Works of Philoxenus of Mabbog: Respect and Distance', *Hugoye*, 7/1, 2004, §§1–34.

Food and the Senses, and One Very Special Taste of Paradise

Danuta Shanzer

Introduction: Yearning and Cogitation

Hell and paradise evoke divergent reactions (fear vs. yearning) and different emotional and intellectual responses (acceptance of some received idea vs. intellection or cogitation, which involve questions).[1] These dichotomies exercise a powerful attraction. The present paper seeks to work at the intersection of pleasure and intellection by re-examining the senses and the problem of their place in the afterlife paradise. I will eventually narrow down to one sense in particular (taste), its concrete object (food), and one lowly foodstuff. I will propose new interpretations of, and connections between, two odd Latin texts, one familiar, one less so.

The Problem of the Senses in the Afterlife

Senses in the afterlife are problematic because they are functions of bodies, yet there will be no fleshly bodies in the afterlife after death, and no bodies (as humans know them) after the resurrection.[2] Here are the canonical five senses in a typical

Author's Note: My thanks to Alessandro Scafi and Markham J. Geller for their invitation to participate in the wonderful Warburg conference on the Cosmography of Paradise and to Alessandro in particular for his editing. My thanks, as always, to Howard Jacobson who helps me with my Hebrew, and who also made me aware of 'roast Leviathan.' We always learn when assembled as a group, and I am most grateful to members of the conference: to Jan N. Bremmer for his offprints and comments, to Dimitris J. Kyrtatas for help locating Pater Muthios's figure, and to Nick Wyatt for Old Testament bibliography on the Psalmic mountain.

Abbreviation used in this essay:

MGH = *Monumenta Germaniae Historica.*

[1] And of course also filling intentional catechetical lacunae using the imagination, as J. Amat, *Songes et visions: l'au-delà dans la littérature latine tardive,* Paris, 1985, p. 401 rightly notes.

[2] Tertullian's more positive evaluation in his *De resurrectione* is a defense of the flesh as God's handiwork as well as local polemic against deniers of the resurrection, not just pagans, but also heretics. Tertullian, *De carnis resurrectione*, 4, *Opera: Pars III*, ed. E. Kroymann, *CSEL* 47, Vienna and Leipzig, 1906, p. 30: 'Quid tum de consequentiis carnis? rursusne omnia necessaria illi, et inprimis pabula atque potacula? Et pulmonibus halandum et intestinis aestuandum et pudendis non pudendum et omnibus membris laborandum? Rursus ulcera et vulnera et febris et podagra? et mors redoptanda?' But Tertullian, ibid., 60, pp. 120–21, does not say that the resurrected body will be just like the human body in its first life nor that it will perform the same functions. The opponents had argued with regard to the mouth: 'Quo enim iam, inquiunt, spelunca haec oris et dentium statio et gulae lapsus et competum stomachi et alvei gurges et intestinorum perplexa proceritas, cum esui et potui locus non erit?' (ibid.). Tertullian argued for the redeployment of the members, not for bodily functions, but for punishment: 'Et ipsa autem liberata tunc vita a necessitatibus, liberabuntur et membra ab officiis, nec ideo non erunt necessaria: licet enim officiis liberentur, sed iudiciis retinentur, ut quisque referat per corpus prout gessit' (ibid.). For a recent treatment of his views, see C.W. Bynum, *The Resurrection of the Body in Western Christianity, 200–1336, Lectures on the History of Religions*, New York, 1995, pp. 37–43. At p. 38 she notes Tertullian's simultaneous extravagantly materialistic view of the resurrected body and his emphasis on radical change and (p. 40) transformation.

(Aristotelian) order: sight, hearing, smell, taste, and touch.[3] Ancient and medieval intellectuals loved to speculate about them and debate their hierarchy.[4] They thought long and hard. Four of the senses were localized in the head alone, whereas the lowest, touch, worked all over the body.[5] Sight, and hearing did not require contact with their objects, as did smell, taste, and touch, but an intervening medium (light, air). Hearing and smell could transcend physical barriers in a way sight could not. Sight was preeminent.[6]

Platonism and Christianity put a low value on the senses, particularly on the lower three. Yet it is through them that men apprehend the world around them – particularly its pleasures and pains. Touch and taste even change our bodies.[7] We are trapped in a sensual framework when describing either bliss or horror. We all know what's so awful about hell (and air travel): 'The noise, my dear, the smell – the *people*', βόρβορος ('mud') and worse. The many sensual delights of heaven are the stuff of dreams. Both, however, are usually apprehended at second hand from others' accounts. This raises the question of how witnesses to other worlds absorbed their sensory data and how conscientious they were in telling us what they experienced and how.

New Heaven and the New Earth

Revelation (c. AD 100/110), a waking vision with sounds and sights, is both foundational and early. In the new heaven and new earth John sees a beatific vision and hears hymns of praise, but no more than that. Noticeable is the lack of smell, touch, or taste – at all, anywhere, either of bad things or of good ones, such as living water or the tree of life.[8] So the question arises: did John have the narrative option of smelling, tasting, or touching in such a text? Or does the *sensorium* of Revelation exhibit a deliberate Christian austerity – unconstrained by medium of focalization – that appealed to hearing and sight alone?[9] Apocryphal texts such as the *Apocalypses of Peter* and *Paul* show greater pleasure in the scents and potential tastes of paradise.[10]

[3] Aristotle, *De anima*, 2.7–11 and *De sensu*, 1.437a and 4.400b (omitting tangibles).

[4] For a lucid modern treatment with limited ancient material, see C. Korsmeyer, *Making Sense of Taste: Food and Philosophy,* Ithaca NY, 1999, pp. 11–37; also the older L. Vinge, *The Five Senses: Studies in a Literary Tradition,* Lund, 1975, pp. 15–46.

[5] Avitus, *De spiritalis historiae gestis*, 1.85–86.

[6] F. Mütherich, 'An Illustration of the Five Sense in Medieval Art', *Journal of the Warburg and Courtauld Institutes*, 18, 1955, pp. 140–41.

[7] C. Korsmeyer, *Making Sense of Taste: Food and Philosophy,* Ithaca NY, 1999, p. 22.

[8] The book of Revelation reads: 'sitienti dabo de fonte aquae vivae gratis' (21:6) and mentions 'lignum vitae' with 'fructus' (22:1), but no smells or tastes.

[9] S. A. Harvey, *Scenting Salvation: Ancient Christianity and the Olfactory Imagination*, Berkeley CA, 2006, p. 235, finds the resurrected *sensorium* essentially mono-sensory.

[10] See *Apocalypse of Peter* 16 (Ethiopic) and 14–16 (Greek), transl. M. R. James and J. K. Elliott, *The Apocryphal New Testament: A Collection of Apocryphal Christian Literature in an English Translation,* Oxford and New York, 1993, pp. 610–11; also *Visio Pauli* 22, 23, 26, and 31 for milk and honey.

How? Enter the Spiritual Senses ...

Origen noticed the problem involving seers' senses and defended voices heard and visions of the heaven opening in his *Contra Celsum*.[11] He had recourse to Old Testament prophets, and people who can 'see things superior to corporeal beings' and 'perceive sounds that have no objective existence in the air', whose taste can 'feed on the living bread ... from heaven', whose sense of smell 'smells spiritual things', and whose touch, 'handled with [his] hands the word of Life'.[12] And from this point came his doctrine of the so-called 'spiritual senses', a mystic gift (θεία αἴσθησις) granted to those with special perception[13] – but also a useful way of enabling full sensory perception of the afterlife and otherworld for dreamers and visionaries.

Lower Sense and Food in the Pagan Afterlife

So much for the higher senses (sight and hearing), which contended for first place. But there were three other senses too, usually ranged in the order smell, taste, and touch. All of these were involved in the consumption of food. In the pagan Greek world there was food of the magical sort in heaven, viz. nectar and ambrosia.[14] Even Epicurean deities ate and drank – for pleasure, not sustenance.[15] One had to watch what one ate when one visited the house of the dead.[16] Consumption of even one pomegranate seed could confine one there forever – as Persephone learned to her cost.[17]

Food and Drink in the Afterlife in the New Testament and Early Passions

But what did Christianity do with these and similar traditions? At the Last Supper Jesus promised that he would drink no wine until he drank in heaven.[18] After Jesus's first

[11] Origen's context was polemical, primarily against pagans, but also against Jews who seem to have sided with pagans in questioning New Testament visions.

[12] Origen, *Contra Celsum*, I.48, transl. H. Chadwick, Cambridge, 1965, p. 44; ed. M. Borret, 4 vols, I, *SC* 132, Paris, 1967, pp. 204 and 206: ὁράσεως πεφυκυίας βλέπειν τὰ κρείττονα σωμάτων πράγματα, ἐν οἷς δηλοῦται τὰ χερουβὶμ ἢ τὰ σεραφίμ, καὶ ἀκοῆς ἀντιλαμβανομένης φωνῶν οὐχὶ ἐν ἀέρι τὴν οὐσίαν ἐχουσῶν, καὶ γεύσεως χρωμένης ἄρτῳ ζῶντι καὶ ἐξ οὐρανοῦ καταβεβηκότι καὶ ζωὴν διδόντι τῷ κόσμῳ, οὕτω δὲ καὶ ὀσφρήσεως ὀσφραινομένης τοιῶνδε, καθὸ «Χριστοῦ εὐωδία» λέγει εἶναι «τῷ θεῷ» Παῦλος, καὶ ἁφῆς, καθ' ἣν Ιωάννης φησὶ ταῖς χερσὶν ἐψηλαφηκέναι «περὶ τοῦ λόγου τῆς ζωῆς»· οἱ μακάριοι προφῆται τὴν θείαν αἴσθησιν εὑρόντες καὶ βλέποντες θείως καὶ ἀκούοντες θείως καὶ γευόμενοι ὁμοίως καὶ ὀσφραινόμενοι, ἵν' οὕτως ὀνομάσω, αἰσθήσει οὐκ αἰσθητῇ καὶ ἁπτόμενοι τοῦ λόγου μετὰ πίστεως, ὥστ' ἀπορροὴν αὐτοῦ ἥκειν εἰς αὐτοὺς θεραπεύσοσαν αὐτούς.

[13] K. Rahner, 'Le Début d'une doctrine des cinq sens spirituels chez Origène', *Revue d'ascétique et de mystique*, 13, 1932, pp. 113–45; J. M. Dillon, 'Aisthesis Noete: A Doctrine of Spiritual Senses in Origen and Plotinus', in *Hellenica et Judaica : hommage à Valentin Nikiprowetzky*, eds V. Nikiprowetzky, M. Hadas-Lebel, J. Riaud and A. Caquot, Leuven, 1986, pp. 443–55.

[14] K. Wernicke, 'Ambrosia,' in *Real-Encyclopädie der classischen Altertumswissenschaft*, eds A. F. Pauli and G. Wissowa, 1/**2**, Stuttgart, 1932, cols 1809–11.

[15] Philodemus, *De dis*, 77.5–6.

[16] See J. G. Frazer, *Apollodorus: The Library,* Cambridge MA, 1961, pp. 39–41 for a long cross-cultural footnote on eating the food of the dead and the food of fairies.

[17] *Hymn to Demeter*, 371–74 and 411–12. Food of the dead – or love magic as per C. A. Faraone, *Ancient Greek Love Magic,* Cambridge MA, 1999, pp. 75–6. See also H. P. Foley, *The Homeric Hymn to Demeter: Translation, Commentary, and Interpretive Essays,* Princeton NJ, 1994, pp. 56–7.

[18] Matthew 26:29; Mark 14:25; Luke 22:18; 22:30 [italics mine]: '*so that you may eat and drink at my table* in my kingdom'. Abraham's bosom (Luke 16:22–23) is a feasting-image. See H. L. Strack and P. Billerbeck,

reappearance on earth he ate some broiled fish to prove the reality of his resurrected body.[19] Imprisoned martyrs had a final *cena libera*.[20] The ἀγάπη was an attested, if still mysterious, early Christian ritual.[21] There was talk of a *convivium* in heaven,[22] yet no serious detailed textual descriptions of it.[23] There are good reasons, as we shall see, why a text like the *Cena Cypriani*, which depicts a heavenly feast in detail, is only possible in a parodic mode.[24]

The Taste of Heaven

Several third-century African *passiones,* however, feature visions that include heavenly food and drink.[25] They exhibit interesting features that have something to contribute, both to understanding Christian conceptions of the senses and to elucidating some arcane lore about a homely food in paradise.

In 203 a young Carthaginian matron, Vibia Perpetua, awaiting trial in prison, dreamt that she had to climb a brazen ladder rising to heaven, covered with instruments of execution. Then she entered a verdant landscape with a Good Shepherd-like figure, who offered her some of the cheese he was milking. She ate. All the company around said 'Amen'. At this, she awoke with the taste of something sweet in her mouth (*commanducans adhuc dulce nescio quid*). She then knew she was going to die.[26] A similar story appears in the *Acts of Marian and James* (*c.* post 6 May 259). Here Marian dreamt that in a lovely countryside he met Cyprian, the recently martyred Bishop of Carthage, who filled a cup with spring-water, drank, and handed it to him. Marian drank with pleasure, and when he said, 'Thanks be to God!' he awoke.[27] In the *Acts of Montanus and Lucius* for the dreamer Quartillosa,

Kommentar zum neuen Testament aus Talmud und Midrasch, II, Munich, 1924, p. 225. The *heiq* (חיק or חוק 'bosom') (the act of lying or sitting in Abraham's bosom) is an expression of the joys of blesssedness. It often appears in the image of a feast or meal, where one lies at breast height with one's neighbour – hence in the bosom of or 'right next to' Abraham. See Matthew 8:11: 'et recumbent cum Abraham et Isaac et Jacob in regno caelorum' for the image of a feast in heaven.

[19] Luke 24:42–3.

[20] *Passio Perpetuae* 17.1.

[21] H. Leclercq, 'Agape', *Dictionnaire d'archéologie chrétienne et de liturgie*, I, Paris, 1907, cols 775–847.

[22] For example the eschatological feasting of Isaiah 25:6 and Luke 22:29–30.

[23] *Passio Mariani and Jacobi* 11.3, ed. H. Musurillo, *The Acts of the Christian Martyrs, Oxford Early Christian Texts*, Oxford and New York, 1972, p. 208: 'Et bene, inquit, Iacobus, ad Agapii ceterorumque martyrum beatisssimorum pergo convivium'; ibid., p. 209: 'James said: "Good! I am on my way to the banquet of Agapius and the other blessed martyrs"'. For a few funerary images from the Roman catacombs, see Leclercq, 'Agape' (n. 21 above).

[24] For the text of the late antique *Cena Cypriani*, see C. Modesto, *Studien zur Cena Cypriani und zu deren Rezeption*, Tübingen, 1992, pp. 14–35, or the more recent medieval *Cena Cypriani, MGH Poetae Latini*, ed. K. Strecker, Munich, 2000. Also the discussion in M. Bayless, *Parody in the Middle Ages: The Latin Tradition,* Ann Arbor MI, 1996, pp. 19–56.

[25] For more on paradise in these texts, see J. N. Bremmer, 'Contextualizing Heaven in Third-Century North Africa', in *Heavenly Realms and Earthly Realities in Late Antique Religions*, eds R. A. S. Boustan and A. Y. Reed, Cambridge and New York, 2004, pp. 159–73.

[26] *Passio Perpetuae* 4.2–10, ed. Musurillo, *The Acts of the Christian Martyrs* (n. 23 above), pp. 110–12.

[27] *Passio Mariani et Jacobi* 6, ed. Musurillo, *The Acts of the Christian Martyrs* (n. 23 above), p. 14. Cyprian seems to have helped make dreams known to encourage his fellow Christians in their confession even when

awaiting trial in prison, two cups of milk carried by a wondrous young man in a vision of heavenly brightness symbolized the previous martyrdoms of her husband and son; a third cup would be her own death.[28]

Haunting Food vs. Transient Cups

Two of these visions show the landscape of paradise to the would-be martyr and leave its haunting taste in his or her mouth. Perpetua's is purely pastoral, and her nourishment solid. The last two visions, however, conflate a rather darker garden, Gethsemane,[29] and the cup of martyrdom that Jesus prayed might be taken away from him.[30] The martyr tastes 'the milk of paradise'[31] as a premonition of his execution.[32]

These three examples from visions all involve the sense of taste and the phenomenon of dream food and drink. Although the food of dreams notoriously fails to nourish,[33] the audience of the *Acta* must have found such visions reassuring.[34] And incentives to let oneself be executed clearly mattered. Yet, at the same time, in one of the same texts there seems to be an implicit criticism of the need for such visions of paradise. One imprisoned martyr, the presbyter Victor saw a child in his dream and demanded to be shown heaven. The answer came: 'Where would your faith be?'[35] This probably alludes to the alternative Christian hierarchy of the senses where true Christians have faith *ex auditu*[36] – not from sight – or worse – touching, like Doubting Thomas! (John 20:24)

One Grand Narrative of the Christian Senses

A broad (and not untrue) story about the history of the five senses in ancient paganism and Christianity would run something like this (according to recent work on smell by

alive. See Cyprian, *Epistula* 38.1, in *Opera Omnia*, ed. W. Hartel, *CSEL* 3/1, Vienna, 1868, p. 582: 'Qui cum consentire dubitaret, ecclesiae ipsius admonitu et hortatu in visione per noctem compulsus est ne negaret.'

28 *Passio Montani et Lucii* 8, ed. Musurillo, *The Acts of the Christian Martyrs* (n. 23 above), p. 220.

29 Pace M. Meslin, 'Vases sacrés et boissons d'éternité dans les visions des martyrs africains', in *Epektasis : Mélanges J. Daniélou*, Paris, 1972, pp. 139–53 discussing the *Passio Maximiani*.

30 Luke 22:41; Matthew 26:39.

31 Samuel Taylor Coleridge, *Kubla Khan* 53–54, in *Christabel: Kubla Khan, a Vision ; The Pains of Sleep*, London, 1816, p. 58: 'For he on honey-dew has fed, And drank the milk of Paradise.'

32 For more, see Pseudo Cyprian, *De laude martyrii*, 21, ed. W. Härtel, *CSEL* 3/3, Vienna, 1871, pp. 43–4, where a *locus amoenus* is described.

33 E.g. Augustine, *Confessions*, 3.6.10, ed. M. Skutella, Stuttgart, 1969, p. 43: 'cibus in somnis simillimus est cibis vigilantium, quo tamen dormientes non aluntur: dormiunt enim'.

34 These are the dreams from which one does not wish to be awakened. See Augustine, *De Genesi ad litteram*, XII.32.61, ed. J. Zycha, *CSEL* 28/3.2, Vienna, Prague and Lepizig, 1894, p. 427: 'Unde quidam in rebus, quas concupiverant, constituti se evigilasse doluerunt'.

35 *Passio Montani et Lucii* 7.5–6, ed. Musurillo, *The Acts of the Christian Martyrs* (n. 23 above), p. 218: 'nam hunc eundem dominum de paradiso interrogavit ubi esset. cui ille ait: "Extra mundum est". "Ostende mihi illum", inquit, et ait illi: "Et ubi erit fides?"'; ibid., p. 219: 'Now this was the Lord from heaven, and Victor asked him where heaven was. "It is beyond the world", said the child. "Show it to me", said Victor. He said to Victor: "Where then would your faith be?"'. Victor felt weak and had need of a sign to tell the others.

36 Romans 10:16–17: 'sed non omnes oboedierunt evangelio. Esaias enim dicit, Domine quis credidit auditui nostro? Ergo fides ex auditu, auditus autem ex verbo Christi.' Lactantius, *Divinae institutiones*, 6.21 promotes hearing in connection with the Psalms alone.

Susan Ashbrook Harvey).[37] Platonism devalued the bodily senses, and a broad pagan tradition regarded them as troublesome at the very least. Christianity inherited this attitude, particularly before the Peace of the Church. Subsequently many Christians reclaimed and revalorized the senses for the liturgy. At the same time, ascetics spouted the 'rhetoric of sensory danger'.[38] Here is a representative example from Ambrose:

> For who amongst all the many passions of this body, amongst all the enticements of this world, can keep his course safe and unpolluted? The eye took a look – and distracted the right thinking of the intellect. The ear heard – and it led the focus astray. Smell inhaled – and impeded thought. The mouth tasted – and related a crime. Touch made contact – and kindled a fire. "Death entered through the window," said the prophet. "Your window is your eye. If you behold a woman to lust after her, death has entered; if you listen to whorish endearments, death has entered; if lechery captures your senses, death has entered."[39]

There were ways out of the difficulty, however, such as Origen's doctrine of the 'spiritual senses' – on which more anon.[40]

Sensory Competition: Getting There With Less

While this outline is essentially correct on the macroscopic level, details matter, and more attention to individual texts reveals an interesting tug-of-war or sensual seesaw in some sources and authors. There is a tension between belief from hearing alone and the human experience of visions to fortify one for martyrdom. These experiences of paradise left evidence of the taste of heaven in the dreamer's mouth. This taste is a sensory version of the 'apports' granted to modern spiritualists and also attested in ancient dreams.[41]

Physical as opposed to gustatory apports

As time went on, and the period of epic passions arrived in the fourth century, we even find legends of fruit deliveries from paradise, both to unbelievers and to monastic

[37] S. A. Harvey, 'Locating the Sensing Body: Perception and Religious Identity in Late Antiquity', in *Religion and the Self in Antiquity*, eds D. Brakke, M. L. Satlow and S. Weitzman, Bloomington, IN, 2005, pp. 140–62; see also Harvey, *Scenting Salvation* (note 9 above).

[38] Nice term from Harvey, 'Locating the Sensing Body' (n. 37 above), p. 146. There is a fine example in Lactantius, *Divinae Institutiones*, 6.20–23.

[39] My translation; Ambrose, *De fuga saeculi*, 3, ed. C. Schenkl, *CSEL* 32/2, Vienna, 1897, p. 164 [my emphasis]: 'Nam quis inter tot passiones huius corporis, inter tot illecebras huius saeculi tutum atque intemeratum servare potest vestigium? Respexit **oculus**, et sensum mentis avertit: audivit **auris**, et intentionem inflexit: inhalavit **odor**, et cogitationem impedivit: **os** libavit, et crimen retulit: **tactus** contigit, et ignem adolevit. Intravit mors per fenestram, dixit Propheta (Jer. 9:21). Fenestra tua est oculus tuus. Si videas mulierem ad concupiscendum, intravit mors: si audias sermones meretricios, intravit mors: si luxuria sensus tuos capiat, penetravit mors.'

[40] Rahner, 'Le Début d'une doctrine des cinq sens spirituels chez Origène' (n. 13above).

[41] E. R. Dodds, *The Greeks and the Irrational,* Berkeley CA, 1971, p. 106 on dreams that demonstrate their objectivity by leaving a physical apport behind. For examples from ancient healing-cults, see O. Weinreich, *Antike Heilungswunder: Untersuchungen zum Wunderglauben der Griechen und Römer,* Giessen, 1909, p. 81.

brethren. The *Acta Dorotheae* provide a delightful elaboration of the theme of paradise.[42] This particular saint preached that Christ invited men 'to come to the paradise of his delights, where the groves are adorned with fruit at all times of year'. As she was being led to execution, one Theophilus, a lawyer, mocked her: 'Hey, you, "bride of Christ", send me apples or roses from the garden of your husband.'[43] Dorothea said that she would certainly do so. Before the axe fell, she prayed. A boy appeared with three fine apples and three roses. She had him take them to Theophilus. The scoffer was instantly converted. 'It was February and bitter cold. Where could they have come from?[44] A pretty little boy came to me – very short, not more than four years old. I thought he couldn't talk, given his age.'[45] When he spoke perfectly, Theophilus knew that an angel had provided fruit from paradise. He was likewise martyred.

A similar apport regaled ascetics. Pater Muthios (Πατερμούθιος), an Egyptian ascetic, was said to have brought back a fig from paradise:

> and he said that he had been carried in the flesh to paradise and had seen a multitude of saints [there]. He said that he had partaken of the fruits of paradise and he pointed to evidence for the statement: he had brought one large choice fig full of fragrance for his disciples, showing them that what he said was true. The one who told us about the fig, the old man Kopres, was young at the time and saw it in the hands of his disciples and kissed it and marveled at its fragrance. 'It remained for many years with his disciples', he said, 'kept for show. It was very large. One had only to smell it, if one was in bad shape and one was instantly cured of one's disease.'[46]

It improved the life of ascetics, healing the sick by its smell alone.[47] Thus there were two ways of perceiving paradise, the ethereal way through sight or hearing alone (with some competition between the two), or using the lower senses also (with competition between taste and touch too). Smell was most easily redeemed.

[42] *Acta Dorotheae*, 3, in *Acta Sanctorum*, 6 February. The following translations and paraphrases are my own. For the Latin text, see *Acta Sanctorum Februarii*, I, eds J. Bollandus and G. Henschenius, Paris, 1863, p. 783.

[43] Ibid., 3.12: 'O tu sponsa Christi, dum perveneris ad paradisum sponsi tui, mitte mihi exinde rosas aut mala.'

[44] Ibid., 3.13: 'At ille, Dicite mihi, inquit, quis mensis nunc est? Dicunt ei: Februarius. Dicit eis Theophilus: Cum totam Cappadociam glaciale frigus tegat, et nulla penitus virgulta iam foliorum suorum tegmine vestiantur; unde istas rosas aut ista mala mirifica cum suis frondibus adesse æstimatis?'

[45] Ibid., 3.13: 'Et ecce repente, posteaquam pro nomine Christi passa est, venit ad me infantulus pulcher, brevissimus, non plus quam quatuor annorum, quem ego putabam loqui non posse præ ipsa infantia.'

[46] My own translation; *Historia Monachorum in Aegypto* 10.21–2, ed. Festugière (n. 8 above), p. 84: καὶ ἀπῆχθαι δὲ αὐτὸν ἐν σαρκὶ ἔφη εἰς τὸν παράδεισον καὶ ἑωρακέναι πλῆθος ἁγίων. μετειληφέναι δὲ αὐτὸν ἔλεγεν ἐκ τῶν καρπῶν τοῦ παραδείσου καὶ μάρτυρα τοῦ πράγματος ἐδείκνυεν· σῦκον γὰρ ἓν μέγα καὶ ἐξαίρετον καὶ εὐωδίας πολλῆς μεστὸν πρὸς τοὺς ἑαυτοῦ μαθητὰς ἀπεκόμισεν ἐπιδεικνύων αὐτοῖς ὅτι ἀληθές ἐστι τὰ παρ' αὐτοῦ λεγόμενα. ὅπερ σῦκον ὁ διηγούμενος ἡμῖν ταῦτα Κόπρης ὁ πρεσβύτερος νεανίας τότε ὑπάρχων ἐν ταῖς χερσὶ τῶν μαθητῶν αὐτοῦ ἐθεάσατο καὶ κατεφίλησεν καὶ τὴν πνοὴν ἐθαύμασεν. «Πολλοῖς γὰρ ἔτεσι, φησί, διέμεινεν παρὰ τοῖς αὐτοῦ μαθηταῖς εἰς ἐπίδειξιν φυλαττόμενον. ἦν γὰρ παμμέγεθες. μόνον γάρ τις ὀσφρήσατο αὐτοῦ τῶν κακουμένων, εὐθέως τῆς νόσου ἀπηλλάττετο»'. See A. J. Festugière, *Les Moines d'Orient: Enquête sur les moines d'Égypte,* 4.1, Paris, 1961, pp. 73–4 for a French translation. The honeycomb produced by the bees of paradise in *Joseph and Aseneth* 16.8 is a Judaeo-Christian variant on the theme.

[47] For a modern version of this legend, see the apple from the Land of Youth (viz. Narnia) that heals Digory's ill mother in C. S. Lewis, *The Magician's Nephew,* New York, 1966, pp. 99 and 162–3.

The treatment of taste, however, deserves more attention. It is dangerous to base an argument on negative evidence and likewise to make general negative statements but – despite Abraham's bosom and New Testament promises of eating and drinking in heaven,[48] there seems to be some reluctance to have recourse to taste or to depict eating in Christian apocalyptic and visions.[49] Eating came with all the sordid baggage of digestion and excretion, and hence required special apology.[50] *Midrashim* provide a contrast, for in them roast Leviathan was a pleasure of paradise.[51] Here is an early pleasance from Pseudo Cyprian:

> For those who always sought God or for those to whom He was known the place belonging to Christ is never forgotten, where grace lies, where the earth revels in green fields and dresses itself in the grass it bears and is fed on scented flowers, where the groves reach a great height, and where the thick foliage of trees clothes whatever the spreading scenic backdrop shadows with its curving branches. All things there are neither cold nor hot, nor, as in autumn, do the fields lie fallow, nor does the fecund earth give birth again in spring. All things have but one season: fruits are produced from one summer. Why, since the moon is then no slave to her monthly cycles, nor does the sun rush through each hour's moments, nor daylight put to flight give way to night, a happy peace enfolds the inhabitants, a calm seat holds them, where a fountain in the middle bubbles up from the bosom of a brimming riverbed and glides in a noisy circle with sinuous windings through the intervening spaces to be split up into the mouths of the rivers that are born to it. From here therefore comes the great praise of the martyrs, from here the noble crown of the victors to whom things yet greater than these are promised, whose prizes are greater. And that the body confronts the beasts or the threatening sword is not feared is a measure of their worth and a manifestation of their status as Chosen. For it would have been inappropriate that He should have held back amidst earthly vices and filth those whom he considered worthy of such a boon.[52]

[48] E.g. Luke 22:29–30.

[49] Warnings about feasting needed to be extended to the baptized. For an interesting example of such admonition to a 'caeleste prandium honestum purum, etc.' with allegorical exegesis of biblical foods, see Zeno of Verona, *Tractatus* 1.24.3, ed. B. Löfstedt, *CCSL* 22, pp. 71–2: 'David regius pastor omnibus momentis lac argenteum subministrat et caseum'. M. Bayless, *Parody in the Middle Ages: The Latin Tradition,* Ann Arbor MI, 1996, p. 33 rightly notes that, 'feasting was only safe when free of earthly taint, when expressed in a purely ethereal sense. To specify which joints of meat the citizens of heaven take, and how they cook them is to degrade the image'.

[50] My thanks to Alessandro Scafi and Pétér Toth for a nice example of the substitution of smell (and a proleptic defense) in the Pseudo Ambrosian, *Acta Sebastiani*, 4.13, *PL* XVII, col. 1027C: 'Odor aeris delectationem per omnia membra diffundit, et esca ibi nulla stercora conficit'.

[51] See M.E.C. Walcott, 'Curiosities from the Talmud', *The Ecclesiastic* 29, 1867, pp. 1–11 (7): 'The second dish is the whale, called Leviathan, or Lifjasan; he forms the course of fish – rather an Esquimaux taste.' The passage is Babylonian Talmud *Baba Bathra* 75a: 'God will provide the righteous with a banquet from the flesh of Leviathan' (transl. Howard Jacobson); The Babylonian Talmud, ed. Isidore Epstein, London, online at: http://www.come-and-hear.com/bababathra/bababathra_75.html.

[52] My translation; Pseudo Cyprian, *De laude martyrii*, 21, ed. Hartel (n. 32 above,) pp. 43–4 [my emphasis]: 'Quibus autem inquisitus semper Deus fuit aut notus, numquam excidit Christi locus; ubi iacet gratia, ubi virentibus campis terra luxurians alumno se induit gramine, et **redolenti pascitur flore** [emphasis added]; ubi altum nemora tolluntur in verticem, et ubi densior arborem coma vestit quicquid curvantibus ramis scena deiacens inumbrarit. Omnia illic non frigoris, nec ardoris, nec ut autumno arva requiescant, aut iterum tellus vere novo fecunda parturiat. Unius cuncta sunt temporis, unius poma feruntur aestatis; quippe cum nec mensibus suis tunc luna deserviat, nec per horarum sol momenta decurrat, aut in noctem lux fugata concedat. Habet populos quies laeta, sedes tenet placida, ubi fons scaturiens medius sinu alvei prorumpentis emergit, et rauco per intervalla circuitu sinuosis flexibus labitur,

The absence of taste is noticeable – except in a synaesthetic form, where the earth (or better 'one' or 'the martyr'?) feeds on the scent of flowers: 'redolenti pascitur flore'. A similar feature appears in a vision in the *Passio Perpetuae*, where Saturus describes being nourished by 'an indescribable scent that left us full'.[53] In the *Acts of Nereus, Achilles, and Domitilla* all celestial food consumption happens through the nostrils.[54] But in addition to broader cultural *sensoria*, there are also the *sensoria* of individual authors and texts, some of which are unusual and may prove influential. So, let's turn to a fourth-century Syriac poet: Ephrem and his *Hymns on Paradise*.[55]

Ephrem's Sensorium

Ephrem is very sensual indeed.[56] The soul cannot enjoy paradise's delights without the resurrected body.[57] He revels consequently in the senses and their properties and objects.[58] And while, yes, his paradise is more fantastic and sensuous than that of Western authors, he often both simultaneously affirms and denies the senses.[59] Nourishment in the afterlife is quite different from that on earth: 'Spiritual food befits the spiritual being'. Hence winds feed the discerning,[60] the perfume of paradise does the duty of bread, and the breath of life acts as drink.[61] Moses ascending Sinai without provender shames the greedy man. In one unusual passage he grossly contrasts paradisiacal and terrene feeding.[62] Key phrases such as 'spiritual being', and 'the greedy man' suggest an audience experiencing difficulty

ut in ora nascentium sibi fluminum dividatur. Hinc ergo magna laus martyrum, hinc nobilis corona victorum, quibus maiora his repromissa sunt, quibus auctiora sunt praemia. Quod vero vel feris corpus obicitur, vel imminens gladius non timetur, dignationis ostenditur ratio, manifestatur electio. Etenim incongruens fuerat ut quem aequum tanti muneris iudicasset, inter terrena vitia sordesque retineret.'

[53] My translation; *Passio Perpetuae* 13.8, ed. Musurillo, *The Acts of the Christian Martyrs* (n. 23 above), p. 122: 'odore inenarrabili alebamur qui nos satiabat'.

[54] Pace Amat, *Songes et visions* (n. 1 above), pp. 396. See *Acta Nerei, Achillis, et Domitillae* 6, *Acta Sanctorum* 12 May, in *Acta Sanctorum Maii*, III eds G. Henschenius and D. Papebrochius, Paris and Rome, 1866, pp. 7–8) [my emphasis]: 'Inter frondes Paradisi et vitæ æternæ immarcessibiles flores, inter nemora angelorum et prata umbrosa et sanctis odoribus exhalantia: ubi **vita æterna naribus assumitur**, ubi aër ipse habet **odoramenti virtutem**, ut qui eum naribus suis assumpserit, ultra tristis esse non possit, ultra nulla in eum possit dominationem exercere infirmitas, nullus dolor, nulla tristitia; sed semper erit gaudens, semper exultans anima, et semper de vitæ suæ perpetutate secura.'

[55] Note that these brief remarks are based on study of the hymns in French and English translation, not in the original Syriac. See Sergey Minov's paper in this volume.

[56] For some discussion, see Harvey, *Scenting Salvation* (n. 9 above), pp. 235–9, who concentrates on the olfactory.

[57] And hence believes in the slumber of the soul before Judgment in a 'pre-paradise', where it is a quasi-embryonic state. See Ephrem, *Hymns on Paradise,* 1.16 and 8.5–6, eds. R. Lavenant and F. Graffin, Paris, 1968, p. 24 . Also the edition by S. P. Brock, Crestwood, NY, 1990, p. 131.

[58] E.g. *Hymn* 6.3–4.

[59] E.g. picking up the image of the joys at the 'Table of the Saints' in *Hymn* 5.15.

[60] *Hymn* 6.3, 7.18: the winds of paradise serve as both food and drink for the righteous.

[61] See *Hymn* 9.8–9, 10, 11, and 17.

[62] *Hymn* 9.23, eds Lavenant and Graffin (n. 57 above), p. 130: 'Car notre nourriture ne demeure qu'ordure; sa souillure nous trouble, son odeur nous égare. Ne se nourrir que peu nous procure bien-être, mais faire excès nous nuit. Si les joies de la terre peuvent render déjà prospère et plantureux, Ah! combien plus encore, dans la joie déferlante, quand toutes les puissances se mettront à sucer le sein de la Sagesse!'.

with gluttony. And special time devoted to that sin and awards awaiting watchers, fasters, and abstainers[63] suggests an ascetic milieu.[64] His treatment of paradise corroborates my working hypothesis that communities that are more ascetic renounce (as best they can) the sense of taste and transmute it into smell. There was continuity on this point through to the late sixth century at least.[65] In Gregory the Great's paradise fragrance obviated the need for taste.[66]

Augustine's Spiritual Sense

I'd like now to introduce a famous mystical passage that that deploys and redeems all five spiritual senses. Augustine is speaking to God:

> Late have I come to love you, beauty so ancient, yet so new; late have I come to love you! And lo! You were inside, and I without, and there I sought you and, lacking beauty myself, rushed to fall upon the beautiful things you have made. You were with me, but I was not with you, and those things kept me distant from you that, had they not had their being in you, would not have existed at all. You called, you cried out, and you shattered my deafness. You sparkled and you shone and you put my blindness to flight; you gave forth a sweet scent, and I breathed it in and

[63] See *Hymn* 6.3: heaven satiates the nostrils and mouth. 'Watchers' and 'Fasters' who in return for their fasting shall graze in its splendid fields.

[64] Although external *testimonia* to Ephrem's life depict him as a monk, e.g. the *Lausiac History* (Syriac version), the picture is probably anachronistic. In his edition of the *Hymns* (n. 57 above), pp. 13–26, S. P. Brock argues instead for a proto-monastic ascetic tradition.

[65] And also beyond. In Jonas's *Vita Columbani* 2, ed. B. Krusch, *MGH, Scriptorum Rerum Merovingicarum* 4, Berlin, 1902, pp. 113–52, appears a long sequence of miracle pericopes concerned, above all, with the deathbeds of the nuns of Faremoutiers. Noticeable is the almost complete absence of taste and eating. Heavenly delights and the good death are marked by mystic singing and the fragrance of sanctity. E.g. Sweet music in ibid., 2.11, 13, 14, 16, 17; Sweet scents in ibid., 2.12, 16, 21. But in 2.22 a sinister story about a girl who was a furtive eater, was granted a punitive vision of the boar that pigged out with her, helps contextualize ladylike nuns's fear of the perils of taste. And Gregory the Great has a cautionary tale for the male secret eater: *Dialogi*, 4.40.11 tells of a monk of Iconium who pretended to fast, while secretly eating. On his deathbed he confessed his sin. His soul was sucked out of his mouth by an infernal snake that inserted its head in his mouth. Bede, no surprise, is more cerebral. Sight and hearing are specifically mentioned: for Bede, *Historia Ecclesiastica*, 3.19, ed. C. Plummer, 2 vols, I, Oxford, 1896, p. 164, relating Fursey's vision ('angelicorum agminum et aspectus intueri et laudes beatas meruit audire'), as in Gregory, fragrance is positive. Ibid., 5.12, ed. Plummer, p. 307, on Drythelm's vision [my emphasis]: 'Cumque procedentes transissemus et has beatorum mansiones spirituum, aspicio ante nos multo maiorem luminis gratiam quam prius, in qua **etiam vocem cantantium dulcissimam audivi, sed et odoris fragrantia miri tanta de loco effundebatur, ut is, quem antea degustans quasi maximum rebar, iam permodicus mihi odor videretur,** sicut etiam lux illa campi florentis eximia, in comparatione eius quae nunc apparuit lucis, tenuissima prorsus videbatur et parva.' But two of his visions imply a memory of what must be touch. Fursey burned after his visit to hell, while Drythelm remembered its chill. The former sported a sinister apport, burns on his shoulders that provided verification through touch and sight. See ibid., 3.19, ed. Plummer, p. 167, on the marks of burning: 'Qui postmodum in corpore restitutus, omni vitae suae tempore signum incendii, quod in anima pertulit, visibile cunctis in humero maxillaque portavit, mirumque in modum quid anima in occulto passa sit, caro palam praemonstrabat.'

[66] Gregory the Great, *Dialogi*, 4.37.8, eds A. De Vogüé and P. Antin, 3 vols, III (Livre IV), *SC* 265, Paris, 1980, p. 130 [my emphasis]: 'Tantusque in eodem loco odor suavitatis inerat, ut ipsa suavitatis **fragrantia** illic deambulantes habitantesque **satiaret**.' The only 'food' experienced in the punitive afterlife is the redemptive mass. See ibid., 4.57.6 and 4.58.1 for the prisoner on earth released by his wife's masses and 4.59.3 for a shipwrecked man rescued from starvation by a mass.

> pant after you; I tasted and I hunger and thirst; you touched me, and I was set on fire by a longing for your peace.[67]

In this sensual extravaganza the five are presented from the top down: hearing first, sight, smell, taste, and touch. First God calls, sparkles, gives out a sweet smell; then Augustine tastes and continues to hunger and thirst, God touches him – and he burns for God's peace. Augustine follows this ecstatic passage with an exploration of the senses in his own personal vale of temptation. Some of what he says is personalized and mundane, for example the discussion of touch mentions the ever-present danger of wet dreams.[68] Smell was not a serious temptation for him.[69] Sight and hearing are both potentially dangerous.[70] But taste is the issue here. Note the sequence of second person singular verbs ('Thou') for each sense – with one exception, taste. No action of God elicits it. Instead it is somehow almost synaesthetically elided into the more ethereal smell. Noticeable too is the way that for these mystical-spiritual senses *in this life* continuing to hunger and thirst is a good.[71] At the Resurrection, Augustine tells us, God will destroy both food and stomach with a wondrous satiety (*mirifica satietas*).[72] At the moment however, one ought to take food like medicine, not for pleasure. Even the transition from hunger to satiety, Epicurean kinetic ἡδονή, is pleasing. So there are manifest hints of the ascetic's daily battle with food and taste.[73]

At this stage it is worth gathering together some preliminary observations based on what we have seen so far. First of all, taste and eating seem to not to be full citizens of the Christian paradise. Instead its ghost foretaste lures only the living to sweeten their deaths. It does not regale the dead. Taste creates the mystic phenomenon, the 'taste of paradise' that Augustine would memorably call a 'loving memory' in one of his early failed mystical visions:

> And [my power of reason] came to That Which Is in a single twinkling of its fearful glance. And then I beheld your invisible things understood through visible things. But I was not able to keep my gaze fixed constantly, and, once my weakness was rebuffed, and I was returned to my usual concerns, I bore off nothing more than a loving memory that yearned, as it were, for what it had smelled, but was not yet able to eat.[74]

[67] Translation mine; Augustine, *Confessions,* 10.27.38, ed. M. Skutella, Stuttgart, 1969, p. 237 [my emphasis]: 'sero te amavi, pulchritudo tam antiqua et tam nova, sero te amavi! et ecce intus eras et ego foris et ibi te quaerebam et in ista formosa, quae fecisti, deformis inruebam. mecum eras, et tecum non eram. ea me tenebant longe a te, quae si in te non essent, non essent. **vocasti** et clamasti et rupisti **surditatem** meam, **coruscasti**, splenduisti et fugasti **caecitatem** meam, **fragrasti**, et duxi spiritum et anhelo tibi, **gustavi** et esurio et sitio, **tetigisti** me, et **exarsi** in pacem tuam.'

[68] Augustine, *Confessions*, 10.31.42.

[69] Ibid., 10.32.48.

[70] Ibid., 10.33.49 and 10.34.51.

[71] An Epicurean would call it kinetic rather than katastematic pleasure.

[72] Augustine, *Confessions*, 10.31.43.

[73] See Cassian, *Collationes*, 5.21 and D. R. Shanzer, 'Bishops, Letters, Fast, Food, and Feast in Later Roman Gaul', in *Culture and Society in Later Roman Gaul*, eds R. W. Mathisen and D. R. Shanzer, 2001, pp. 221, 215–36.

[74] Translation mine; Augustine, *Confessions,* 7.17.23, ed. M. Skutella, Stuttgart, 1969, p. 146: 'Et pervenit ad id quod est in ictu trepidantis aspectus. Tunc vero invisibilia tua per ea quae facta sunt intellecta conspexi. sed aciem figere non evalui et repercussa infirmitate redditus solitis non mecum ferebam nisi amantem memoriam et quasi olefacta desiderantem quae comedere nondum possem.'

This is a rebuff, not a lure – hence the opposite of the martyr's encounters. And Augustine cannot *taste* in this vision, for he is unbaptized, cannot eat the right (Eucharistic) food yet, and has not come to know Jesus.[75] He must have recourse to the *scent* of what he cannot yet eat. But the basic concept is identical. Second a synaesthetic substitution of smell for taste occurs in the description of the sensory experience of the dead in paradise. Third, some texts substitute physical apports for the haunting taste or 'loving memory', but these are grosser and more naïve: they smack of fairy tale. Fourth, these latter cater instead to those lesser souls who demanded signs.

Why? Not-Eden?

But why should this be the case? Two possible answers require further research. First, it seems obvious that taste may have become somehow tainted by the form of the Edenic Fall. Eve's fruit was after all both attractive and tasty: In the words of Genesis 3:6, 'the woman saw that the tree was good for food, and that it was a delight to the eyes'.[76] And she sinned by consumption.

Not-Cockaigne?

The second possible answer points in a new direction, namely to the medieval Land of Cockaigne, or the modern American *Schlaraffenland*, the hobo's 'Big Rock Candy Mountain'.[77] These are the vulgar fantastic vernacular versions of the *paradisus voluptatum*. Such traditions existed in the ancient world as parodies of the Golden Age.[78] And their vulgarly pagan and carnal emphasis on food and drink and preposterous automation may have exercised a repellent effect on serious Christians' sober ideas of paradise – on which more shortly.

Perpetua's Cheese or 'Blessed are the Cheese-Makers'

But it is time to revisit and perhaps even to connect two odd Latin texts. The first, *Passio Perpetuae* 4.9: 'de caseo quod mulgebat dedit mihi quasi buccellam', we have seen already. 'And he gave me as it were a mouthful[79] of the cheese he was milking.'

The controversy surrounding Perpetua's cheese is considerable. Is it really cheese, or merely soft curds,[80] or is it to be read as milk?[81] While cheese is obviously related to milk,

[75] These implications are clear from *Confessions*, 7.18.24.

[76] In the Latin Vulgate: 'vidit ergo mulier quod bonum esset lignum ad vescendum et pulchrum oculis aspectuque delectabile'.

[77] Song first recorded in 1928 by Harry McClintock. For a commonly available recent recording, see the soundtrack of Joel and Ethan Coen's *O Brother, Where Art Thou?* (2000).

[78] W. F. Hansen, *Ariadne's Thread: a Guide to International Tales Found in Classical Literature,* Ithaca NY, 2002, pp. 378–92.

[79] Or 'morsel' or 'small lump'. See *Thesaurus linguae latinae*, I, Leipzig, 1900, s.v. 'buccella', [col. 2227].

[80] Pace Meslin, 'Vases sacrés (n. 29 above), p. 150: 'une bouchée du caillé qu'il est en train de faire'. Perhaps a typo for 'traire?'

[81] The conclusion of E. Zocca, 'Un passo controverso della Passio Perpetuae IV,9 , De caseo quod mulgebat dedit mihi quasi bucellam', *Studi e Materiali per la Storia delle Religioni*, 50, 1984, pp. 147–54.

it is highly 'marked' in this passage and hence significant in some way that the banal milk would not have been.

Milking cheese (as opposed to milk) has been seen as a surreal chronological compression from an authentic dream.[82] Or as having a latent sexual meaning.[83] Even in antiquity it was connected with the weaning of Perpetua's own child.[84] Dronke saw it as embryological, signifying Perpetua's own celestial birth.[85] Amat pointed to the pagan initiate who like the kid has 'fallen into milk.'[86] Some have seen a Montanist Artotyrite communion in bread and cheese rather than in bread and wine.[87] Or is there some other way to read it?

It is difficult to evaluate any of the first three possibilities, and only Perpetua could confirm the second and third. Dronke's embryological interpretation seems over-intellectual, requiring a knowledge of the book of Job that is unlikely for a recent convert like Perpetua. The Artotyrite interpretation has, I think, rightly been rejected, even though the context is clearly liturgical and Eucharistic.[88] But we may be able to get further by looking at the distinctive feature from a different angle. Most scholars see the substance eaten as a liquid (i.e. as a milk-equivalent of some sort).[89] But Perpetua uses the rare and almost entirely Christian word *buccella* ('morsel', 'mouthful')[90] and emphasizes eating ('chewing' *commanducans*).[91] So she saw it as a *solid* – specifically cheese. She thus may

[82] E. R. Dodds, *Pagan and Christian in an Age of Anxiety: Some Aspects of Religious Experience from Marcus Aurelius to Constantine,* New York and London, 1970, p. 51.

[83] Ibid. p. 51; even more weirdly Meslin, 'Vases sacrés (n. 29 above), p. 152: 'Cette sensation gustative est un symptôme de *jouissance érotique bien connu* et propre à certains états extatiques.' [Italics mine!]

[84] At least one ancient reader thought so. See Quodvultdeus, *Sermo De tempore barbarico*, I.5, ed. R. Braun, *CCSL* 60, pp. 430–31: 'Una earum erat praegnans, alia lactans. Sed tamdiu haec Perpetua lactavit, quamdiu acciperet ab illo pastore simul et patre buccellam lactis: qua accepta dulcedo felicitatis perpetuae eam fecit contemnere filium, spernere patrem, non inhaerere mundo, perdere animam pro Christo.'

[85] P. Dronke, *Women Writers of the Middle Ages: A Critical Study of Texts from Perpetua (†203) to Marguerite Porete (†1310),* Cambridge, 1984, pp. 6–7 denies that the cheese could be Eucharistic, and at 9 suggests that its coagulation is embryonic, and that Perpetua is given her own celestial birth.

[86] Amat, *Songes et visions* (n. 1 above), p. 75.

[87] Epiphanius, *Panarion Her. 34–64*, ed. K. Holl, *Die Griechischen christlichen Schriftsteller der ersten Jahrhunderte: Epiphanius*, II, Berlin, 1980, 49.2.6, p. 243; *Panarion*, II.4.2.6, transl. F. Williams, Leiden and New York, 2 vols, 1987–94, II (1994), p. 22: 'they call them Artotyrites because they set bread and cheese on the altar in their mysteries'. The Artotyrites derived their name from the Greek terms *ἄρτος* (*artos*, 'loaf') and *τυρός* (*turos*, 'cheese'), as they added cheese, instead of wine, to the bread in the Eucharist.

[88] Dodds, *Pagan and Christian in an Age of Anxiety* (n. 82 above), p. 51; also earlier L. Gatti, 'La Passio SS Perpetuae et Felicitatis', *Didaskaleion* 1, 1923, pp. 31–43 (37–8) ; Meslin, 'Vases sacrés' (n. 29 above), p. 151, who notes the lack of bread. Zocca, 'Un passo controverso' (n. 81 above), p 147, n. 3; A. B. McGowan, *Ascetic Eucharists: Food and Drink in Early Christian Ritual Meals,* Oxford and New York, 1999, p. 102 notes the silence in Tertullian about Montanism. In this case a strong argument.

[89] E.g. Meslin, 'Vases sacrés' (n. 29 above), pp. 150–51; also McGowan, *Ascetic Eucharists* (n. 88 above), p. 107: 'The importance of cheese is likely, then, to be the importance of milk in another form.'

[90] It is very interesting to see that the word appears in the Old Latin Bible versions of John 13:26, where it signifies the morsel Judas took at the Last Supper that caused the devil to enter in to him.

[91] Lucian, *True Histories*, 2.5, transl. A. M. Harmon, in *Lucian*, 8 vols, I, *Loeb Classical Library*, London and New York, 1913, p. 308: ἡσθέντες δὲ τῇ ὀσμῇ.

have seen herself not as the New Testament infant who can only drink milk,[92] but as one of the Pauline *perfecti* whose food was solid.[93] But there may be more.

'The Island is Made of Green Cheese'

While milk and honey are familiar features of many paradises,[94] cheese had a place there too. In Lucian's *Vera Historia*, the narrator visits an island, prelude to the Isles of the Blessed, called *Tyroessa*, 'the Cheesy Island':

> The island was a great solid cheese, as we afterwards learned by tasting it. It was twenty-five furlongs in circumference. The vines were full of grapes, but the liquid which we squeezed from them and drank was milk instead of wine. A temple had been constructed in the middle of the island in honour of Galatea the Nereid … . All the time that we stopped in the island the earth was our bread and meat and the milk from the grapes our drink. The ruler of that region was said to be Tyro, daughter of Salmoneus.[95]

No serious paradise this, but a comic pastoralists' Cockaigne, where you eat the soil as cheese and squeeze the grapes for milk – automation, excess, and nourishment from one substance in two forms. No Cyclops, here, but Tyro, an etymologically correct goddess and queen.[96] When the crew arrives near the Isles of the Blessed fragrance wafts towards them at sea,[97] and they eventually join the Symposium of the Blessed in the Elysian Fields, a glutton-and-brawlers' paradise, if there ever was one.[98]

[92] I Peter 2:2: 'sicut modo geniti infantes rationale sine dolo lac concupiscite'.

[93] Hebrews 5:13–14: 'omnis enim qui lactis est particeps expers est sermonis iustitiae parvulus enim est. perfectorum autem est solidus cibus …'

[94] H. Usener, 'Milch und Honig', *Rheinisches Museum fürPhilologie*, 57, 1902, pp. 177–95; K. Wyss, *Die Milch im Kultus der Griechen und Römer,* Giessen, 1914, pp. 52–61. For the Old Testament, see (for example) Exodus 3:8, Joel 3:18. *Oracula Sibyllina* 2.318 features milk, honey, and wine. See J. L. Lightfoot, *The Sibylline Oracles: with introduction, translation, and commentary on the first and second books,* Oxford and New York, 2007, pp. 528–9 for commentary. Tertullian saw the Eucharist on earth as a foretaste of paradise; see Tertullian, *De corona*, 3.2–3, ed. J. Fontaine, Paris, 1966, pp. 64–7 [my emphasis]: 'Denique, ut a baptismate ingrediar, aquam adituri ibidem, sed et aliquanto prius in ecclesia, sub antistitis manu, contestamur nos renuntiare diabolo et pompae et angelis eius. Dehinc ter mergitamur amplius aliquid respondentes quam Dominus in evangelio determinavit. Inde suscepti, **lactis et mellis concordiam praegustamus**, exque ea die lavacro quotidiano per totam ebdomadem abstinemus. Eucharistiae sacramentum, et in tempore victus et omnibus mandatum a Domino, etiam antelucanis coetibus nec de aliorum manu quam praesidentium sumimus.'

[95] Lucian, *True Histories*, 2.3, transl. Harmon (n. 91 above), pp. 306–7: μετ' οὐ πολὺ δὲ εἰς πέλαγος ἐμβαίνομεν, οὐχ ὕδατος, ἀλλὰ γάλακτος· καὶ νῆσος ἐν αὐτῷ ἐφαίνετο λευκὴ πλήρης ἀμπέλων. ἦν δὲ ἡ νῆσος τυρὸς μέγιστος συμπεπηγώς, ὡς ὕστερον ἐμφαγόντες ἐμάθομεν, σταδίων εἴκοσι πέντε τὸ περίμετρον· αἱ δὲ ἄμπελοι βοτρύων πλήρεις, οὐ μέντοι οἶνον, ἀλλὰ γάλα ἐξ αὐτῶν ἀποθλίβοντες ἐπίνομεν. *ἱερὸν δὲ ἐν μέσῃ τῇ νήσῳ* ἀνῳκοδόμητο Γαλατείας τῆς Νηρηίδος … ὅσον δ' οὖν χρόνον ἐκεῖ ἐμείναμεν, ὄψον μὲν ἡμῖν καὶ σιτίον ἡ γῆ ὑπῆρχεν, ποτὸν δὲ τὸ γάλα τὸ ἐκ τῶν βοτρύων. βασιλεύειν δὲ τῶν χωρίων τούτων ἐλέγετο Τυρὼ ἡ Σαλμωνέως.

[96] P. v. Möllendorff, *Auf der Suche nach der verlogenen Wahrheit: Lukians Wahre Geschichten,* 21, Tübingen, 2000, pp. 272–4; also M. Paschalis, 'Tityrus and Galatea (Virgil, Eclogue 1): An Expected Relationship', *Dictynna*, 5, 2008, pp. 153–69 (160–61).

[97] Isles of the Blessed Fragrance wafts from the island, Lucian, *True Histories,* 2.5, transl. Harmon (n. 91 above), p. 308: ἡσθέντες δὲ τῇ ὀσμῇ.

[98] Ibid., 2.11, p. 312: Τοὐντεῦθεν αὐτομάτων ἡμῖν τῶν στεφάνων περιρρυέντων ἐλελύμεθα καὶ εἰς τὴν πόλιν ἠγόμεθα

One could read Perpetua's cheese-milking as an image of magic *automation*, like Vergil's vermilion and purple dyed rams.[99] This sort of *sponte sua* (with cheese, not the normal Old Testament milk) might owe something to pagan comic utopias. One might speculate about a certain amount of pagan-Christian descriptive competition, dialogue, and parody in the late second century. The *Vera Historia* with its clear take-off on the New Jerusalem is a rare surviving example.[100] So this might be one side of geographical cheese-features near paradise.

Verecundus's Repose on the Psalmic Cheese Mountain

The second (and last) text that needs to be brought into the discussion is Augustine's epitaph of Verecundus:

> Thanks be to you, our God! We are yours. Your acts of encouragement and consolation make it clear. Faithful to your promises, in return for his country estate at Cassiciacum, where we rested in you from the heat of the world, you requite Verecundus with the beauty of your paradise that is perpetually green, for you forgave him his earthly sins in the mountain of cheese, your mountain, the plentiful mountain.[101]

In Book IX of the *Confessions* Augustine gave thanks to the Lord for his friend Verecundus's baptism, and ended with a prayer: that he grant Verecundus rest in paradise in return for the rest Augustine had enjoyed at Cassiciacum. The sentence ends with a citation of Psalm 67.17: 'in monte incaseato, monte tuo, monte uberi'. The literal meaning suggests that, despite its position, the quotation should be in apposition to 'virentis paradisi tui'.[102] But the line is problematic: what is this mysterious cheese-mountain, the *mons incaseatus*?

The difficulty goes back to the text of Psalm 67.16–17. Here below the Greek Septuagint, followed by Jerome's Vulgate [my emphasis]:

> ὄρος τοῦ θεοῦ ὄρος πῖον, ὄρος **τετυρωμένον**, ὄρος πῖον. ἵνα τί ὑπολαμβάνετε, ὄρη **τετυρωμένα**, τὸ ὄρος, ὃ εὐδόκησεν ὁ θεὸς κατοικεῖν ἐν αὐτῷ; καὶ γὰρ ὁ κύριος κατασκηνώσει εἰς τέλος.

καὶ εἰς τὸ τῶν Μακάρων συμπόσιον; see also ibid., 2.14, p. 316: Τὸ δὲ συμπόσιον ἔξω τῆς πόλεως πεποίηνται ἐν τῷ Ἠλυσίῳ καλουμένῳ πεδίῳ.

[99] *Bucolic*, 4.42–5. A. B. McGowan, *Ascetic Eucharist* (n. 88 above), p. 101 refers to 'instant cheese', but with no implication of magic automation.

[100] Lucian, *True Histories*, 2.11. See P. v. Möllendorff, *Auf der Suche nach der verlogenen Wahrheit: Lukians Wahre Geschichten,* 21, Tübingen, 2000, pp. 17–22; idem, 'Christliche Apokalypsen und ihr mimetisches Potential in der paganen Bildungskultur: Ein Beitrag zu Lukians Wahren Geschichten', in *Die Bibel im Dialog der Schriften*, eds S. Alkier and R. B. Hays, Tübingen and Basel, 2005, pp. 181, 184–6; and J. N. Bremmer, 'Peregrinus' Christian Career', in *Flores Florentino. Dead Sea Scrolls and Other Early Jewish Studies in Honour of Florentino García Martínez*, eds A. Hilhorst. e. al, Leiden, 2007, pp. 729–47 for Lucian and Christianity.

[101] My translation; Augustine, *Confessions,* 9.3.5, ed. Skutella (n. 33 above), p. 183: 'gratias tibi, deus noster! tui sumus. indicant hortationes et consolationes tuae: fidelis promissor reddis Verecundo pro rure illo eius Cassiciaco, ubi ab aestu saeculi requievimus in te, amoenitatem sempiterne virentis paradisi tui, quoniam dimisisti ei peccata super terram *in monte incaseato, monte tuo, monte uberi* (Ps. 67.16).'

[102] But it probably is not. Instead it is parallel to *in te* (God the Father).

mons dei mons pinguis **mons coagulatus** mons pinguis ut quid suspicamini **montes coagulatos** mons in quo beneplacitum est Deo habitare in eo.

The meaning of the Hebrew word, גבננם, translated by Jerome as *coagulatus,* is unknown.[103] The translators of the Septuagint guessed that it had something to do with cheese from a similar word (גבינה): hence τετυρωμένον 'turned into cheese'. The lines were rarely discussed and cause an exegetic bifurcation.[104] The first Westerner to mention them before Augustine was Hilary of Poitiers, one of the witnesses for the Old Latin Bible text that read *montes coagulatos* (accusative). He had a pejorative interpretation: the mountains were diabolical powers because cheese was corrupted milk.[105]

For Augustine, however, the *mons* was a happy place with paradisiacal connotations. He would later elucidate it as Christ, for he fed the young on milk and appeared on a mountain.[106] His positive paradisiacal exegesis here might seem in the first instance to be motivated by a clear pun between '*Cas*siciacum' and 'in*cas*eatus'.[107] To achieve this Augustine used an Old Latin Bible text of limited attestation that had the reading 'caseatus' for Hilary's 'coagulatus'.[108] *Incaseatus* seems to be Augustine's *unicum*. Verecundus lent Augustine his estate at Cassiciacum for a period of philosophical *otium*, and the Lord is asked to grant Verecundus rest in return in (or through) Christ the Cheese-Mountain.

[103] See F. Brown, E. Robinson, S. R. Driver, C. A. Briggs, and W. Gesenius, *A Hebrew and English Lexicon of the Old Testament*, repr. with corrections, Oxford, 1966, p. 148 for גבן. Many translators of the Hebrew Bible connect it with 'giben,' which means 'hump-backed'. For more on this mountain in its Ur-context, see M. Dijkstra, 'The Weather God on Two Mountains', *Ugaritische Forschungen*, 23, 1991, pp. 127–40 and N. Wyatt, *The Mythic Mind: Essays on Cosmology and Religion in Ugaritic and Old Testament Literature,* London and Oakville CT, 2005, pp. 209–12.

[104] Not to mention serious syntactic problems in line 17. Was ὄρη τετυρωμένα vocative or accusative? I am deliberately avoiding discussing v. 17, which raises a different set of difficulties.

[105] Hilary, *In Psalmum 143 Tractatus*, 9, ed. A. Zingerle, *CSEL* 22, Prague, Vienna and Leipzig, 1891, p. 820 [my emphasis]: 'Ut quid suscepistis montes coagulatos? et quantum intelligi conceditur, montes interdum etiam diabolicas potestates, ut nunc in hoc loco, significari arbitramur: qui constitutiones divinae legis transgressi, tamquam **ex lactis sinceritate, vitiosa coaguli corruptione coagulaverint**. Et hos esse potentissimos inimicos suos psalmus anterior locutus est. (an in malo interpretation).' The *in malo* interpretation is also to be found in Cassiodorus, *Expositio in Psalmum* 118, ed. M. Adriaen, *CCSL* 98, Turnhout, 1958, p. 1087, line 1186.

[106] Augustine, *Ennarrationes in Psalmos*, 67.22, eds E. Dekkers and J. Fraipoint , *CCSL* 39, Turnhout, 1956, p. 885 [my emphasis]: 'Sed quem montem intelligere debemus montem Dei, montem uberem, montem **incaseatum**, nisi **eumdem Dominum Christum**, de quo et alius propheta dicit: Erit in novissimis temporibus manifestus mons Domini, paratus in cacumine montium)? Ipse est **mons incaseatus, propter parvulos gratia tanquam lacte nutriendos**; mons uber, ad roborandos atque ditandos donorum excellentia: nam et ipsum lac, **unde fit caseus**, miro modo significat gratiam; manat quippe ex abundantia viscerum maternorum, et misericordia delectabili parvulis gratis infunditur.'

[107] G. N. Knauer, *Psalmenzitate in Augustins Konfessionen,* Göttingen, 1955, pp. 122–3 finds the similarity in the sounds quite a plausible motivation for Augustine's citing the verse here.

[108] It is to be found in Apponius, *In Canticum Canticorum expositio*, 3.525, eds B. de Vregille and L. Neyrand, *CCSL* 19, Turnhout 1986, p. 78: 'Est "mons caseatus" firmioribus namque solidum cibum tribuendo' and in Jerome, *Epistula* 66.,10, ed. I. Hilberg, *CSEL* 54, Vienna, 1996, p. 660: 'qui inter caseatos nutritus est montes'. Compare the text of the *Psalterium Gallicanum iuxta LXX* [my emphasis]: 'is mons dei, mons pinguis, mons **coagulatus**, mons pinguis. Ut quid suspicamini montes coagulatos with Vulg. Iuxta Hebr. Is mons dei, mons pinguis, mons **excelsus**, mons pinguis. Quare contenditis montes excelsi adversum montem quem dilexit deus ut habitaret in eo.'

Augustine's exegesis looks much less odd, however, in an Eastern context, where the Septuagint was the definitive text. Here, when exegetes had to make sense of the 'coagulated' or 'cheesy' mountain in v. 16, the Christological solution was current. Gregory of Nyssa provides the first identification of Christ with the coagulated mountain.[109] For Athanasius, however, the coagulated mountain was the Church, full of milk, which is identified as simple speech.[110] According to him the Psalmist challenged those who suspected that churches of the heretics were full of milk.[111] Asterius of Emesa read the question as addressed by Christ to those who identified him with Moses or Elijah.[112]

It is Didymus the Blind's exegesis, however, that is most relevant.[113] He saw Jesus as the mountain of God, for he was the 'fat', 'cheesy', or 'coagulated' mountain that gave milk (in the Pauline sense, viz. doctrine) to some to drink and solid food (viz. cheese) to others:

> Quite rightly is the Savior the mountain of God, for it is said of him, 'The mountain of the Lord will appear at the end of days' and again, 'Onto the lofty mountain let there go up the one bringing good news to Zion.' The ascent is through development. The Logos of God since it is wisdom and life and sanctification is fat. It does not receive fatness from outside, but provides it to those who partake of it. This mountain, in addition to being fat, is also coagulated/cheesy, giving some people milk to drink and imparting solid food to others that is none other than milk, for, when the latter is solidified, it has become the food of the perfect.
>
> The following will be said about its coagulation. The fat mountain, being God, became Man, taking on the flesh from the Virgin through which it is called 'coagulated'. His birth as a man was made evident in this passage, as was clear when Job said to God, 'Remember that you fashioned me as mud', and subsequently, 'did you not milk me like milk, and coagulate me as cheese?' 'You have clothed me with skin and flesh' and so forth. If after having been coagulated, he preserves his godhead unharmed and, after having been turned into cheese, he is fat, as the text itself attests, saying, 'Coagulated mountain, fat mountain'. He is fat both before and after being coagulated.[114]

[109] Gregory of Nyssa, *In Cantica Canticorum oratio 1*, in *Opera*, VI, ed. H. Langerbeck, Leiden, 1960, p. 11.15. Eusebius, *In Psalmos* (*PG* XXIII, col. 700) has nothing on the term.

[110] Athanasius, *Expositiones In Psalmos,*. 67.16, *PG* XXVII, col. 297: Ορος μὲν τὴν Εκκλησίαν ὀνομάζει, πῖον δὲ, ὅτι εὐτραφεῖς καὶ λιπαρὰς τῶν ἐν αὐτῇ τὰς ψυχὰς ἀπεργάζεται ... Καὶ μὴν καὶ τετυρωμένον, τουτέστι γάλακτος μεστὸν, ἁπλουστέρου δηλαδὴ λόγου, κατὰτό· *Γάλα ὑμᾶς ἐπότισα, οὐ βρῶμα.*

[111] Ibid.: Επιτιμᾷ οὖν τοῖς τὰς τῶν αἱρετικῶν ἐκκλησίας ὑπολαμβάνουσιν εἶναι τετυρωμένας. Οὐδὲν γὰρ ἐν αὐταῖς τὸ δυνάμενον τρέφειν εἰς ἕξιν πνευματικήν.

[112] Asterius of Emesa, *Commentarii in Psalmos Homilia* 18, ed. M. Richard, Oslo, 1956, Section 15. Line 28: Πρὸς τοὺς λέγοντας αὐτὸν Μωυσέα καὶ Ηλίαν· Ινα τί ὑπολαμβάνετε, ὄρη τετυρωμένα, τὸ ὄρος ὃ ηὐδόκησεν ὁ θεὸς κατοικεῖν ἐν αὐτῷ.

[113] His Psalm-commentary belongs to the 370's. See R. A. Layton, *Didymus the Blind and his Circle in Late-antique Alexandria: Virtue and Narrative in Biblical Scholarship,* Urbana IL, 2004, p. 6.

[114] My translation; Didymus the Blind, *In Psalmum 67.16*, Fr. 692A, in E. Mühlenberg, *Psalmenkommentare aus der Katenenüberlieferung,* Berlin and New York, 1975–77, pp. 68–69: Κυρίως ὄρος θεοῦ ὁ σωτὴρ ὑπάρχει· εἴρηται γὰρ περὶ αὐτοῦ Εσται ἐπ' ἐσχάτων τῶν ἡμερῶν ἐμφανὲς τὸ ὄρος κυρίου· <καὶ πάλιν Επ' ὄρος ὑψηλὸν ἀνάβηθι ὁ εὐαγγελιζόμενος τῇ Σιών. ἡ δὲ ἀνάβασις διὰ προκοπῆς.> ὁ τοίνυν θεὸς λόγος σοφία τε καὶ ζωὴ καὶ ἁγιασμὸς ὑπάρχων πίων ἐστίν, οὐκ ἔξωθεν δεχόμενος πιότητα ἀλλὰ παρέχων αὐτὴν τοῖς μετέχουσιν αὐτοῦ. τοῦτο τὸ ὄρος πρὸς τῷ πῖον εἶναι καὶ τετυρωμένον ἐστίν, τοὺς μὲν γάλα ποτίζων, τοὺς δὲ στερεᾶς τροφῆς μεταδιδούς, οὐχ ἑτέρας οὔσης ταύτης παρὰ τὸ γάλα· αὐτὸ γὰρ παγὲν τελείων γέγονε τροφή. Λελέξεται δ' εἰς τὸ τετυρωμένον καὶ οὕτως· τὸ πῖον ὄρος θεὸς ὢν ἐνηνθρώπησεν, ἐκ παρθένου ἀναλαβὼν σάρκα δι' ἣν τετυρωμένον κέκληται, δηλουμένης τῆς ἀνθρωπίνης γενέσεως τῇ λέξει ταύτῃ, ὡς ἐκ τοῦ Ιὼβ φανερὸν λέγοντος πρὸς θεὸν <Μνήσθητι ὅτι πηλόν με ἔπλασας, καὶ ἑξῆς·> πότερον οὐχ ὥσπερ γάλα με ἤμελξας,

Job 10:10–11 ('Did you not pour me out like milk and curdle me like cheese? You clothed me with skin and flesh, and knit me together with bones and sinews.')[115] was adduced to create an embryological reading of τετυρωμένον ('coagulatus'). Jesus before the Incarnation was milk; when he was incarnate, he became cheese. It is unclear how precisely Didymus's exegesis made its way to Augustine. But Jerome might be a missing link, for the latter was a student of Didymus and used his work extensively.[116] Butter was cousin to cheese, so Jerome merged the Didyman exegesis (*mons caseastus*) with Isaiah 7:15 ('butyrum et mel comedet') and 7:22 ('et prae ubertate lactis comedet butyrum, butyrum enim et mel manducabit') (on Emmanuel, who will eat butter and honey). This odd passage in the consolatory letter on Paulina's death recommends that Pammachius (transgendered in the *persona* of the bride of the Song of Songs) should seek Jesus (the small boy fed on butter and honey):

> And when you find him and hold him, do not let go. And if for a while he should slip from your hands as you nod off, do not immediately despair. Go out into the streets! Beseech the daughters of Jerusalem! You will find him reclining at noon, tired, drunk, soaked with the dew of night, amongst the bands of his companions, amidst many types of perfumes, amidst the fruits of paradise. There give him your breasts! Let him suckle at your learned breast and let him take his ease amidst the clergy. His feathers are like those of a silver-gilt dove and his viscera gleam like gold. This child, this boy who is replete with butter and honey, who was raised amongst the mountains of cheese, swiftly grows into a youth, swiftly despoils your enemies, and in due course preys on Damascus and conquers the Assyrian king.[117]

Thus by 396/410 Western writers had access to a positive image of the Christological, cheese mountain in Psalm 67:16.

Conclusion

Taste was one of the lower senses, and, as a consequence, despite the existence of a doctrine of the spiritual senses, its delights were not explored in most early Christian depictions of paradise. This may seem odd in the presence of 'pull' from texts such as

ἐτύρωσας δέ με ἴσα τυρῷ; σάρκα καὶ δέρμα με ἐνέδυσας, καὶ τὰ ἑξῆς. εἶτ' ἐπεὶ τετυρωμένος γεγονὼς ἀβλαβῆ φυλάττει τὴν θεότητα, καὶ μετὰ τὸ τετυρῶσθαι πίων ἐστίν, ὡς αὐτὴ ἡ λέξις παρίστησι λέγουσα Ὄρος τετυρωμένον, ὄρος πῖον. πρὸ γὰρ τοῦ τετυρῶσθαι καὶ μετὰ τοῦτο πῖόν ἐστιν.

[115] Vulgate: 'nonne sicut lac mulsisti me et sicut caseum me coagulasti. pelle et carnibus vestisti me et ossibus et nervis conpegisti me?'.

[116] Jerome translated Didymus's *De sancto spiritu* and commissioned Didymus to dictate commentaries on Hosea and Zechariah. He spent a month with him in Alexandria in 386. See J. N. D. Kelly, *Jerome: His Life, Writings, and Controversies,* London, 1975, pp. 85 and 124–6.

[117] My translation; Jerome, *Epistula* 66.10, ed. I. Hilberg, *Pars I, Epistulae I-LXX*, *CSEL* 54, Vienna and Leipzig 1910, pp. 660–61 [italics mine]: 'cumque inveneris eum et tenueris, ne dimittas. et si pauxillulum dormitanti elapsus fuerit e manibus, noli protinus desperare. egredere in plateas, adiura filias Hierusalem, repperies eum cubantem in meridie, lassum, ebrium, noctis rore madefactum, inter greges sodalium, in aromatum varietatibus, inter poma paradisi. ibi ei da mamillas tuas, sugat de erudito pectore et requiescat inter medios cleros, pinnae deargentatae columbae et interiora eius in fulgore auri. paruulus iste et puer, qui butyro et melle saginatur, *qui inter caseatos nutritus est montes*, cito crescit in iuvenem, velociter in te hostes spoliat, mature praedatur Damascum et regem vincit Assyrium.'

Psalm 33:9 ('Gustate et videte quoniam suavis est Dominus') and I Peter 2:3 ('si gustastis quoniam dulcis Dominus'). But there may also have been 'push', from the pagan philosophical tradition, from anxiety about an Original Sin that evolved from eating, and perhaps even from contemporary trauma about consumption of sacrificial food.[118] Instead there was synaesthetic recourse to the more refined sense, smell, in paradise. The magical taste of paradise was experienced, however, by living martyrs alone – in anticipation of their deaths.

In the second part of the paper I discussed one of the odder of these tastes of paradise, Perpetua's cheese. I then attempted to show by comparing Lucian that there were close-to-contemporary ancient folk-traditions of Cockaignian cheese-islands, bastard cousins of paradise.[119] Working backwards from Augustine's *mons incaseatus*, we find a Judaeo-Christian exegetical cheese mountain: the incarnate Christ. The Septuagint translators put this non-place on the map, when they eliminated the historical Bashan from Psalm 68:16, substituting a 'fat' mountain.[120] In the subsequent verse a problematic Hebrew word that looked a bit like 'cheese', resulted in a 'cheese mountain' that would rightly trouble exegetes to this day. It is unclear whether Perpetua's taste of the sweetness of paradise in the form of cheese could be attributed to contamination from paganism, or to her sense that she was ready for Pauline solid food, or to an exegetic connection between Christ and cheese first explicitly attested rather later in fourth-century exegesis.[121] Thanks to Didymus the Blind, the image was available for Augustine's Christian use in a highly poetic epitaphic context, where the mountain was the afterlife paradise. It is even possible (though impossible to prove) that the Cockaignian cheese-lands *enabled* the mutation in the Septuagint. But that is just a guess. An interesting chain of exegetic and epitaphic improvisation gives cheese a place in paradise – even if one cannot know how it tastes. Getting to paradise is all about connecting points.

[118] In which I locate the origins of the medieval tasting ordeals.

[119] Not, I think, any kindred of the mythic cosmogonical cheese discussed in C. Ginzburg, *The Cheese and the Worms: The Cosmos of a Sixteenth-Century Miller,* New York, 1989, p. 58.

[120] ὄρος πῖον, as Howard Jacobson suggests, could have been derived either from a text that read 'dashan' ('fertile') instead of 'Bashan', or it could be an interpretation of the place Bashan, which was known for its fertility. See L. Köhler and W. Baumgartner, *Lexicon in Veteris Testamenti libros,* Leiden, 1958, p. 158.

[121] The significance would then be clearly Eucharistic. There is of course a gap of well over a century to account for. One wonders, for example, what role Origen might have played.

‘The Heavens Declare the Glory of God’: Mapping Cosmos and Activating Heaven through Holy Icons

Veronica della Dora

1. Introduction

> *Let the heavens rejoice and the earth be exceedingly glad, for the Lord has shown the power of His reign: He has conquered death by death, and become the first born of the dead. He has delivered us from the depths of Hades, and has granted the world great mercy*
> (Resurrectional *Apolytikion,* Third Tone).

In the Orthodox Church, heaven descends on earth every time a divine liturgy is celebrated. As the priest prepares Holy Communion, so we are told by St John Chrysostom, the invisible presences of thousands of angels and archangels fill the space of the church.[1] Heaven is activated in a strong, ontological sense. It becomes ‘here’ and ‘now’. And the faithful taking part to the holy mystery can get a glimpse of it. ‘Through the divine Eucharist we are transported to the heavenly kingdom’ – we are transposed to what Orthodox theologian Georgios Metallinos called ‘meta-history’.[2]

Theologically, heaven is not a physical place. It is a disposition, a state of the soul attainable in this life and made eternal after death. It is uncreated light at the sight of which righteous souls will shine and sinful ones will burn. Geographically, heaven encompasses what the ancients called κόσμος (*kosmos*), an ordered creation comprising all things visible and invisible. Aristotle, after Plato, conceived of the cosmos as a closed spherical shape. At its centre was a stationary and corruptible earth surrounded by the mutable concentric spheres of water, air, and fire. These were in turn materially enclosed within seven concentric crystalline planetary spheres separated by a fifth (and this time incorruptible) element called ether. The sphere of the fixed stars contained all of these,

[1] *The Divine Liturgy of Saint John Chrysostom*, transl. Holy Cross Orthodox School of Theology, Brookline MA, 1985, pp. 70–71: Εὐχαριστοῦμεν σοί καὶ ὑπὲρ τῆς λειτουργίας ταύτης, ἥν ἐκ τῶν χειρῶν ἡμῶν δέξασθαι καταξίωσας, καίτοι σοί παρεστήκασι χιλιάδες Αρχαγγέλων καὶ μυριάδες Αγγέλων, τὰ Χερουβεὶμ καὶ τὰ Σεραφεὶμ, ἑξαπτέρυγα, πολυόμματα, μετάρσια, πτερωτά. ‘We also thank You for this liturgy, which You have pleased to accept from our hands, even though You are surrounded by thousands of Archangels and tens thousands of Angels, by the Cherubim and Seraphim, six winged, many-eyed, soaring with their wings’. See also the priest’s prayer before the Small Entrance, ibid., pp. 44–45: Δέσποτα Κύριε ὁ Θεὸς ἡμῶν, ὁ καταστήσας ἐν οὐρανοῖς τάγματα καὶ στρατιὰς ἀγγέλων καὶ ἀρχαγγέλων εἰς λειτουργίαν τῆς σῆς δόξης, ποίησον σὺν τῇ εἰσόδῳ ἡμῶν εἴσοδον ἁγίων ἀγγέλων γενέσθαι, συλλειτουργούντων ἡμῖν καὶ συνδοξολογούντων τὴν σὴν ἀγαθότητα. ‘Master and Lord our God, You have established in heaven the orders and hosts of angels and archangels to minister to Your glory. Grant that the holy angels may enter with us that together we may serve and glorify Your goodness.’

[2] G. Metallinos, ‘Paradise and Hell According to Orthodox Tradition.’ From: http://www.oodegr.co/oode/esxata/kol_par1.htm (retrieved on 1/November/2008).

with nothing beyond it except the *primum mobile*.[3] Philo of Alexandria, the first-century AD Hellenistic Jewish philosopher who contributed so much to the incorporation of Platonism into Christianity, declared God to be a 'hypercosmic star', but only in an allegorical sense, for God (and thus paradise) was not spatial.[4] The opposition between a finite and transitory cosmos and an eternal Logos beyond time and space is evocatively expressed in the book of Psalms: 'In the beginning, Lord, you founded the earth, and the heavens are the work of your hands; they will perish but you remain; they will all wear out like clothing; like a cloak you will roll them up, and like clothing they will be changed. But you are the same, and your years will never fail.'[5]

On the one hand, heaven in Orthodoxy can thus be thought through glimpses of divine grace, as an interior and more-than-representational spiritual state, as 'the way we individually experience the sight of Christ, depending on the condition of our heart'.[6] On the other, we can think of heaven in relation to the cosmos, a mathematical construct we need to visualize, to map, in order to 'grasp' with our mind's eye. On the one hand is the micro-scale of the fleshly human body enshrining an immortal soul, catching glimpses of paradise through liturgical mysteries and prayer; on the other is the macro-scale of a timeless realm encompassing perfect celestial spheres, in turn enclosing a corruptible earth and its fallen human inhabitants. Materiality clothes the invisible. The invisible clothes materiality. How do these two conceptions of heaven come together? How were they visualized and represented by the Byzantines? This essay explores these questions through a paradoxical object that lies at the heart of Byzantine (and Orthodox) spirituality: the icon.

Icons are 'paradoxical' because they problematize, or rather collapse, the binary distinctions through which Western thought has been traditionally articulated. They blur the boundaries between referent and symbol, same and otherness, temporal and eternal, immanent and transcendent. They do so not through philosophical effort, but by godly initiative, through the incarnation of the eternal Logos. For the faithful, icons are statements and embodiments of the truth, vehicles of Grace which make the intangible tangible, the invisible visible, divinity accessible to humans, as the incarnated Christ did.[7]

[3] See D. Cosgrove, *Apollo's Eye,* Baltimore MD, 2001, p. 36.

[4] Philo of Alexandria, *On the Creation of the Cosmos According to Moses*, 4.17, transl. D. T. Runia, Leiden, Boston and Cologne, 2001, p. 50. On the non-spatiality of God, see K. Hutchinson, 'The Cosmology of the Globe-Mandorla', unpublished paper. A distinction needs to be made between terrestrial paradise, the Garden of Eden, lost to mankind after Adam's fall and the Heavenly Jerusalem – a condition attained by righteous souls after death. On Byzantine writings on Eden, see H. Maguire, 'Paradise Withdrawn,' in *Byzantine Garden Culture,* eds A. Littlewood, H. Maguire and J. Wolschke-Bulmahn, Dumbarton Oaks DC, 2002, pp. 23–35 and Dimitris J. Kyrtatas's paper in this volume.

[5] Psalm 102:25–7; also quoted in Hebrews 1:10.

[6] Metallinos, 'Paradise and Hell' (n. 2 above): Παράδεισος και κόλαση δεν είναι απλώς ανταμοιβή και τιμωρία (καταδίκη), αλλά ο τρόπος με τον οποίο βιώνουμε καθένας μας τη θέα του Χριστού, ανάλογα με την κατάσταση της καρδιάς μας.

[7] A. Kartsonis, 'The Responding Icon,' in *Heaven on Earth: Art and the Church in Byzantium*, ed. L. Safran, University Park PA, 2002, p. 58; L. Ouspensky, *Theology of the Icon,* Crestwood NY, 1982, p. 35.

Icons are symbols in a strong sense. While for most people today the term 'symbol' tends to imply something essentially different from the thing it stands for, in the Patristic tradition, a symbol 'actually participates in some sense with the spiritual reality it symbolizes'.[8] It is not completely separate from its referent. 'It contains in itself the presence of what is symbolized'.[9] Through symbols, icons seek 'to incarnate the visible of the invisible, God beyond everything and through everything'.[10] If properly engaged through prayer, they make traces of immortality manifest in the life of mortals. They are royal doors to heaven.

My approach to icons and heaven is that of the geographer *and* of the faithful. I am interested in the way in which heaven is both mapped *on* icons and how it is activated *through* icons. In this essay I shall thus consider icons both as maps of the cosmos *and* as doors to heaven, which is, as images open to visual analysis and as more-than-representational objects configuring sacred space through their physical interaction with the faithful; as 'visual texts' and as sacred objects spatially embedded in the microcosm of the church, since, it has been observed, the only authentic approach to the icon is through prayer, and 'an icon divorced from a place and act of worship would be a contradiction in terms'.[11] My selection of examples is purposely disparate, encompassing medieval Byzantine and post-Byzantine icons on different media and from different centuries and places. While temporally and stylistically diverse these examples are nevertheless connected by similar iconographies. In particular, they are connected by shared topographic and cosmographic features including visible natural elements such as mountains and caves, as well as supernatural elements (manifestations of the divine Logos, usually represented as a series of concentric circles and mandorlas of light).

In the aftermath of iconoclasm these elements were systematized in consistent symbolic patterns and, albeit with stylistic variations, they continued to repeat themselves throughout the centuries.[12] Henry Maguire described Byzantine pictorial cycles and the churches they adorned as a 'set of variations on a theme'. Unlike Western art, Byzantine church decoration was 'subject to conventions of decorum beyond which artists rarely strayed'.[13] Rather than focusing on stylistic differences and change, I use the iconographic

[8] B. Foltz, 'Nature Godly and Beautiful: The Iconic Earth,' *Phenomenology*, I, 2001, p. 128. 'The world is one ... for the spiritual world in its totality is manifested in the totality of the perceptible world, mystically expressed in symbolic pictures for those who have eyes to see. And the perceptible world in its entirety is secretly fathomable by the spiritual world in its entirety, when it has been simplified and amalgamated by means of the spiritual realities. The former is embodied in the latter through the realities; the latter in the former through the symbols. The operation of the two is one' (St. Maximos the Confessor, *Mystagogia*, quoted in ibid.).

[9] Quoted in Foltz, 'Nature Godly and Beautiful' (n. 8 above), p. 133.

[10] My own translation; M. Zibawi, *L'Icône: sens et histoire,* Paris, 1993, p. 13: 'Tout au long de sa traversée, l'art sacré cherche à incarner le visible de l'invisible, "Dieu au-delà de tout, à travers tout et en tout"'.

[11] K. Ware, *Philip Sherrard: A Prophet of our Time*, Oxford, 2008, p. 17.

[12] L. Brubaker, *Vision and Meaning in Ninth-century Byzantium: Image as Exegesis in the Homilies of Gregory of Nazianzus*, Cambridge, 1999, pp. 33–4.

[13] H. Maguire, 'The Cycle of Images in the Church', in *Heaven on Earth*, ed. L. Safran (n. 7 above), p. 121.

continuity of these background elements as a reflection of deeper theological continuities in approaching the *kosmos* and mapping heaven in the Christian East.

The essay falls into three main parts. It opens with a few general words on icons' 'micro-geographies', which is, the symbolism and role of these 'standard' iconographic topographic and cosmographic features, and the sacred space constructed between the icon and the gazing faithful. The following section focuses on representations of mandorlas in icons of the Transfiguration, Ascension and Majestas Domini as cosmic maps and situates these representations within Byzantine cosmographic traditions that were, in turn, influenced by ancient Greek understandings of creation. The essay concludes with a discussion on the liturgical space of the medieval Byzantine church as both a map and a re-enactment of a transfigured cosmos.

2. Mapping Icons's Iconographies

The word 'icon' is the transcription of the ancient Greek εἰκών (*eikón*), which indicated a 'likeness, image or picture', but it could also refer to 'a living image, an image in a mirror, a phantom, or an image of the mind'.[14] In the Byzantine Middle Ages a 'holy' *eikón* designated an image used for Christian purposes, such as providing a focus for the veneration of Christ and saintly figures who offered models for a perfect Christian life, and as an instrument for participation in a transcendent reality.[15] These images could be produced on a variety of media, including wooden tablets, frescoes and mosaics (hence my use of the word 'icon' here applies to all these genres). Sacred space was reconfigured and enacted through performance: in between the viewer and the gazing icon; between the faithful venerating, 'bowing down before the icon' (προσκυνητής, *proskynētēs*) and the venerated icon.

According to John of Damascus, the eye did not merely 'see' holy images, but 'embraced' them and 'kissed' them. Because, as Patriarch Photius explained in the ninth century, the optical rays emanated by the eyes were believed to touch the object seen, vision was in fact deemed to be 'haptic, as well as optic, tactile as well as visual'.[16] Much of the charge was thus placed on the beholder, for it was him who bridged the ontological gap between image and prototype. 'Because God having been made flesh in Jesus Christ, humans are able to glimpse the very face of God in matter itself'.[17] Icons were thus not holy in themselves, but holy *by participation*. They demanded personal engagement in the act of perceiving. Viewers 'conversed with holy persons; [they] entered the holy places:

[14] A. Kartsonis, 'The Responding Icon' (n. 7 above), p. 58.

[15] Icons are thus holy by participation, rather than holy in themselves. See P. Evdokimov, *The Art of the Icon: A Theology of Beauty,* Redondo Beach CA, 1990, p. 178.

[16] R. Nelson, 'To Say and to See: Ekphrasis and Vision in Byzantium', in *Later Byzantine Painting: Art, Agency and Appreciation*, Burlington VT, 2007, p. 153.

[17] B. Lane, *The Solace of Fierce Landscapes: Exploring Desert and Mountain Spirituality*, New York and Oxford, 1998, p. 125. See also Kartsonis, 'The Responding Icon' (n. 7 above), p. 58; Ouspensky, *Theology of the Icon* (n. 7 above), p. 35.

Bethlehem, Jerusalem, Golgotha'.[18] In this sense, the icon was epiphanic – it was created precisely to involve, or rather, address the viewer; to 'invade' his or her own physical and emotional space.[19]

While we tend to focus our attention on the appearance of an image, Byzantines tended to focus on its prototype. Icons were not concerned with imitation of the material world, but with its transfiguration. What was important was not the faithfulness of the image to historical events and places or to holy figures, but the holy events and holy figures themselves – and their effect on the beholder.[20] In this sense the 'icon' endured as 'a form of art which both promoted and supported Christian faith and worship, and which communicated the ways in which believers may understand their world':[21] as a sacred object, and as a visual image.

Basil the Great conceived of the earth as a vast reservoir in which the Creator had left 'everywhere memorials of His wonders'.[22] On medieval Byzantine icons these 'reminders' of God's creative power feature as 'standard' background elements of narrative scenes from the Scriptures, especially from the life of Christ. These scenes usually come in cycles of icons located on the church's iconostasis, or in cycles of frescoes painted on the church's walls. As a sequence, they serve as powerful *loci memoriae* of historical happenings; as stations of the Life of Christ (the Annunciation, the Nativity, the Baptism in the Jordan, the Transfiguration, the Crucifixion, the Resurrection, and so on), they serve as icons of feasts embedded in the cyclical time of the liturgical calendar, and in the eternal sacred time of divine liturgy.

These images are not 'worlds by themselves' but, as we have seen, they are related to the beholder. They must therefore be 'visible, easy to recognize and to interpret', hence the 'standard' coding of their compositional elements.[23] Landscape features in these scenes are not meant to portray the actual aspect of physical places, but the spiritual reality behind them. Landscape is therefore not realistic, but symbolic; its elements contribute to the ultimate goal of the icon as a 'mnemonic prompt' and window on eternity.

In medieval Byzantine cycles earthly and heavenly natural elements typically give the scenes narrative continuity. For example, on icons of the Nativity, the infant Christ is usually portrayed in a cave, indicating the earth itself taking the glory of divinity through Incarnation (Fig. 1). Likewise, on icons of the Theophany, the banks of the river Jordan generally feature in a cave-like shape (Fig. 2) and are sometimes enclosed by symmetrical rocky pinnacles bending towards the centre. The baptismal water of the river topped by the Holy Spirit echoes the first creation's imaging of 'the Spirit moving

[18] O. Demus, *Byzantine Mosaic Decoration: Aspects of Monumental Art in Byzantium*, London, 1948, p. 6.

[19] Foltz, 'Nature Godly and Beautiful' (n. 8 above), p. 135.

[20] Brubaker, *Vision and Meaning* (n. 12 above), p. 22.

[21] R. Cormack, *Icons*, London, 2007, p. 8.

[22] Literally, 'the reminders of the benefactor [i.e. of God]' [τοῦ εὐεργέτου τὰ ὑπομνήματα], Basil of Casearea, *Homélies sur l'hexaéméron*, ed. S. Giet, Paris, 1968, p. 242.

[23] Demus, *Byzantine Mosaic Decoration* (n. 18 above), p. 6.

upon the face of the waters',[24] and thus signifies the restoration of paradise. In later scenes of the Crucifixion a small cave under the cross on Golgotha often shelters the skull of Adam whose bones are still part of the dust of the earth from which they were formed, yet whose fall is now being redeemed.[25] Finally, in the scene of the Resurrection, Christ is usually shown in a vast cave deep within the earth, triumphantly rescuing Adam and Eve from hell (Fig. 3).[26] 'From a cave, You shone forth the world, into a cave You sink from the world. Then a cave was found when there was no room at the inn. And now you are lodged in another cave where You lack a tomb', we read in a sermon composed by Germanos II, Patriarch of Constantinople in the thirteenth century.[27]

In later Byzantine cycles rock assumes a more prominent and dramatic presence, speaking to the beholder with an almost explosive violence (Figs 4 and 5). Yet, besides the earthly, corruptible, rocky element, the iconic cycle of the Life of Christ is also signposted by theophanies, by divine manifestations, by encounters between heaven and earth. In these scenes, the Father manifests Himself through the concentric heavenly spheres of the cosmos held together and ordered by the divine Logos. The word λόγος (*logos*) comes from the verb λέγειν (*legein*), literally 'to say', but also 'to order', 'to gather', 'to select' (from which we have inherited words as disparate as 'legion', an ordered army, and 'elegant', a person who keeps herself in order and tastefully selects).[28] The Logos is the ordering principle of the cosmos. Εν ἀρχῇ ἦν ὁ Λόγος καὶ ὁ Λόγος ἦν πρὸς τὸν Θεόν, καὶ Θεὸς ἦν ὁ Λόγος ...; 'In the beginning was the Word, and the Word was with God, and the Word was God ... Everything was made through the Logos and nothing that exists was made without it', we read at the opening of the Gospel of St John.[29] As stated at the beginning of the Old Testament, the cosmos was created through the separation and ordering of its elements operated by the Logos.[30] The Byzantine mosaics on the first dome in the narthex of Saint Mark's basilica in Venice (thirteenth century) illustrate this process (Fig. 6). Following the scenes in an anticlockwise sense, day is separated from night (at 6 o'clock), water from the heavens, the land from the seas, and so on.[31] On the fourth day, the cosmos is completed, and the heavens are beautified with the creation of lights (the stars). The completed starry cosmos is represented in the form of five concentric circles painted in different gradations of blue – the same (though enlarged) circles that surround the Spirit 'moving upon the face of the waters' portrayed in the first

[24] Genesis 1:2.
[25] Maguire, 'The Cycle of Images' (n. 13 above), p. 130.
[26] Foltz, 'Nature Godly and Beautiful' (n. 8 above), p. 138.
[27] Quoted in Maguire, 'The Cycle of Images' (n. 13 above), p. 122.
[28] See F. Farinelli, *L'invenzione della Terra*, Palermo, 2007, p. 24.
[29] John 1:1.
[30] Genesis 1:1.
[31] See W. Dorigo, 'I mosaici del nartece di San Marco nel paesaggio fra la cultura figurativa mediobizantina e la cultura musiva veneziana,' in *La Basilica di San Marco: arte e simbologia*, ed. B. Bertoli, Venice, 1999, pp. 47–71.

scene of creation. The cosmos thus appears as nothing but a reflection, or the visible manifestation, of the divine Logos.

As with the rocky element, the same concentric circles usually appear in most scenes of the life of Christ. The different gradations of tone are interpreted theologically as the experience of the uncreated light and the 'divine darkness' inside which God dwells – φῶς οἰκῶν ἀπρόσιτον.[32] The central dark blue circle represents this divine darkness, the totally unapproachable light, whereas the exterior light blue (sometimes white) circle symbolizes that which can be perceived by pure-hearted humans.[33] In the 'standard' Byzantine iconography of the scenes of the Annunciation and of the Nativity (Fig. 1) we often find partial views of these circles in the upper part of the icon, radiating a beam of light, signifying the descent of Logos and its incarnation. In the Theophany (Fig. 2) the mystery of Holy Trinity is mapped out through a vertical axis: Father – Holy Spirit – Son (the first commonly still symbolized by the concentric circles and the last as a dove sometimes enclosed within smaller circles). This axis is described in the *troparion* (or hymn) of the feast:

> When Thou, O Lord, wast baptized in the Jordan the worship of the Trinity was made manifest! For *the voice of the Father* bare witness to Thee, calling Thee His *Beloved Son*. And *the Spirit*, in the form of a dove, confirmed the truthfulness of His Word.[34]

In these initial scenes of the Life of Christ the concentric circles are partial and relegated to the upper part of the icon. The Logos, the voice of the Father, is heard from above, often from within a cloud, as it was heard by Moses on Mount Sinai.[35] It is only on the top of Mount Tabor that Christ fully reveals His divine nature to Peter, James, and John, that He literally opens heaven for them to see, an idea consistent with patristic tradition.[36] Here past, present, and eternity converge in a single moment and the cosmographic and topographic scales converge in a single scene. In late Byzantine icons, such as Theophanes the Greek's icon of the Transfiguration (Fig. 7), Mount Tabor is usually transfigured into the mountains of the Old Testament's revelations: Mount Sinai (topped by Moses) and Mount Horeb (by Elijah). In the upper part of the icon the two

[32] 'Dwelling in the light which no man can approach', I Timothy 6:6.

[33] Nunnery of St Stephen, *The Depiction of the Second Coming or the Last Judgement*, Holy Meteora, 2010, pp. 35–7.

[34] *Menaion*, January 6. My own translation. *Apostolike Diakonia Ekklesias Hellados, Menaia,* Athens, 1970, available on line: Εν Ιορδάνῃ βαπτιζομένου σου Κύριε, ἡ τῆς Τριάδος ἐφανερώθη προσκύνησις· τοῦ γὰρ Γεννήτορος ἡ φωνὴ προσεμαρτύρει σοι, ἀγαπητόν σε Υἱὸν ὀνομάζουσα· καὶ τὸ Πνεῦμα ἐν εἴδει περιστερᾶς, ἐβεβαίου τοῦ λόγου τὸ ἀσφαλές. Ο ἐπιφανεὶς Χριστὲ ὁ Θεός, καὶ τὸν κόσμον φωτίσας δόξα σοι.

[35] Exodus 19. We encounter similar iconographic motifs (of the Logos speaking from above represented as concentric semicircles, often enclosing the blessing hand of God) also on icons of saints, indicating the revelations they had.

[36] A. Andreopoulos, *Metamorphosis: The Transfiguration in Byzantine Theology and Iconography*, Crestwood NY, 2005, p. 240.

prophets are set against an abstract golden background. Their poses mirror each other as they engage a timeless dialogue with Christ. On the lower part of the icon time continues to flow and three apostles below (sometimes set against a flowering terrain, indicative of movement) are each in a different stage of awakening.[37]

Eastern Christian doctrine is at once apophatic and kataphatic: 'beyond' (ἀπό, *apó*) every 'speech' (φάσις, *phasis*) and, therefore, image one might use to speak of God; and yet at the same time operating 'according to' (κατὰ, *katà*) the image, acknowledging the metaphorical character of human thought – a tension encapsulated in the above-mentioned representations of the 'dark light'.[38] Patristic theology rests on this tension: absence and presence, image and darkness, language and silence. 'The knowledge of God', Gregory of Nyssa argued in the fourth century, 'is a mountain steep indeed and difficult to climb.' The one 'who is going to associate intimately with God must go beyond all that is visible and (lifting up his own mind, as to a mountaintop, to the invisible and the incomprehensible) believe that the divine is there where the understanding does not reach'.[39] Gregory employed the image of Moses approaching and ascending Mount Sinai to map out spiritual progress towards *theosis*.[40] The union with God was achieved in darkness, for 'when Moses grew in knowledge, he declared that he had seen God in the darkness, that is, that he had then come to know that what is divine is beyond all knowledge and comprehension'.[41]

For Gregory, Sinai thus symbolizes the aniconic power of apophatic tradition – the dark inner circle of the divine light. On Sinai God reveals Himself in absence of clarity, in a dense cloud of smoke – ἐν νεφέλῃ πυκνῇ (Exodus 19:9). Byzantines saw in the mountain of Moses a prefiguration of Tabor, where God revealed Himself in a sharpness of lucidity of image: 'the iconic, imaginative power of the kataphatic tradition' – the external white, approachable circle.[42] On Tabor the transfigured Christ appears to the disciples ἐν νεφέλῃ φωτεινῇ (Matthew 17:5), within a luminous cloud radiating divine Grace:

> The mystery of the incarnation is disclosed in Jesus of Nazareth, his clothes glistening and intensely white. The disciples know themselves to have encountered the living God in human flesh. What was not given to see and know to Moses, becomes manifest in the New Testament.[43]

[37] Maguire, 'The Cycle of Images' (n. 13 above), p. 143.

[38] See Lane, *The Solace* (n. 17 above), pp. 62–4.

[39] Gregory of Nyssa, *The Life of Moses,* transl. A. J. Malherbe and E. Ferguson, New York, Ramsey and Toronto, 1978, pp. 43 and 95.

[40] Lane, *The Solace* (n. 17 above), pp. 101–2.

[41] Gregory of Nyssa, *The Life of Moses* (n. 39 above), p. 164.

[42] Ibid.; Mount Tabor is the object of poetical comparisons on the part of the Psalmist (Psalm 88:13, Jeremiah 46:18).

[43] Lane, *The Solace* (n. 17 above).

In Byzantine iconography, the transfigured Lord and Creator of heaven and earth is embedded within a *mandorla* of light. He is the ἥλιος τῆς δικαιοσύνης, a sun bestowing the uncreated light of divine grace.[44]

This mountain prefiguration is celebrated through Byzantine liturgical performance. In the vespers of the Feast of the Transfiguration, Sinai and Tabor are continually juxtaposed. Readings from Exodus about the revelation to Moses and from Kings about the revelation to Elijah anticipate the Transfiguration of Christ, which finds its fulfilment in the Divine Liturgy.[45] Throughout these services Moses and Elijah are described not only as the greatest figures of the Old Testament dialoguing with the transfigured Christ and worshipping him, but as allegories (or embodiments) of the Old Testament itself: Moses stands for the Law (which he holds in his hands), Elijah for the Prophets, and Christ is the fulfilment of both the Law and the Prophets.[46] As in the Theophany, all the persons of the Trinity are revealed: the voice of the Father speaking to the disciples; the Holy Spirit appearing as the dazzling Uncreated Light; and the transfigured Son on the top of the mountain. Tabor, the 'high mountain', therefore becomes a point of encounter and revelation; it suddenly manifests itself a cosmic *axis mundi* joining heaven and earth:[47]

> Going up with the disciples into the mountain, Thou hast shone forth with the glory of the Father. Moses and Elijah stood at Thy side, for the Law and the prophets minister to Thee as God. And the Father, acknowledging Thy natural Sonship, called Thee Son. We praise Him in song together with Thee and the Spirit.[48]

3. *Mapping Heaven*

On early representations of the Transfiguration from the monastery of Saint Catherine we are offered glimpses of heaven through oval *mandorlas* of light, 'cuttings' through the golden backdrop, passages to eternity, as if the viewer were enabled to pierce into the upper chambers of creation for a moment. In late- and post-Byzantine representations, however, the Taborean mandorla was more often represented in a circular shape. In these mandorlas Andreas Andreopoulos envisaged Christian mandalas, or sacred maps of the cosmos 'based on the essential representation of the earth around its sacred axis, which in this case is none other than Christ'; maps of the cosmos the Lord disclosed for His disciples to see.[49]

[44] 'His face shone like the sun' (Matthew 17:2).

[45] Exodus 24:12–18, 33:11–23, 34:4, 34:6, 34:8; I Kings 19:3–9, 19:11–13, 19:15–16.

[46] Matthew 5:17.

[47] The original Hebrew word *TaBoR* itself means 'navel' (the *omphalos* of the ancient Greeks), the centre of the cosmos. The idea of a cosmic axis centred on an *omphalos* (marked by a tree or mountain reaching heaven, around which creation was arranged and which was perhaps the first part of creation) is a very ancient one and is discussed by K. Hutchinson, 'The Cosmology of the Globe-Mandorla' (n. 4 above). See also M. Eliade, 'Centre du monde, temple, maison', in *Le Symbolisme cosmique des monuments religieux*, Rome, 1957, pp. 57–82.

[48] My own translation, Kathisma, 4th tone, Matins of the Transfiguration, *Menaion*, 6 August. *Apostolike Diakonia Ekklesias Hellados, Menaia.* Athens, 1970, available on line: Ο ἀνελθὼν σὺν Μαθηταῖς ἐν τῷ ὄρει, καὶ ἐν τῇ δόξῃ τοῦ Πατρὸς ἀπαστράψας, σὺν Μωϋσῇ Ηλίας σοι παρίστανται, *Νόμος καὶ Προφῆται* γὰρ ὡς Θεῷ λειτουργούσιν, ᾧ καὶ τὴν Υἱότητα τὴν φυσικὴν ὁ Γεννήτωρ, ὁμολογὼν ἐκάλεσεν Υἱόν, ὅν ἀνυμνούμεν σὺν σοὶ καὶ τῷ Πνεύματι.

[49] Andreopoulos, *Metamorphosis* (n. 36 above), p. 240.

The cosmos of the Byzantines was predominantly the Aristotelian cosmos, characterized by its spherical concentric pattern.[50] In scientific geography such a cosmos was given visual form through the work of second-century Alexandrian geographer and astronomer Claudius Ptolemy.[51] In the *Almagest* Ptolemy explained how the Aristotelian cosmos – both corruptible elemental earth and the incorruptible, ethereal heaven – could be mathematically coordinated and mapped.[52] In southern Italy the Christian Cassiodorus (*c.* AD 487–583) suggested that young monks should learn geography and cosmography through Dionysius's map and Ptolemy's *Geography*; whereas in Alexandria, the greatest intellectual centre of the late Roman world, Johannes Philoponus (*c.* 490–570), another Christian, commented on Aristotle's works and taught, like him, that the earth was spherical and that it was located at the centre of a celestial sphere.[53] 'Cosmic mandorlas' reflect this concentric pattern, with Christ becoming the centre of the universe. Since the fourteenth century other shapes were often superimposed on the concentric circles of the Taborean mandorla, including stars reminiscent of the seal of Solomon (Fig. 7), which was itself understood as an amulet embedding all the elements of the universe and summarizing the unity of the cosmos (Fig. 8).[54] Andreopoulos linked the emergence of this new 'cosmic mandorla' to the triumph of Saint Gregory Palamas and the canonization of hesychasm in the 1340s. This was a monastic tradition of contemplative prayer that began roughly in the fifth century on Mount Sinai. Hesychastic contemplation was sometimes accompanied by visions of divine light, the same uncreated light the disciples saw on Tabor and theologians identified with heaven.[55]

[50] This influence is seen, for example, in the church of St George in Madaba, Jordan, whose 6th-century mosaic, known as the Madaba map, presents a cosmological view according to Ptolemy's principles (see ibid., p. 236).

[51] O. A. W. Dilke, 'Cartography in the Byzantine Empire', in *History of Cartography,* I, eds. J. B. Harley and D. Woodward, Chicago IL and London, 1987, p. 268; and on line at http://www.geography.wisc.edu/histcart/.

[52] D. Cosgrove, 'Images of Renaissance Cosmography, 1450–1650', in *The History of Cartography*, III, ed. D. Woodward, Chicago IL and London, 2007, p. 55; and on line at http://www.geography.wisc.edu/histcart/. In the 14th century Byzantine scholar Maximos Planoudes revived and collected many precious manuscripts for the library of the monastery at Chora, including Ptolemy's *Alamagest*, and designed a map to accompany his copy of Ptolemy's *Geography* which was admired by Emperor Palaiologos II, in whose court Saint Gregory Palamas was educated. See Andreopoulos, *Metamorphosis* (n. 36 above), p. 236.

[53] Dilke, 'Cartography in the Byzantine Empire' (n. 51 above), p. 261.

[54] M. Chelli, *Manuale dei simboli nell'arte: l'era paleocristiana e bizantina*, Rome, 2008, pp. 85–6.

[55] The word *hesychasm* itself derives from ἡσυχία (*hesychia*) which means 'stillness', or 'quiet'. See I Thessalonians 4:1; II Thessalonians 3:12; I Timothy 2:2; I Peter 3:4. On hesychasm, see, for example, Gregory Palamas, *Triads*, New York, 1983; J. Meyendorff, *Byzantine Hesychasm: Historical, Theological and Social Problems,* London, 1973. An example of experience of the uncreated light is recounted by Symeon the New Theologian (949–1022): 'One day, as he stood and recited, "Lord, have mercy upon me, a sinner", uttering it with his mind rather than with his mouth, suddenly a flood of divine radiance appeared from above and filled all the room. As this happened, the young man lost all awareness (of his surroundings) and forgot that he was in a house or that he was under a roof. He saw nothing but light all around him and did not know if he was standing on the ground. ... Instead, he was wholly in the presence of immaterial light and seemed himself to have turned into light. Obvious of all the world he was filled with tears and ineffable joy and gladness' (Symeon the New Theologian, *The Discourses*, quoted in S. Nes, *The Uncreated Light*, Eerdmans, 2007, p. 98).

In the fourteenth century hesychasm was practiced by the monks of Mount Athos, but it was attacked by Barlaam, a Greek monk from Calabria who represented the humanist views of the time. Palamas came to the defence of the monks of Athos, and the hesychastic doctrine was officially adopted by the Orthodox Church in the councils of AD 1341, 1347 and 1351.[56] By insisting that humans only relate truly to God as integrated entities of soul and body, the Hesychasts managed to emphasize that *theosis* encompasses the whole person. Whether a painted icon, or a saint's shining face, matter could therefore receive and communicate what was genuinely spiritual – a glimpse of heaven.[57] The depiction of the hesychastic mandorla, with its superimposed geometrical forms, is to be found for the first time in an illumination in the manuscript of John IV Kantakouzenos (1375), the emperor who openly supported Palamas and voluntarily abdicated to become a monk and dedicate the rest of his life to prayer and to the defence of hesychasm.[58]

The full concentric circles are not unique to icons of the Transfiguration. They also appear in various forms around the resurrected Christ in the scene of the Ascension, where He is elevated in His heavenly throne and exalted in all His Glory above His Mother and disciples, as well on icons of the Majestas Domini. In both scenes two ribbons are occasionally added to the concentric circles, making for more sophisticated 'mappings'. Unlike the geometrical shapes in the Taborean mandorlas, however, the appearance of these ribbons is not confined to a specific period, but spans from the early Byzantine to the post-Byzantine era. For example, in the seventh-century Sinai icon of Christ as the Ancient of Days, the Lord sits on one ribbon and leans His feet on the other ribbon, echoing Isaiah's verse ὁ οὐρανός μοι θρόνος, καὶ ἡ γῆ ὑποπόδιον τῶν ποδῶν μου [the heaven is my throne and the earth my footstool], a motive that re-emerges in Matthew's Gospel.[59] In the Marciana z540 Gospel manuscript (mid-twelfth century) the Majestas Domini is once again enthroned in heaven in the same fashion (Fig. 9). This time it is also framed by images of revelation (the prophets Elijah and Ezekiel proclaiming their visions of God, presented here as the youthful Christ Emmanuel, the 'God with us' of His incarnation and eschatological return) and of transmission (through the four Evangelists, portrayed through their fixed signs of the zodiac: for St Mark a lion, St Luke a bull, St Matthew a man, and for St John an eagle).[60] This combination of images invites the reader to 'see' the revelation of the Logos as it literally takes flesh in the parchment of the book.[61] The two ribbons re-appear on the mosaics of the central dome of the St Mark's Basilica in Venice (thirteenth century) (Fig. 10), as well as in a number of other icons of the Ascension and Majestas Domini dating from post-Byzantine times (see, for

[56] Andreopoulos, *Metamorphosis* (n. 36 above), pp. 214–15.
[57] Nes, *The Uncreated Light* (n. 55 above), p. 116.
[58] Andreopoulos, *Metamorphosis* (n. 36 above), p. 228.
[59] Isaiah 66:1; Matthew 5:35.
[60] A. T. Mann, *Sacred Architecture*, Shaftesbury, 1993, p. 148.
[61] A. W. Carr, entry 223 in *Byzantium 330–1453*, eds R. Cormack and M. Vassilaki, London, 2008, p. 436.

example, Fig. 11). On these icons Christ is revealed to the viewer as the Creator and the centre of the cosmos; as the Alpha and the Omega. Like the circle, His kingdom has no end.

4. The Church as Cosmos

'Cosmos' means both creation and 'ornament'. Cosmos therefore implies beauty and geometrical perfection. It also implies the harmonious integration of its parts. While such geometrical perfection is visually expressed in the above-discussed cosmic mandorla, integration finds its most evocative and sophisticated expression in the three-dimensional space of the Byzantine church. The Byzantine καθολικόν (*katholikón)* summarizes the harmonious totality of the cosmos, the beauty and the order of creation (as the word *kath-olikón* itself suggests). Otto Demus defined the medieval Byzantine church as a 'grand icon', an 'image of the cosmos symbolizing heaven, paradise (or the Holy Land) and the terrestrial world in an ordered hierarchy, descending from the sphere of the cupolas, which represent heaven, to the earthly zone of the lower parts'.[62] This pattern, fixed and popularized between the ninth and eleventh centuries, drew on the conceptualization of the cosmos formulated by Cosmas Indicopleustes, a sixth-century traveller and geographer from Alexandria who became a monk later in his life and deemed the spherical Aristotelian cosmos and any other inheritance from the pagan world inappropriate for a Christian audience; a truly Christian cosmography, he maintained, should only come from Scripture.[63]

According to Cosmas, on Sinai God revealed to Moses not only the archetypal pattern of the tabernacle of the Temple of Jerusalem, but also the whole structure of the universe. He made manifest to him 'an interlocking correspondence of spiritual and terrestrial geographies'.[64] In his *Topographia Christiana*, Cosmas thus envisaged the cosmos in the form of the tabernacle; as a large rectangular box with a vaulted top (Fig. 12).[65] As described in Exodus and in Leviticus, the tabernacle had two precincts: an outer court where the priests executed the sacrifices, and an inner Holy of Holies to which only the high priest had access. On Cosmas's representation the two spaces corresponded respectively to the terrestrial and celestial realms and, by extension, to the present and future

[62] Demus, *Byzantine Mosaic Decoration* (n. 18 above), pp. 13, 15.

[63] A. Scafi, *Mapping Paradise: A History of Heaven on Earth,* London and Chicago IL, 2006, pp. 160–62. While the authenticity of this figure has recently been much debated his work, the *Topographia Christiana*, is one of the very few Byzantine mappings that have survived. Some have recently attributed the *Cosmographia* to Constantine of Antioch; see H. Kessler, *Spiritual Seeing: Picturing God's Invisibility in Medieval Art,* Philadelphia PA, 2000. God says to Moses 'And let them make me a sanctuary; that I may dwell among them. According to all that I shew thee, after the pattern of the tabernacle, and the pattern of all the instruments thereof, even so shall ye make it' (Exodus 25:8–9). On the genealogy of the *Topographia*, see M. Kominko, 'The Map of Cosmas, the Albi Map, and the Tradition of Ancient Geography', *Mediterranean Historical Review*, 20, 2001, pp. 163–86.

[64] Lane, *The Solace* (n. 17 above), p. 142.

[65] Andreopoulos, *Metamorphosis* (n. 36 above), p. 236.

conditions.[66] The upper part, a vaulted chamber separated from the lower section by the screen of the firmament, was occupied by Christ Pantokrator and the Kingdom of Heaven, or the invisible world. The rectangular lower prism (or terrestrial section) contained the οἰκουμένη·(*ecumene,* or 'inhabited world'), entirely dominated by a huge mountain stretching to heaven – 'as high as the breadth of the land towards the northern and western regions'.[67] The eastern part of the lower prism was occupied by terrestrial paradise (the Garden of Eden lost to mankind after Adam's fall).[68] In a way Cosmas's map presented the unceasingly transfigured cosmos where 'God shall be all in all', as announced by the apostle Paul.[69]

This vertical division between heaven and earth was mapped on the three zones of what Demus called the 'ideal scheme' of the middle Byzantine church: the cupolas and high vaults, including the conch of the apse; the pendentives and upper parts of the vaults; and the lower parts of the vaults and of the walls. In this tripartite space the higher an image featured, the more sacred it was held to be. Thus, the first and uppermost zone, 'the celestial sphere of the microcosm of the church', was occupied only by representations of Christ, the Virgin and Angels, and of scenes taking place in heaven. In later churches the Christ Pantokrator and the hierarchies of celestial forces surrounding him (featuring in the main dome) were sometimes encircled by the sun, the moon and the signs of the zodiac, a circle of the animal forms of the powers of the monthly phases of the yearly cycle (which include the symbols of the four evangelists, as we have seen) (Fig. 13).[70] From the main dome usually hung the πολυέλαιον (*polyéleon*), a large circular bronze chandelier symbolizing the stars of the celestial vault. The second zone was dedicated to the life of Christ, therefore featuring scenes from the New Testament. The third and lowest zone contained a hierarchy of icons of the apostles, saints, martyrs, prophets and patriarchs.[71]

This vertical cosmic order permeated not only 'poetry seen', but also 'poetry pronounced' during church services.[72] 'Every creature which you have made offers you

[66] Kessler, *Spiritual Seeing* (n. 63 above), p. 91; see also W. Wolska, *La Topographie Chrétienne de Cosmas Indicopleustés: théologie et science au VI siècle,* Paris, 1962.

[67] *Topographia Christiana*, 2.31; text available at: http://users.uoa.gr/~nektar/history/tributes/byzantine_historians/cosmas_indicopleustes_topographia_christiana.htm: Ὅσον πλάτος ἔχει ἡ γῆ ἀνεπαισθήτως χθαμαλῶς κειμένη, τοσοῦτον εὑρίσκεται ἔχουσα ὕψος εἰς τὰ βόρεια καὶ δυτικά.

[68] Scafi, *Mapping Paradise* (n. 63 above), pp. 160–62: 'In his Christian topography Cosmas intended to prove that any inheritance from the pagan ancients was inappropriate for a Christian audience and that a true Christian teaching about the shape of the universe should come only from Scripture. He thus explained that the earth could not be spherical, as those who only pretended to be Christians claimed, but was structured like a tabernacle. The rectangular inhabited land was surrounded by the ocean, and the terrestrial paradise lay to the east, beyond the ocean, on "a land of the beyond". The "land of the beyond" was another earth surrounding the post-diluvian world, and separated from it by the impassable ocean. Its extremities were attached to the extremities of heaven, and prior to the Flood, mankind had been living here after the expulsion from paradise.'

[69] I Corinthians 15:28.

[70] Mann, *Sacred Architecture* (n. 60 above), p. 147.

[71] Demus, *Byzantine Mosaic Decoration* (n. 18 above), pp. 15–26.

[72] B. Pentcheva, *The Sensual Icon: Space, Ritual and Senses in Byzantium,* University Park PA, 2010, p. 4.

thanks,' a hymn composed by Patriarch Anatolius (fifth century) and chanted in the Vespers of Nativity goes,

> The Angels offer you a song
> The heavens offer you their star
> The wise men offer their gifts
> The shepherds offer their wonder
> The earth offers a cave
> The desert offers a manger
> And we offer you a virgin mother.[73]

The hierarchy of heavenly, human, and earthly elements found their harmonious fulfilment in the birth of Christ. Through His incarnation Christ made heaven manifest on earth: he re-mapped heaven at the centre of the cosmos.

The space of the Byzantine church was structured not only through a vertical axis, but also through a horizontal one, producing a parallel longitudinal hierarchy. Both the Byzantine liturgical texts and the mystagogical tradition relate altar to nave as heaven to earth. A later commentator, Symeon of Thessalonika (1375–1429), however, offered a more complex symbolic interpretation. For example, he associated the three major divisions of the space of the church (narthex, nave and sanctuary) with the tripartite division of the cosmos ('earth, heavens, places beyond heavens') and with the three regions of the 'visible world' (earth, terrestrial paradise, visible heaven). He also associated the same tri-partition to the different categories of people occupying each of these spaces of the church: the clergy, the faithful and 'those in repentance'. 'Because the three distinct spaces of the narthex, nave and sanctuary [were] contained within a single architectural unity, the church pre-eminently signif[ied] the multiplicity within unity of the Holy Trinity'.[74]

These three spaces were separated by vertical surfaces, including the icon screen dividing lay people from the space of the altar (restricted to clergy). The iconostasis symbolized the division between earth and heaven, yet at the same time it constituted 'the paradoxical place where those two worlds communicate'.[75] The flat surfaces of the holy icons embedded in the screen indeed acted as thresholds to heaven, as channels of communication between the faithful and God and His saints. 'As symbolic thresholds that so conspicuously marked the boundary between the sensible and the "intelligible"', Nicholas Constas comments, 'the icon screen effectively realized the unique Byzantine understanding of the Incarnation

[73] My own translation; *Sticheron idiomelon*, 6th tone, Christmas vespers, *Menaion*, 25 December:
Τὶ σοι προσενέγκωμεν Χριστέ, ὅτι ὤφθης ἐπὶ γῆς ὡς ἄνθρωπος δι' ἡμᾶς; ἕκαστον γὰρ τῶν ὑπό σου γενομένων κτισμάτων, τὴν εὐχαριστίαν σοί προσάγει, οἱ Αγγελοι τὸν ὕμνον, οἱ οὐρανοὶ τὸν Αστέρα, οἱ Μάγοι τὰ δῶρα, οἱ Ποιμένες τὸ θαῦμα, ἡ γῆ τὸ σπήλαιον, ἡ ἔρημος τὴν φάτνην, ἡμεῖς δὲ Μητέρα Παρθένον, ὁ πρὸ αἰώνων Θεὸς ἐλέησον ἡμᾶς.

[74] N. Constas, 'Symeon of Thessalonica and the theology of the icon screen', in *Thresholds of the Sacred*, ed. S. Gerstel, Dumbarton Oaks DC, 2006, p. 166.

[75] Eliade quoted in S. Gerstel, 'Introduction', in *Thresholds of the Sacred* (ibid.), p. 3.

as a paradoxical dialectic of revelation and concealment'.[76] In other words, the iconostasis was the bridge that united altar and nave, even while separating them:

> Dominated by the icon of Christ himself, who is joined by the *Theotokos*, John the Baptist, and the saints, the iconostasis situates the boundary between heaven and earth in the event of the Incarnation. What the iconostasis actually and dramatically signifies is that the union of heaven and earth is not merely the absorption of one by the other, nor a *tertium quid*, but a 'Passover' in Christ. The faithful face the iconostasis not as if it were a wall that blocks them from the altar, but as a sign that their Passover into heaven is accomplished in and through Jesus Christ.[77]

The vertical surface of the wall dividing the narthex from the nave separated penitents who had committed great sins and catechumens from other faithful. Since the ninth century this surface was often decorated with representations of the Last Judgement, dramatic maps of the afterlife featuring topographies of salvation and damnation possibly inspired by a new genre of edificatory accounts of symbolic journeys to afterlife meant to be visualized with the mind's eye.[78] The emergence of these pictorial and textual representations can be linked to the post-iconoclastic reaffirmation of the necessity of symbolic images to attain spiritual contemplation. Rather than maps of physical places, representations of the Second Coming are thus to be regarded as maps of states of the soul; maps which would have in turn impacted the souls of their viewers by prompting a fear of eternal damnation (and would have thus proved instrumental for their repentance and conversion).

Finally, a third vertical surface, the church walls, separated the microcosm of the church from the outer macrocosm. Yet, as with the iconostasis, it acted as a liminal threshold, rather than as an impenetrable barrier. The liturgical cycle occurring in the interior of the church was deeply connected to the cycles of the outer macrocosm. The Byzantine church faced eastwards, towards the rising sun and the rising God, and changes in natural light signposted daily church services (Fig. 14).[79] As a microcosm the church

[76] Constas, 'Symeon of Thessalonica' (n. 74 above), p. 163.

[77] K. Anatolios, 'Heaven and Earth in Byzantine Liturgy', *Antiphon*, 5(3), 2000, p. 3; ibid., p. 5: 'At the end of the Little Entrance with the Gospel book, the celebrant prays that "with our entrance there may be an entrance of holy angels serving with us and glorifying thy goodness," and the congregation chants the "Trisagion" hymn, in evocation of the angelic acclamation of "holy, holy, holy." The medieval Byzantine theologian Nicholas Cabasilas comments that it is shown thereby "that angels and men form one Church, a single choir, because of the coming of Christ who was both of heaven and earth"'.

[78] See S. Gerstel, 'The Sins of the Farmer: Illustrating Village Life (and Death) in Medieval Byzantium', in *Word, Image, Number: Communication in the Middle Ages*, eds J. Contreni and S. Casciani, Florence, 2002, pp. 209–17. See for example, the account of mystic and theologian Niketas Stethatos (1000–1090).

[79] The orientation is to the east because scripture says that Christ's Second Coming will be from east, and the central feature of the sanctuary, the altar, was viewed as the throne prepared for him, and like a throne, it was covered by a ciborium or canopy. The 8th-century Patriarch Germanos gives the following explanation of the symbolism of the various parts of the church: 'The conch [of the apse] is after the manner of the cave of Bethlehem, where Christ was born, and that of the Cave where He was buried. ... The holy table [the altar] is the place where Christ was buried, and on which is set forth the true bread from heaven, the mystic and bloodless sacrifice, i.e. Christ. ... It is also the throne upon which God, who is borne up by the cherubim, has rested. At this table, too, He sat down at His last supper' (R. Ousterhout, 'The Holy Space: Architecture and the Liturgy', in *Heaven on Earth* [n. 7 above], pp. 84, 99).

encompassed all the Aristotelian elements: the earth upon which it rested, the water of baptism, the fire of candles, the air and light filtering through the door and windows. As Robert Ousterhout commented, the Byzantine church is insistently three-dimensional. It does not just house events, but it 'amplifies, sanctifies, comments upon, and interacts with the functions it houses, and both form and function can be empowered by the interaction'.[80]

Byzantine liturgical performance exalted all the senses: through the modulation of sound and the smell of incense, through the taste of bread and blessed grapes, through the physical interaction with icons and the scripted use of candles.[81] Theophanic experience was mediated directly through the senses and Divine Grace experienced through participation, rather than the mental elaboration of abstract concepts, or general propositions; it did not speak of things, but *from* things.[82] To our days, as Psalms 135 and 136 are chanted in the Orthodox Church at the matins or all-night vigils of high-ranking feast days, all of the candles of the *polyeleos* are lit and the crown of the chandelier is pushed with a rod so that it circles back and forth. During the singing the space of the church is thus inscribed with joyful orbits and the cycle of creation. The microcosm of the church and the macrocosm of creation overlap in this circular dance (Fig. 15):

> O give thanks unto the Lord; for he is good: for his mercy endureth for ever.
> ...
> To him that by wisdom made the heavens: for his mercy endureth for ever.
> To him that stretched out the earth above the waters: for his mercy endureth for ever.
> To him that made great lights: for his mercy endureth for ever.
> The sun to rule by day: for his mercy endureth for ever.
> The moon and stars to rule by night: for his mercy endureth for ever ...[83]

* * *

Denis Cosgrove showed how, in the Renaissance West, cosmography, the science which describes the general features of the visible universe (both the heavens and the earth), implied detachment, distancing. It implied a third orb: that of the cosmographer's rational, disembodied eye (as represented in Appian's *Cosmographicus Liber*, 1524, for example).[84] On Western representations of the Transfiguration the 'heavenly spheres' do not encircle the Transfigured Christ, but remain above, in the upper part of the icon, for, as Andreopoulos observed, 'the West stressed the historical descent of Christ'.[85] Byzantines, by contrast, stressed the experiential ascent through

[80] Ibid., p. 81.

[81] On the way liturgical performance and changing atmospheric conditions brought icons to life, affecting the way the scenes appeared to the beholder, see Pentcheva, *The Sensual Icon* (n. 72 above).

[82] Ware, *Philip Sherrard* (n. 11 above), p. 19.

[83] Psalm 136:1, 5–9.

[84] Cosgrove, 'Images of Renaissance Cosmography' (n. 52 above), pp. 56–7.

[85] Andreopoulos, *Metamorphosis* (n. 36 above), p. 208.

asceticism and the unity of the incarnated Logos with the cosmos – God in everything and through everything.

In the liturgical space of the Byzantine church distance was similarly annihilated. In the seventh century Maximus the Confessor wrote that God

> encompassed the extremes of the whole creation through the means, as His own parts, and He joined them around Himself, each with the other, tightly and indissolubly: paradise and inhabited world, heaven and earth, the sensible and the intelligible ... and He recapitulated in Himself, in a manner appropriate to God, all things, showing that the whole creation is one.[86]

If the liturgy was celebrated by a mixed choir of humans and angels that was because Christ, 'by passing with his soul and body, that is, with the whole of our nature, through all the divine and intelligible ranks of heaven [has] united the sensible and the intelligible and showed the convergence of the whole creation'.[87] Through Byzantine liturgical performance the topographic and cosmographic scales thus collapsed into one another as they did when Christ was transfigured on Tabor; then, Saint Gregory the Sinaite (*c.* 1265–1346) comments,

> Even the mountains rejoiced and were glad – as is said, 'Tabor and Hermon rejoice in Thy name' – and heaven exulted and all the earth leapt with joy, seeing on the mountain their own master shining more than the sun both sensibly and intelligibly, and illuminating and sanctifying all things. ... The powers, looking down from above, shuddered; and the earth, from below, trembled with joy and fear. Thus all creation, all nature, 'everything that hath breath' praised, glorified and magnified Him, seeing its master and king, who had shown through previously as founder, as creator, now transfigured and transformed on the mountain and resplendent beyond the sun as God.[88]

In this essay I have tried to show how Byzantine representations of heaven and creation cannot be separated from their referents. As Saint John of Damascus wrote in the eighth century in his defence against iconoclasm, 'it is impossible for us to think without using physical images ... through sight we attain spiritual contemplation. Thus Christ took both soul and body, for man is made of both.'[89] Holy icons portray natural objects and theophanies; they make invisibilities visible through symbols (some of which can be connected to pre-Christian cosmographies). Icons link the prototype, its image and its beholder, enabling access and participation to the realm of the divine.[90] As images, they

[86] Maximus the Confessor, *Ambiguum* 41, 1312A, in *Ambigua*, II, transl. N. Constas, Cambridge MA, 2014, p. 115. Καὶ σὺν ἡμῖν καὶ δι' ἡμᾶς τὴν ἅπασαν κτίσιν διὰ τῶν μέσων, ὡς μερῶν ἰδίων, τὰ ἄκρα περιλαβὼν καὶ περὶ ἑαυτὸν ἀλύτως ἀλλήλοις διασφίξας παράδεισον καὶ οἰκουμένην, οὐρανὸν καὶ γῆν, αἰσθητὰ καὶ νοητά, ὡς σῶμα καὶ αἴσθησιν καὶ ψυχὴν καθ' ἡμᾶς ἔχων καὶ νοῦν, οἷς ὡς μέρεσι καθ' ἑκάστω καθόλου συγγενὲς οἰκειωσάμενος ἄκρον κατὰ τὸν προαποδοθέντα τρόπον θεοπρεπῶς τὰ πάντα εἰς ἑαυτὸν ἀνεκεφαλαιώσατο.

[87] Anatolios, 'Heaven and Earth' (n. 77 above), p. 4.

[88] Gregory the Sinaite, *Discourse on the Transfiguration*, transl. D. Balfour, Athens, 1982, p. 49.

[89] Quoted in J. Herrin, *Byzantium: The Surprising Life of a Medieval Empire*, London, 2007, p. 143.

help visualize, and thus map the unmappable; at the same time, however, they also help enact and achieve what cannot be represented – here and now. In other words, they depict the cosmos and, if properly engaged, offer a glimpse of heaven.

This dual function is magnified in the liturgical space of the church. The eighth-century Patriarch Germanos defined the church as 'a heaven on earth wherein the heavenly God "dwells and walks". It typifies the Crucifixion, the Burial and the Resurrection of Christ. It is glorified above Moses's tabernacle of testimony. ... It was prefigured by the Patriarchs, foretold by the Prophets, founded by the Apostles, and adorned by the Hierarchs.'[91] The church's symbolism actively participates in the ontological re-enactment of heaven on earth, the transfiguration of the cosmos through liturgical performance. Paradise is not a physical place; it is a condition. But this condition is achieved through our terrestrial (and thus spatialized) life, for a pure heart, a heart shining in the Taborean light of paradise, is a heart that is also in harmony with the totality of creation.

An Elder was once asked, 'What is a compassionate heart?' He replied: 'It is heart on fire for the whole creation, for humanity, for the birds, for the animals, for the demons, and for all that exists. At the recollection and at the sight of them such a person's eyes overflow with tears owing to the vehemence of the compassion which grips his heart; as a result of his deep mercy his heart shrinks and cannot bear to hear or look on any injury or the slightest suffering of anything in creation.[92]

[90] A. Kartsonis, 'The Responding Icon' (n. 7 above), pp. 58, 71.

[91] Patriarch Germanos I, *Historia Mystagogica*, quoted in C. Mango, *Art of the Byzantine Empire (312–1453)*, Englewood Cliffs NJ, 1972, pp. 141–3.

[92] St Isaac of Syria, quoted in Foltz, 'Nature Godly and Beautiful' (n. 8 above), p. 114.

Paradise in Western Medieval Tradition

Rudolf Simek

As the topic of 'Paradise in Western Medieval Tradition' covers a vast area itself, I shall refrain from tracing the tradition back to the Judaic tradition and the two creation stories of the Old Testament (Genesis 1:26–31; 2.8), but rather concentrate on what became of these accounts in the Western Middle Ages, in the widest sense of the word, a development based on patristic interpretation of the biblical sources.

I shall also refrain, for pragmatic reasons, from combining the discussion of paradise with the motif of the Heavenly Jerusalem, as far as this is possible. But even if we limit our discussion to the notion of paradise itself, we come across (at least) three distinct textual and iconographical traditions of the Middle Ages, which are linked to notions of the respective functions of paradise, namely:

1stly: paradise as the place of man's creation and fall, always mentioned as such in the abundant corpus of medieval commentaries on the Hexaemeron.
2ndly: paradise as the place of future promise, in which sense paradise gains both a physical location in the cosmography of this world and a spiritual location in the other world, called in medieval writings *paradisus terrestris* and *paradisus coelestis* respectively.
3rdly: paradise as the final abode of the blessed after the Last Judgement; in this meaning, paradise is the eschatological aim of human salvation history.

Medieval terminology is even richer than this and by no means limited to this threefold categorization. There is, however, always a danger of becoming inexact by either identifying the place of future promise, in the spiritual sense, with the Heavenly Jerusalem, or the eschatological place with the *coelum spiritualis* (or *coelum intellectualis*) – which is actually the abode of God, but becomes merged with the final abode of the blessed achieving their goal, a place which Augustine calls *paradisus paradisorum*.[1]

Single writers and thinkers were of course much clearer in their distinctions, but if we look at texts outside theology and philosophy proper, confusion prevails, even without entering the world of vernacular literature, where the terminology becomes even less exact.

It must be noted, however, that from a strictly theological point of view, there is of course only one paradise, but in practice this serves the three functions mentioned above and therefore tends to fill, in the mentality of medieval authors, at least these three intentions.

[1] See R. R. Grimm, *Paradisus coelestis, paradisus terrestris: Zur Auslegungsgeschichte des Paradieses im Abendland bis um 1200*, Munich, 1977, pp. 67–71.

The Cosmography of Paradise: The Other World from Ancient Mesopotamia to Medieval Europe, Warburg Institute Colloquia 27, 2016

Despite this functional division reflected in literature and iconography, the theological and biblical unity of paradise is reflected by the fact that all descriptions of paradise go back to the same handful of patristic and early medieval sources, a theological corpus that is distinct from the actual biblical sources. These fundamental Christian texts are: the works of Augustine (354–430), Isidore of Seville (*c.* 570–636), John of Damascus (*c.* 650–750), Bede (672–735) and the twelfth-century compilers of the *Glossa ordinaria*.[2] All there is to be known about paradise can be found in these texts, and through them and from them the descriptions are taken throughout the Middle Ages and then modified according to the function assigned to paradise in the texts and pictures in question.

The three functions ennumerated above have of course more than one aspect to each of them when it comes to their utilization in literature and art, depending not only on the intentions of the author, but also on the necessities imposed by genre.

a) Thus, the first and possibly most important aspects of the protological and cosmological function of paradise as the Garden of Eden is the historical aspect, by giving an actual place to the history of mankind (and thus of salvation history). This is most obvious in the historical literature of chronicles of the history of the world, but also for those innumerous compilations combining biblical and chronological history (in German, *Historienbibeln*). Both genres share the interest in the creation of man, and here, of course, the description of paradise usually closely follows the account given in the second chapter of Genesis.

But quite apart from these historical works, there is another wide-ranging genre of texts that takes a particular interest in the protological paradise, the protoplasts and their immediate fate, namely the literature concentrating on Adam, Seth and the invention of the wood for the cross, that is the Legends of the Holy Rood. Here, special attention is given to the trees of paradise as well as the location of Adam's grave, from which that tree will spring that finally will produce the wood for the Holy Rood. As journeys (particularly Seth's) back and forth to and from paradise are an important element of the story, albeit at a time close to the protoplasts, these are the earliest 'travellers' tales' to paradise. However, this legendary genre strictly belongs to hagiography and thus these journeys have little to do in practice with later attempts to reach paradise in the travel literature of the later Middle Ages.

[2] Augustine, *De Genesi ad litteram*, VIII.1–7, ed. J. Zycha, *CSEL* 28/1, Prague-Vienna-Leipzig, 1894, pp. 229–42 and *De gratia Christi et de peccato originali*, II.23.27, ed. C. F. Urba and J. Zycha, *CSEL* 42, Prague-Vienna-Leipzig, 1902, pp. 185–6; Isidore of Seville, *Etymologiae*, XIV.3, ed. W. M. Lindsay, 2 vols, Oxford, 1911; John of Damascus, *De fide orthodoxa*, II.11, *PG* XCIV, cols 910D–914A; Bede, *In Genesim*, II.8, ed. Ch. W. Jones, *CCSL* 118 A, Turnhout, 1967, pp. 45–6; *Glossa ordinaria*: *ad Genesim* II.8, *PL* CXIII, col. 86C. These references have been conveniently collected by Heinrich Köster, 'Urstand, Fall und Erbsünde in der Scholastik', in *Handbuch der Dogmengeschichte*, eds M. Schmaus, J. Geiselmann, et al, II/3b, Fribourg, Basel, Vienna, 1979, pp. 48–50.

A second aspect of the protological function is an aestethic one: as the descriptions in Genesis already describe an ideal garden with plants and animals, this is naturally the model for an ideal medieval garden. The walls around it, which are a major feature of paradise since its pre-Judaic Persian beginnings, serve the function of keeping humans out (rather than the original one of keeping animals in) since the Fall, became canonical in the Middle Ages and also became the norm in any ideal medieval garden, whether a real pleasure garden attached to feudal manors, the cloister garden of monastic establishments[3] or simply the ideal literary garden representing the *locus amoenus*[4] in medieval works of fiction. Medieval depictions both of gardens (and usually of any *locus amoenus*) therefore owe many features to the descriptions of paradise: the wall, the trees, the well, grass and flowers.

A third aspect, but one that is limited to the visual arts, is the erotic one: here in Eden, of all places, is the chance of depicting the naked human form. While attitudes to chastity in the depictions of the two protoplasts may vary, it is clear that in some cases the illustrators (and their authors?) took a delight in depicting beautiful humans. That the erotic connotations were not lost on them is shown by illustrations in a Middle High German metrical life of Adam and Eve (actually called *Eva und Adam*) where two very pretty and courtly protoplasts are repeatedly shown naked, even after the Fall, and in one scene Adam lovingly cups Eve's shapely breast with his hand.[5]

b) The second function is of paradise as the place of future promise, but as yet (normally) inaccessible to man, and therefore only to be experienced in visions or visionary journeys.

The two main aspects of this function are the exhortatory one, as shown by the contrasts between purgatory, hell and paradise (or even heaven itself) of the visions, and the geographical one. About both I shall talk in more detail below (section III).

c) The third function is the eschatological. This has, of course, foremost a moral aspect, insofar as to achieve access to it man has to follow the steep narrow path of virtue and God's law. But it also has a soteriological aspect: because of Christ's redemption of man, paradise is actually open to man, which is indicated by the fact that medieval iconography frequently shows Christ the Risen, advancing in front of the crowds of the blessed, to the enclosure of paradise which Saint Peter is about to unlock for them: this rather stereotype depiction is the one presented by scenes of the Last Judgement frequently on late Romanesque tympana as well as paintings on church walls and thus must have been ever present in medieval man's mind.

[3] However, a perfect monastic establishment, as represented by the setout on the famous St Gall monastery plan, shows the place of a garden not in the sheltered surroundings of the cloisters, but rather outside, possibly because of the actual location of a herb- and fruitgarden in practice.

[4] Cf. D. Thos, *Studien zum locus amoenus im Mittelalter*, Vienna, 1972.

[5] Vienna, Österreichische National Bibliothek, MS 2980, fol. 43r; cf. M.-B. Halford, *Illustrations and Text in Lutwin's Eva und Adam*, Göppingen, 1980, Table 14.

Possibly directly connected is the consolatory aspect of it: paradise, in these and other representations, is shown with Abraham, Isaac and Jacob awaiting the souls of the saved, and the artists even make a point of Abraham cradling one or more of them in his lap.[6] This possible return to 'the bosom of Abraham' must be seen as a major consolatory factor of eschatology, judging by the frequency of its depiction, with an even higher number of incidents in the iconography of the Eastern Church.

To return to the earthly paradise which is physically existant, but only theoretically accessible by humans, its cosmography and geography may be established through two genres of medieval literature, with both resulting in a rich iconographic genre of book illustrations. These are, firstly, journeys to the other world (in visionary literature) which mention, apart from rather detailed descriptions of the torments of purgatory and glimpses of hell, also paradise – unfortunately in far less graphic detail than the netherworld. Secondly, it is the medieval world maps which tend to show the location of paradise and do not need to claim personal acquaintance for that as they have biblical authority for its depiction. When and if medieval travellers' tales also mention paradise (such as Mandeville's travels or the Old Norse saga I shall mention below), they fuel their description with both these genres.

There is no need, here, to enumerate the works belonging to the genre of medieval visionary literature and its journeys to the other world. I just want to stress the fact that, despite this being a 'typically Western medieval' genre, its origins go back a long way, normally traced to the *Visio Pauli* from the third century. Its main texts, however, and those that have most impact on literature, ecclesiastical art and thus on mentality, date to the twelfth century and all show a connection with a physical place in Northern Ireland.

This is the so-called *Purgatorium Sancti Patricii* located on Station Island in Lough Derg (Co. Donegal), which is already mentioned in the *Topographia Hibernica* of the Welshman Giraldus Cambrensis in 1186.[7] Here, after due religious preparation, pilgrims could enter a cave where they experienced, during a trance-like sleep, an otherworldy journey taking them to the various places of punishment and torment, but also allowed them glimpses of paradise.[8] The literary results of several of these 'journeys' circulated all over Europe and were translated into several languages, not only for the admonitory character – several of those pilgrims experiencing the vision, like one Knight George from Hungary, had major conversion experiences – but also for the sheer descriptions of the other world contained in them.

[6] See J. Baschet, *Le Sein du Père: Abraham et la paternité dans l'Occident médiévale*, Paris, 2000.

[7] Giraldus Cambrensis, *Topographia Hibernica*, II.5, in *Opera*, V, ed. J. F. Dimock, London, 1867, pp. 82–3.

[8] Marie de France, *L'Espurgatoire seint Patriz*, ed. Y. de Pontfarcy, Louvain, 1995; *St. Patrick's Purgatory*, ed. R. Easting, Oxford, 1991.

Despite a wealth of such visionary tales, it was mainly the *Visio Tnugdali* (Latin 1148) that was disseminated and translated all over Europe and is preserved in a number of illustrated (vernacular) manuscripts.[9] The most fascinating aspect of these descriptions and depictions is not the rather archetypical description of the paradisical landscape, but rather the fact that this paradise is empty – apart from occasional angels volunteering information to the other world traveller. The reason for this emptyness is of course the fact that humans, after the Fall, have lost their right to enter paradise before their salvation. But another, more intricate problem seems to be at the bottom of the crass dichotomy between the utterly deserted paradise of visionary literature and the overcrowded paradise of eschatological iconography, where the *turba beatorum* appears as a densely packed mass of the blessed crowding into paradise. This problem concerns the two basically incompatible concepts of the judgement. In medieval Christian thinking, and in Catholic thinking right up to the present, there are two concepts of the judgement of the souls after death: that of Particular Judgement (of every individual immediately after death) versus Universal Judgement (of all the dead on doomsday).[10] There are, of course, logical as well as theological problems involved, but in medieval theology it seems that both types of judgement were thought to occur consecutively. This means that every soul is judged according to its merits right after death (and should therefore go either to hell, purgatory or heaven/paradise), and then is judged again on doomsday. According to medieval theology, the individual soul would either, even if judged to be deserving in its own personal judgement, have to wait until doomsday to reach eternal bliss and paradise (this position was held by Pope John XXII in 1331/32 according to his sermons[11]) or else could reach eternal bliss directly after death, this only to be confirmed on doomsday. This latter position was finally upheld by pope Benedict XII in his constitution *Benedictus Deus*, dated 1336.[12] According to this, the resurrection of the body on doomsday only completes the eternal bliss, but the perception of God and the joys of paradise could be achieved directly after death. Medieval popular thinking, however, seems to have thought otherwise: the stress in salvation history seems to lie on the Harrowing of Hell after the death of Christ, in which the souls of those who had departed before Christ's redemption were freed from hell (but were not yet allowed access into paradise?), and the resurrection of the body at the final judgement: in between, only very few, if any, souls were allowed access to paradise if we look at the descriptions of it in visionary literature.

[9] See, for example, *L'Au-delà au Moyen Age: 'Les visions du chevalier Tondal' de David Aubert et sa source la 'Visio Tnugdali' de Marcus*, ed. and transl. Y. de Pontfarcy, Bern-Oxford, 2010; Albericus, *La visione di Tungdal*, ed. M. Lecco, Alessandria, 1998; Th. Kren, R. S. Wieck, *The Vision of Tondal from the Library of Margaret of York,* Malibu CA, 1990; N. F. Palmer, *'Visio Tnugdali', the German and Dutch Translations and their Circulation in the Later Middle Ages*, Munich, 1982; *Tondolus der Ritter: die von J. und C. Hist gedruckte Fassung*, ed. N. F. Palmer, Munich, 1980.

[10] See J. Baschet, 'Jugement de l'âme, jugement dernier: Contradiction, complémentarité, chevauchement?', *Revue Mabillon*, 6, 1995, pp. 159–203.

[11] N. Wicki, 'Seligkeit, Ewige', in *Lexikon des Mittelalters*, VII, Munich, 2000, cols 1734–5.

[12] H. Denzinger and A. Schönmetzer, *Enchiridion Symbolorum definitionum et declarationum de rebus fidei et morum*, Freiburg, 1963, Nr. 1000, pp. 296–7.

In our second genre, cosmography and maps, these arguments played hardly any role at all: here, authors and artists were concerned with the physical aspects of paradise, both to inform the readers of these qualities or else to use them, in abbreviated form, to make the place depicted instantly recognizable.[13]

All our maps and cosmographical texts depend heavily on the patristic sources which summarized the biblical knowledge on paradise. Therefore, in listing the aspects of paradise as shown on medieval *mappae mundi*, the reference to the Church Fathers and early encyclopedists is never far off.

The most consistent of all qualities of paradise is its location 'furthest east'; although this is not actually found in the *Vulgata*, but only in the *Vetus Latina* version of the Old Testament, this became common basic knowledge from Late Antiquity all through the Middle Ages. Isidore of Seville is probably the most influential of these authors in his etymological development of the theme:

> [Asia] has many provinces and regions, whose names and locations I will briefly explain, beginning with Paradise. 2. Paradise is located in the east. Its name, translated from Greek into Latin, means 'garden'. In Hebrew in turn it is called Eden, which in our language means 'delights.' The combination of both names gives us the expression 'garden of delights,' for every kind of fruit tree and non-fruit bearing tree is found in this place, including the tree of life. It does not grow cold or hot there, but the air is always temperate. 3. A spring which bursts forth in the center irrigates the whole grove and it is divided into the headwaters of four rivers. Access to this location was blocked off after the fall of humankind, for it is fenced in on all sides by a flaming sword. That is, encircled by a wall of fire, so that the flames almost reach the sky. 4. Also the Cherubim, that is, a garrison of angels, have been drawn up above the flaming sword to prevent evil spirits from approaching, so that the flames drive off human beings, and angels drive off the wicked angels, in order that access to Paradise may not lie open either to flesh or to spirits that have transgressed.[14]

Apart from section 4, this is a wholly physical description of paradise, and moreover one which despite the linguistic turn of section 2 can for the greater part be easily depicted in the iconographic legends of a medieval map.

[13] For a more comprehensive treatment of the following notes, see A. Scafi, *Mapping Paradise: A History of Heaven on Earth*, London and Chicago IL, 2006, who traces the individual traditions in much closer detail.

[14] *The Etymologies of Isidore of Seville,* transl. S. A. Barney, W. J. Lewis, et al, Cambridge, 2006, p. 285; Isidore, *Etymologiae*, XIV. 3.1–4 (n. 2 above): '[Asia] Habet autem provincias multas et regiones, quarum breviter nomina et situs expediam, sumpto initio a Paradiso. [2] Paradisus est locus in orientis partibus constitutus, cuius vocabulum ex Graeco in Latinum vertitur hortus: porro Hebraice Eden dicitur, quod in nostra lingua deliciae interpretatur. Quod utrumque iunctum facit hortum deliciarum; est enim omni genere ligni et pomiferarum arborum consitus, habens etiam et lignum vitae: non ibi frigus, non aestus, sed perpetua aeris temperies. [3] E cuius medio fons prorumpens totum nemus inrigat, dividiturque in quattuor nascentia flumina. Cuius loci post peccatum hominis aditus interclusus est; septus est enim undique romphea flammea, id est muro igneo accinctus, ita ut eius cum caelo pene iungat incendium. [4] Cherubin quoque, id est angelorum praesidium, arcendis spiritibus malis super rompheae flagrantiam ordinatum est, ut homines flammae, angelos vero malos angeli submoveant, ne cui carni vel spiritui transgressionis aditus Paradisi pateat.'

From the logic of medieval map making, which would also have to include the zonal and climatic maps[15] beside the usual *mappae mundi*, it followed that if paradise was really furthest east, then it would have to be on the equatorial section of the inhabited portion of Asia. But another reason for the equatorial positioning of paradise lay in the (assumed) equal lengths of day and night in paradise, something obviously desirable. In the writings of Thomas of Aquinas[16] this is stated as a fact, and later authors followed suit.

Interestingly, this fact became also relevant for quite critical astronomers, such as John of Gmunden in the early decades of the fifteenth century, who revised the Alphonsine tables, wrote commentaries on these and several astronomical instruments as well as devising the first calendar to be used by laymen – and the first to get printed in Europe several years after his death in 1442.[17] John of Gmunden included the location of paradise in a famous table of latitudes and longitudes, which lists several European university towns, but also several other well-known towns as well as islands.[18] The first item on the table, however, is paradise, given with a latitude of 0° N and a longitude of 180° E.

What these and other passages do not answer is the vertical position of paradise: Because Isidore describes paradise as the first – and foremost – province of Asia, it is clear that he understood it to be a place actually on the surface of this earth. Other authors, however, adhered to the view of John of Damascus[19] that it was situated aloft; in this John of Damascus was followed and actually surpasssed by the writings of Bede[20] as well as the *Glossa ordinaria*,[21] Thomas Aquinas[22] and a tract by a (Pseudo-) Albert the Great who claimed to have read the fact *in quibus libris antiquissimis* (written by the Apostle Thomas), namely that paradise bordered on the lunar sphere.[23] This was one bit of information that was not easily transferable onto the map on parchment, but not for lack of trying, as is testified by the 1448 map of the Salzburg Benedictine monk Andreas Walsperger, who shows paradise in a extremely elevated position (see below).[24]

One possibilty was to show paradise outside the circle of the oikomene, by placing it as a geographical unit into the border of the map, but clearly not as an island in the ocean,

[15] Cf. R. Simek, *Heaven and Earth in the Middle Ages*, Woodbridge, 1996, pp. 25–8.

[16] Especially in his commentary on the *Sententiae* by Peter Lombard: *Scriptum super libros Sententiarum magistri Petri Lombardi Episcopi Parisiensis*, ed. P. F. Mandonnet, 2 vols, Paris, 1929, II, d. 17, q. 3, a. 2, II, pp. 438–9.

[17] On John of Gmunden, see *Johannes von Gmunden (ca. 1384–1442): Astronom und Mathematiker*, eds R. Simek and K. Chlench, Vienna, 2006.

[18] London, British Library, Add. MS 24070, fol. 74v, written by Georg Prunner, the librarian of the Augustinian foundation at Klosterneuburg, similar to the manuscript in Vienna, Österreichische National Bibliothek, MS 2332, where the table is also found. For the table in the London manuscript, cf. Scafi, *Mapping Paradise* (n. 13 above), pp. 231–3.

[19] John of Damascus, *De fide orthodoxa*, II.11 (n. 2 above) , cols 910D–914A.

[20] Bede, *In Genesim*, II.8 (n. 2 above), pp. 45–6.

[21] *Glossa ordinaria*: *ad Genesim* II.8 (n. 2 above) , col. 86C.

[22] Thomas Aquinas, *Scriptum super libros Sententiarum* (n. 16 above), II, d. 17, q. 3, a. 2, II, p. 439.

[23] Albert the Great, *Summa theologiae, pars secunda*, 13, q. 79, in *Opera omnia*, ed. Auguste C. Borgnet, 38 vols, Paris, 1890–99, XXXIII, 1895, pp. 110–13.

[24] Cf. Scafi, *Mapping Paradise* (n. 13 above), p. 233.

but as a square symbol of an 'extraterrestrial place': this path was chosen by a *mappa mundi* by Ranulf Higden (*c.* 1285–1364/5), included in his *Polychronicon* (dated between 1320 and 1363). Different versions of this world map are found in different manuscripts, or even within the same manuscript, as is shown by the two *mappae mundi* inserted at the beginning of the *Polychronicon* in a manuscript kept in the British Library.[25] In the second version, paradise appears beyond the river Indus, with the typical scene of Adam and Eve and the rivers only sketched. The first version shows paradise as an open space (possibly not fully executed) but still clearly recognizable by the four rivers, three springing directly from the garden and joining the Indus, the fourth possibly being the Nile depicted in Africa. The 'unearthly' rectangularity of the place as well as its stunning emptiness in an otherwise crowded map lift paradise from being an earthly territory to belonging to a transcedental dimension, even though Ranulf – like Isidore long before him and John of Gmunden two generations later – lists the place as the first and foremost region in Asia.

A different and more effective way of showing the elevated position of paradise has been chosen by Andreas Walsperger on his large *mappa mundi* already mentioned: here, paradise is a many-turreted castle on top of a high mountain, with the four rivers springing from the mountainsides, where also the two trees of the sun and the moon grow.[26]

The location of paradise both inside this earth and at the same time above it, as well as theoretically and practically inaccessible to humans but repeatedly visited by travellers according to medieval legends show the complexity and the paradoxical nature of medieval thinking about it. The last step in disentangling the earthly paradise from the forces of gravity was taken by Fra Mauro on his famous map executed at Venice *c.* 1450. Here, on a *mappa mundi* measuring 223 cm in diameter, paradise is no longer integrated into the depiction of Asia, but is also not lost from the map. Instead, Fra Mauro included an idealized circular depiction in one of the four corners left by the main round map on its square background.[27] Here, the mountainous (!) landscape of the *paradisus terrestris* is connected, by the four rivers, to the depiction of the inhabited earth of the type already found since *c.* 1350 in astronomical drawings as a pictorial marker for the earth.[28] 'In short, Fra Mauro had not only found a way to map the unlocatable Garden of Eden as an eartly region, but

[25] London, British Library, Royal MS 14, C.IX, fols 1v-2r and 2v.

[26] For these trees, cf. Scafi, *Mapping Paradise* (n. 13 above), p. 134 and n. 39.

[27] On Fra Mauro and the 'cornered' paradise, see ibid., pp. 235–8.

[28] Cf. the manuscripts of Konrad of Megenberg's translation of the *Liber de sphaera* by John of Sacrobosco: R. Simek, 'Die mittelhochdeutschen Übertragungen von Johannes von Sacroboscos Liber de sphaera', *Codices manuscripti,* 13, 1987, pp. 57–76, and F. B. Brévart, 'Johannes de Sacrobosco', in *Die Deutsche Literatur des Mittelalters. Verfasserlexikon*, 2nd edn, IV, Berlin-New York, 1983, p. 733, and the earliest prints of Konrad's *Deutsche Sphaera*.

he had also shown an acute awareness of the temporal chasm decisively separating the lost paradise of delights from the human realm.'[29]

As opposed to the flat landscape, which seems to have lost its former desirability during the fifteenth century, the aspect of the temperate climate was an item that could be far more easily conveyed in the illuminations of manuscripts of visionary texts than on maps; it is thus one of the items of the prose descriptions of paradise that we will look for in vain on the maps. The same goes for the supposed eternal spring, which is more of a symbolic than a climatic value and anyway in Isidore more of a comparison than an actual statement of everlasting spring.

Standard elements of paradise are of course the Tree of Life and the Tree of Knowledge of Good and Evil. These could easily be depicted on maps, as *mappae mundi* from the earliest Beatus maps all the way to the maps of Walsperger and Fra Mauro show. Yet, these trees are not really cosmographical items, but rather belong, with the apple, the (crowned) snake and the protoplasts themselves, as well as occasionally the hand or face of God above them, to the hagiographical inventory of paradise, taken both from the patristic descriptions as well as the legendary texts dealing with paradise. As such, they are the paratexts of cosmographical representations of paradise.

Alessandro Scafi has shown in his book on the mapping of paradise how these elements can still be found in such an extremely late medieval text as the *Dialogus de somnio quodam* by Aeneas Silvius Piccolomini (or Pope Pius II, after 1458),[30] and this could of course be taken further still in late fifteenth-century writings. However, I would like to close with a text of much earlier date and of surprising similarity to maps and continental traveller's tales, considering it comes from the very fringe of medieval Europe, namely from Iceland. This short saga cannot be dated with absolute certainty, but is preserved in over fifty manuscripts, including some from the fourteenth century. Despite its title, *Eiríks saga víðförla*[31] ('The Saga of Erik the Far-Travelled'), which purports to be one of the typical Scandinavian adventure stories set in the Viking Age or even before, the text presents hardly more than a slight frame of that type and in reality is a didactic text on paradise, partly in the form of a dialogue between the Norwegian protagonist and the Byzantine emperor, a dialogue which is closely modelled on the Latin *Elucidarius* (*c.* 1120–30). This gives us a rough date for the saga of *c.* 1150–1300, but the most interesting part is the description of paradise, because the Norwegian hero Erik wants to find the 'Land of the Undying' (*Ódáinsákr*).

When he finally reaches it, only he and a close companion dare to enter the dragon mouth that turns out to be the bridge to the other world, and enter a type of proto-paradise. Once he has seen it and heard the angel's explanation, he travelles back to Norway and has only just time to tell of his discovery before he dies; as a compilation of

[29] Scafi, *Mapping Paradise* (n. 13 above), p. 240.

[30] Ibid., pp. 195–8.

[31] *Eiríks saga víðförla*, ed. H. Jensen, Copenhagen, 1984, *Editiones Arnamagaeanae* B 25.

elements from descriptions of paradise, the saga includes elements not very common elsewhere, and therefore I shall quote it more extensively:

> After the Norwegian Eirek and his companion had jumped into the dragon's mouth, it seemed to them as if they were wading through smoke, and when they came out of the smoke they saw a fair land with grass as soft as purple cloth with sweet scent and many blossoms. There were streams of honey running everywhere in this land. This land was big and flat. There was such sunlight that it was never dark and there was never any shade. The air was calm, but there was a gentle breeze just above ground level so that one scent was stronger than the other.
>
> They walked for a long time and wondered if they would find any settlement or inhabited area, and they wanted to know the size of this land. Then they saw something like a tower suspended in the air without pillars. When they got nearer they saw the tower hanging in the air without pillars. On the southside, a ladder led up to the tower. They saw that it was decorated with extremely beautiful silken stuff. There was a well-set table with a silver salver on it, on which all sorts of delicacies were laid out, and laden with sweet smelling white bread. There was a golden jug covered with gems on it and also a cup of wine. There were well-made beds spread with gold-striped covers and silken stuff.
>
> But while Eirek slept, a young man, beautiful and shining, appeared to him. He said to Eirek: '[...] The place you see here is but a desert compared with paradise, which is not far from here and slopes from there to the place you have seen. No living man will ever get there, but the souls of the just will live in it. The place you have visited is called the Land of the Living. Before you arrived, God had this place guarded by us, and let us show you the Land of the Living and receive you and repay you for your trouble.'[32]

[32] Ibid., pp. 84–92, my translation from the Old Norse. For the various elements of this description and their literary sources, see R. Simek, 'Die Quellen der Eiríks saga víðförla', *Skandinavistik*, 14, 1984, pp. 109–14.

Paradise in the Islamic Religious Imagination

Christian Lange

It has been stressed by historians of Islam that the medieval Muslim cartographers, as a rule, did not see fit to mark the location of paradise on their maps of the world.[1] At best, as in the case of the world map of al-Kāshgharī (Baghdad, eleventh century) discussed by Emilie Savage-Smith in this volume, they indicated the spot were Adam fell to the earth after his expulsion from the Garden of Eden.[2] From this eloquent gap in Muslim cartography, it seems reasonable to infer that paradise was generally conceived to be removed from this world, so remote that humans could under no circumstances aspire to enter it during their lifetime. Such an inference indeed resonates with the idea that Islamic theology stresses the remoteness and otherness of God, the lord of paradise. 'Islam', in the words of one of its greatest living interpreters,

> does not know the idea of a mediating instance [between God and creation]; Muḥammad brings salvation only in the sense of delivering information. In other respects, too, all bridges are torn down: there are no sacraments, no images to be worshipped, no church music. God is transcendent.[3]

In this sublunar world there is no place, then, for paradise; it simply is, and must remain, out of reach, both physically and conceptually. According to a report (*ḥadīth*) transmitted by the famous collector al-Bukhārī (d. 870), the Prophet Muḥammad declared (echoing a notion from biblical literature) that what God has in store for Muslims is 'that which no eye has ever seen, and no ear has ever heard, and no human heart has ever imagined'.[4] In the judgement of the celebrated al-Ghazālī (d. 1111), known in European scholasticism as Algazel, no better definition of paradise could be given.[5]

Author's Note: The author wishes to acknowledge the support of the European Research Council (Project no. 263308: "The here and the hereafter in Islamic traditions") during the research for, and writing of, this essay.

Abbreviations used in this essay:

*EI*² = *The Encyclopaedia of Islam*, eds H. A. R. Gibb, B. Lewis, C. Pellat, C. Bosworth et al, 2nd edn, 11 vols, Leiden, 1960–2002.

EQ = *Encyclopaedia of the Qur'ān*, eds J. D. McAuliffe et al, 6 vols, Leiden, 2001–6.

[1] J. van Ess, *Theologie und Gesellschaft im 2. und 3. Jahrhundert Hidschra. Eine Geschichte des religiösen Denkens im frühen Islam*, IV, Berlin, 1997, p. 552.

[2] Often this place was identified with *Sarandīb*, or Sri Lanka. Thus in Ibn Qutayba (d. 889), *K. al-Maʿārif*, ed. S. Okacha, Cairo, 1960, p. 15.

[3] J. van Ess (with H. Küng), *Christentum und Weltreligionen: Islam*, Munich and Zurich, 1994, p. 110. The translation from German is mine.

[4] Bukhārī, *Ṣaḥīḥ*, ed. M.D. al-Bughā, Beirut, 1987, VI, p. 2723. In Paul (I Corinthians 2:9) the eschatological content of the adage seems only alluded to, but the Babylonian Talmud (Sanhedrin 99a) refers it explicitly to the world to come. Cf. Isaiah 52:15, 64:4.

[5] Ghazālī, *al-Maqṣad al-asnā*, Cairo, s.d., p. 21. Also in Ghazālī, *Mīzān al-ʿamal*, transl. ʿA. Elschazli, *Das Kriterium des Handelns*, Darmstadt, 2006, p. 87.

And yet, the idea of an inaccessibly removed paradise in the Islamic tradition deserves to be nuanced and qualified in important respects. 'It only sounds *to us* as if God is remote or impersonal', writes Josef van Ess. For Christians, 'God shows Himself in several persons, and monotheism is a rather philosophical idea. For Muslims, on the other hand, God is not only the One but also the Only-One, the one towards whom everything is directed and who takes care of everything; He is the Lord and the All-Merciful.'[6] In this perspective, paradise moves back a degree or two closer to the world of the here-and-now. In fact, as another Prophetic *ḥadīth*, again transmitted by al-Bukhārī, tells us, 'the Garden is closer to you than the strap of your sandal'.[7] In this vivid image, paradise cuts through earthly reality, in the same way in which the strap of a sandal penetrates the cavity between the two toes of the foot. The horizontal and the vertical become fused. There is immanence of the divine in creation.

We are thus confronted with a paradox. Paradise in the Islamic tradition has aspects of being both remote and proximate, outside and inside the world. In apparent contradiction to traditions stressing the spatial, temporal and conceptual gulf separating this world from the otherworld there are important strands of thought in Islamic culture that betray a feeling of nearness of the two realms. This calls for the obvious, namely, to introduce a measure of differentiation into the picture. While traditional Muslim cartography may offer us no examples of this sense of nearness of paradise, other Islamic traditions of Islamic literature, such as eschatological *ḥadīth* and local legends, do. It is with these traditions of the Islamic religious imagination that this essay is primarily concerned.

Doors to the otherworld in the Qurʾān[8]

First, however, let us take a look at the Qurʾān. Arguably, and in contrast to the dominant Western (post-Augustinian) Christian tradition, one does not find a developed sense of 'paradise lost' in the Qurʾān, nor is the idea of original sin present. According to the Qurʾān, Adam and Eve were expelled from paradise, but they received God's full forgiveness immediately after their fall. In its retelling of Genesis 3, the Qurʾān states:

> Adam disobeyed his Lord, and so he erred. Thereafter his Lord chose him, and turned again unto him, and He guided him. Said He, 'Get you down, both of you together, out of it [the Garden]... but if there comes to you from Me guidance, then whosoever follows My guidance shall not go astray.'[9]

[6] Van Ess, *Christentum und Weltreligionen* (n. 3 above), pp. 110–11.

[7] Bukhārī, *Ṣaḥīḥ* (n. 4 above), V, p. 2380. Also in Aḥmad b. Ḥanbal (d. 855), *Musnad*, Cairo, s.d., I, pp. 287, 413, 442. The tradition continues: '... and so is hell'. Here, however, I am only interested in exploring the interpretive horizon opened up by the first half of this well-known saying. For a full textual and cultural history of the Islamic paradise and hell, see my *Paradise and Hell in Islamic Traditions*, Cambridge, 2016.

[8] The following section is partly based on my inaugural lecture, *The Discovery of Paradise in Islam*, Utrecht, 2012.

[9] Qurʾān 20:121–3. Quotations, with minor emendations, from the Qurʾān follow the classic translation of Arthur Arberry (1955).

As people unto whom God 'turned again' and to whom He gave guidance after their error, Adam and Eve did not pass on any essential, inherited human depravity to their descendants.[10] The expulsion from paradise in the Qurʾān, therefore, does not signify an ontological 'fall' from a state of grace to one of sin and damnation but rather, a momentary and reversible loss of divine favour. In a sense, then, the Qurʾān carries the promise that paradise remains within sight of humans. To be sure, such a reading does not coincide with canonical or 'orthodox' interpretations of the Qurʾān such as the majority of medieval Muslim exegetes would have practiced; but nonetheless, it may be worthwhile to point out that the sense of the temporal, spatial and conceptual nearness between this world and paradise is a clearly identifiable feature of the Qurʾān. Let us take a closer look at these three dimensions of the relationship between this world and the otherworld in the Qurʾān.

Little in the Qurʾān suggests that paradise will be created or populated only at the end of time. The synchronicity of life on earth and in the otherworld is stressed. Adam and Eve were in the Garden of Eden, after all, and the notion that God would destroy this garden and then recreate it at the end of time would seem patently absurd. This argument assumes, of course, that the garden inhabited by Adam and Eve is identical with the eschatological garden. This was sometimes denied by Muslim theologians (see below). However, the Qurʾān itself makes no such distinction: it refers to both the primordial and the eschatological abode of the blessed as '*the* Garden' (*al-janna*, cf. 2:35 and 2:82), the usual Qurʾānic term for paradise.

Some verses in the Qurʾān seem to express the idea of synchronicity directly. For example, the Qurʾān enjoins the believers to 'hasten to a garden that has been prepared for the godfearing' (3:133). This was usually understood to mean that paradise 'has *already* been created' (*qad khuliqat*), that is, that it is already in place, waiting to receive the believers.[11] A verse that was revealed after the battle of Badr, in which several of Muḥammad's companions were killed, reassures the survivors that those who have fallen are not dead but rather, that 'they are living with their Lord, well provided for' (3:169). Presumably, this indicates that paradise is entered immediately after death, in the same way in which Jesus, in Luke 23:42–3, promises the thief on the cross next to him that they will be together in paradise 'to-day'.[12]

Descriptions of paradise in the Qurʾān are generally formulated in the present tense. The Qurʾān states, for example, that 'the godfearing are in gardens and bliss, they rejoice in what their Lord has given them; their Lord has guarded them against the chastisement of hell' (52:17–18). It is true that Arabic, like other Semitic languages, does not distinguish

[10] As the Qur'an (6:164) states, 'every soul only bears its own burden'.

[11] Ashʿarī (d. 935 or 6), *Maqālāt al-islāmiyyīn*, ed. H. Ritter, Instanbul, 1929–30 [2nd edn, Wiesbaden, 1963], p. 475; Pazdawī (d. 1100), *Uṣūl al-dīn*, ed. H.-P. Lins, Cairo, 1963, p. 170.

[12] Note, too, that according to the Qurʾān, unbelieving sinners, such as the wives of Noah and Lot (66:1) and 'the family of Pharao' (3:10–11), are said to have entered hell after their death.

precisely between present and future tense (it only knows two aspects of time, complete and incomplete).[13] However, there are certain extra particles in Arabic, such as the prefixes *sa-* and *sawfa*, which, if they are added to verbs, unequivocally refer to future events, and it is striking that these particles are *not* used, as a general rule, in Qurʾānic descriptions of the otherworld. It is as if the line between the time of the here and the time of the hereafter is intentionally blurred.[14] This observation squares with the fact that the Qurʾān challenges the fatalistic conception, common among pre-Islamic Arabs, of time (*dahr*) as an irreversible power and process to which all humans must yield. By contrast, the Qurʾān disempowers chronological time and subjects it to God's sovereignty.[15]

According to the Qurʾān, the Garden of Eden is located somewhere above the earth: Adam and Eve are told to 'go down' (2:36, 7:24) from it, as if from a mountain. The Qurʾān here reprises the ancient Judeo-Christian motif of a paradise mountain,[16] thus positing a physical connection between paradise and earth. Another verse, however, widens the gap between this world and the otherworld, by telling us that paradise is accessed through the 'gates of heaven' (7:40). One ascends towards these gates (or descends from them, as in the case of Adam and Eve) over stairways (70:4) or ropes (40:37). These ropes – 'sky-ropes' they have been termed[17] – hold the edifice of the cosmos in place, like a tent.

This is a cosmological metaphor that must have made sense in the context of seventh-century Arabia, heavily influenced as it was by bedouin culture. Indeed, when God created the world, according to the Qurʾān, he set up mountains like tent poles (78:6–7) to support the heavenly dome, and then spread out the earth like a rug (2:22) to cover the inside of the cosmic tent. The 'sky-ropes' literally offer an escape through gates in the roof of the tent (21:32), toward the other side of the firmament. Particularly prophets and angels travel along them, while evil spirits and demons with sky-walking pretensions who try to breach these gates are repelled by star-hurling angels (37:6–10). Paradise, 'a great kingdom' (76:20), stretches out over the entire width of the cosmic roof: it is 'a garden the breadth of which is as the breadth of heaven and earth' (57:21).

However, even though according to this cosmology, paradise is situated on the other side of the roof of the world, a sense of proximity and immediateness persists. The Qurʾān

[13] Cf. W. Reuschel, *Aspekt und Tempus in der Sprache des Korans*, Frankfurt, 1996, pp. 211–17.

[14] In the words of Michael Sells, the overlap between the tenses in the Qurʾān is what gives it its 'depth, psychological subtlety, texture, and tone'. See M. Sells, *Approaching the Qurʾān*, 2nd edn, Ashland OR, 2007, p. 24.

[15] I borrow this phrase from G. Tamer, *Zeit und Gott. Hellenistische Zeitvorstellungen in der altarabischen Dichtung und im Koran*, Berlin, 2008, pp. 187–214. Tamer speaks of the 'Entmachtung der Zeit' in the Qurʾān (197–205).

[16] See, for example, *1* [*Ethiopic Apocalypse of*] *Enoch* (2nd century BC – 1st century AD) 24:1–6; Ephrem (*fl.* late 4th century), *Hymns on paradise*, transl. S. Brock, Crestwood NY, 1990, 2:10. Cf. G. Anderson, 'The Cosmic Mountain: Eden and its Interpreters in Syriac Christianity', in *Genesis 1–3 in the History of Exegesis*, ed. G. A. Robbins, Lewiston NY, 1988, pp. 187–223, and the essay by Sergey Minov in this volume.

[17] K. van Bladel, 'Heavenly Cords and Prophetic Authority in the Qur'an and its Late Antique Context', *Bulletin of the School of Oriental and African Studies*, 70/2, 2007, pp. 223–46 (231).

is known for its powerful evocations of the havoc wrought on earth during the apocalypse. Creation is undone: the mountains are set moving and become like tufts of wool, the sun disk is rolled up, the stars are thrown down, heaven will be rent asunder, split open and full of gaping holes, the earth empties its bowels, ejecting the dead from their graves.[18] What, then, happens to paradise? The Qur᾽ān suggests that, on the Day of Resurrection, paradise and hell do not perish together with 'whosoever is on the earth' (55:26). Instead, they survive the re-ordering of the cosmos in their original form. As the Qur᾽ān puts it, they are 'brought near' (26:90–91) to the place where the resurrected are gathered. On the Day of Resurrection, in other words, it is as if the sticks on which the cosmic tent rests are knocked out, so that the roof collapses onto the tent's foundation. The world that humankind inhabits is squeezed in between heaven and hell, while all three realms are an integrated whole: they equal reality *in toto*. They form, to use a term from rhetoric, a *merismos*.[19]

As for the physical contents of the Qur᾽ānic paradise, suffice it to highlight two areas in which the conceptual slippage between paradise and this world is manifest. The flora and topography of the Qur᾽ānic paradise establish a firm link between this world and the otherworld. One finds palms and pomegranate trees (55:68), as well as acacias and a thornless version of the evergreen jujube tree (56:28) in the Qur᾽ānic paradise. The Western botanical name of the jujube tree (Arab. *sidr*) is *ziziphus spina christi*, 'Christ's Thorn Jujube'. This is a plant that is found from the lands of Syria to the Yemen and beyond. Its fruit is an important source for honey-making bees, which may explain its association with paradise.[20] In later tradition, one regularly comes across encomia of earthly fruits as deriving directly from paradise. For example, according to a *ḥadīth*, every pomegranate has one seed from paradise in it.[21] Plants in paradise are watered by a plethora of fountains, wells and rivers. In one verse (47:15), the Qur᾽ān reprises the biblical idea of the four rivers of paradise (Genesis 2:10). Later authors elaborated that these four rivers flow uninterruptedly from paradise into this world, usually equating them with the Nile, Euphrates, and the two eastern Turkish rivers Sayḥān and Jayḥān, which were all said to spring from the al-Kawthar well in paradise.[22] In general, the geomorphization of paradise (as, for that matter, of hell[23]) is a striking feature of Qur᾽ānic and later Islamic eschatology.

[18] Qur᾽ān 81:1–4, 99:1–2, 101:1–5 and *passim*. Cf. F. Leemhuis, 'Apocalypse', in *EQ*, I, pp. 111–14.

[19] A. Neuwirth, 'Spatial relations', in *EQ*, V, pp. 104–8.

[20] As its name indicates, in the Christian tradition the tree is associated with the passion and resurrection of Christ. Cf. A. Dafni, Sh. Levy and E. Lev, 'The Ethnobotany of Christ's Thorn Jujube (*Ziziphus Spina-Christi*) in Israel', *Journal of Ethnobiology and Ethnomedicine*, 1/8, 2005.

[21] Ṭabarānī (d. 971), *al-Mu῾jam al-kabīr*, ed. Ḥ. al-Salafī, 2nd edn, Mosul, 1983, X, p. 263.

[22] Qurṭubī (d. 1272), *al-Tadhkira fī aḥwāl al-mawtā wa-umūr al-ākhira*, 2 vols in 1, Cairo, 1980, II, pp. 167–8.

[23] On hell in Islam, cf. now C. Lange, 'Hell', *Encyclopaedia of Islam*, 3rd edn, eds G. Krämer et al., Leiden, 2010/2, pp. 143–8. See also *Locating Hell in Islamic Traditions*, ed. C. Lange (Leiden, 2016).

The material culture of the Qurʾānic paradise is ripe with the kind of mundane luxury items, including imports from Persian lands, which a rich inhabitant of seventh-century Ḥijāz would have encountered in the travelling canteens that pitched their tents on the great fairs of pagan Arabia.[24] In pre-Islamic and early Islamic poetry one reads about visitors to these taverns 'reclining on carpets over which are spread embroidered rugs most sumptuous', who are served 'wine cooled by mixing with water' in 'full flagons overflowing with foam atop' by 'a boy, girt up, with two pearls in his ears' and by 'white ones, with beautiful eyes' who carry around 'great cups filled full with wine'.[25] These images reappear, almost *verbatim*, in Qurʾānic descriptions of the paradisiacal banquet.[26]

It is unmistakable, then, that the Qurʾānic paradise is derived from the observation of this-worldly phenomena. Western polemicists of all ages have used this fact to attack the perceived exaggerated sensuality of the Qurʾānic imagination.[27] It is possible, however, to see in it a deliberate strategy to highlight how thin, how exquisitely permeable, the line is that separates humans from the blissful state of an ideal, otherworldly existence. Seen in this light, the Qurʾānic paradise is not a form of *dilectatio morosa* but a creative act of re-ordering the cosmos and filling it with meaning. It is an assemblage of worldly images used to build a new deep structure of meaning, a mosaic in which all those dimensions of human life that are experienced as transcending time are combined into one glorious picture. The sensuality of the Qurʾānic paradise does not result, in other words, from a bedouin's vision of a decadent life filled with wine, women and poetry. Rather, it evokes an ideal, a perfectly structured and ideally harmonious world, a world that humans, in the happiest moments of their life on earth, can already see before them.

In sum, in a certain reading the Qurʾān outlines a world in which salvation of humankind toward a better world is a dominant theme, but not in the sense of salvation from an innate state of sin and impurity to a far-removed and utterly transcendent afterworld. Rather, we are talking about a recovery from a momentary lapse of focus, about a switch from ordinary life to the simultaneous, accessible reality of paradisiacal space and a state of being in which all material and spiritual potentialities are realized and lived to the full. In Eden, Adam and Eve existed in a state of perfect purity: they did not urinate or defecate, they did not even sleep. According to the Muslim scholars, urination, defecation and sleep are the three main types of impurities that the Islamic ritual of ablution (*wuḍūʿ*) does away with.[28] The state of purity achieved by the Islamic ablution ritual, in other

[24] On Persian terms in the Qurʾān, see W. Eilers, 'Iranisches Lehngut im arabischen Lexikon: Über einige Berufsnamen und Titel', *Indo-Iranian Journal*, 5/3, 1962, pp. 203–32 (203–7).

[25] See J. Horovitz, 'Das koranische Paradies', in *Scripta Universitatis atque Bibliothecae Hierosolymitanarum*, Jerusalem, 1923, repr. in *Der Koran*, ed. R. Paret, Darmstadt, 1975, pp. 53–75 (64–73); C. Wendell, 'The denizens of paradise', *Humaniora Islamica*, 2, 1974, pp. 29–59; A. Neuwirth, *Der Koran als Text der Spätantike. Ein europäischer Zugang*, Berlin, 2010, pp. 221–2, 424–32. The examples are taken from Wendell, 'The denizens of paradise', pp. 35, 43.

[26] Cf. L. Kinberg, 'Paradise', in *EQ*, IV, pp. 12–19.

[27] The classic account of the history of Western polemics against Islam is still N. Daniel, *Islam and the West: The Making of an Image*, Edinburgh, 1960, rev. edn, Oxford, 1993. For the Islamic paradise, see especially pp. 172–6.

words, enacts the state of Adam and Eve in the Garden of Eden. Arguably, such notions are anchored in that peculiar Qurʾānic *Weltanschauung* according to which life in this world is lived with a door open to the otherworld.

*The protological and the eschatological garden in Muslim scholastic theology (*kalām*)*

In the developing discourse of post-Qurʾānic theology, there was disagreement as to whether paradise was located on earth or in the heaven. The Qurʾān, as we have seen, provides no clear answer to the question; rather, it seems to place the garden somewhere at the nexus of the sublunar world and the heavenly spheres, where heaven and earth meet. The Damascene Ibn Qayyim (d. 1350), in his great eschatological *summa* entitled 'Urging Souls Forward Toward the Lands of Happiness' (*Ḥādī l-arwāḥ ilā bilād al-afrāḥ*), provides glimpses of the debate. The famous al-Ḥasan al-Baṣrī (d. 728), the early authority Abū Ḥanīfa (d. 767) 'and his followers' as well as the celebrated al-Jubbāʾī (d. *c.* 915) from Basra, for example, reportedly opined that paradise was in heaven and thus unconnected to the earth.[29] However, two younger contemporaries of al-Jubbāʾī, Abū l-Qāsim al-Balkhī (d. 931) and Abū Muslim al-Iṣfahānī (d. 934), proposed that paradise was on earth. They explained that, when in the Qurʾān Adam is commanded to 'go down' (*ihbiṭ*) from the garden, this does not indicate a vertical movement, or 'fall' from paradise. Rather, the verbal root *h-b-ṭ* refers to horizontal dislocation (*intiqāl*), as in another Qurʾānic verse in which Moses, annoyed at the Israelites' disobedience, tells them to 'go down to Egypt' (*ihbiṭū miṣr*, 2:61).

A century before al-Balkhī's and al-Iṣfahānī's intervention, the Baghdadi *littérateur* and collector of religious lore Ibn Qutayba (d. 889), in his 'Book of Knowledge' (*K. al-Maʿārif*), had gathered a number of Adamic legends from which later theologians inferred 'proofs' for the earthly location of Adam's garden. According to these narratives, when Adam and Eve were created in the garden they were told to 'multiply and fill the earth' – not the heavenly paradise; from paradise spring the four rivers Sayḥūn, Jayḥūn, Tigris and Euphrates – which suggests that there is a seamless connection of paradise with the inhabited world; the snake in the garden was 'the biggest riding animal on earth' – an appellation that would make no sense if paradise were in heaven; Adam and Eve's banishment from the garden was from Eden (Arab. *ʿadn*) to the east of India – that is, a horizontal move; when Adam came to die, he asked his sons to fetch him some grapes from the garden – which therefore must have been 'within reach' on the surface of the earth.[30]

[28] Scholars differ as to whether the human body in Islam is considered 'pure' by default (and only temporarily 'contaminated' by ritual impurity) or whether it is by default 'impure' (and only temporarily 'decontaminated' by ritual washing). The first position is outlined by Kevin Reinhart in his ground-breaking article, 'Impurity/No Danger,' *History of Religions*, 30/1, 1990, pp. 1–24. More recently B. Wheeler has challenged Reinhart's interpretation and defended the second position in his *Mecca and Eden: Ritual, Relics, and Territoriality in Islam*, Chicago IL, 2006, pp. 47–70.

[29] Ibn Qayyim, *Ḥādī al-arwāḥ*, ed. Z. ʿUmayrāt, Beirut, 2007, pp. 24–5.

[30] Ibn Qutayba, *K. al-Maʿārif* (n. 2 above), pp. 11–15, 19.

Ibn Qutayba relates many of these traditions on the authority of the Yemeni scholar Wahb b. Munabbih (d. *c.* 730), a judge in Ṣanʿā celebrated for his knowledge of the Bible and of Judeo-Christian scholarship, who is often credited with introducing Judeo-Christian narratives, so-called *Isrāʾīliyyāt*, into the growing corpus of Islamic religious literature.[31] Ibn Qayyim, who summarizes the stories preserved by Ibn Qutayba, draws attention to this genealogy by pointing out that ʿAdn (Aden) is in Yemen, as if to suggest that Wahb's stories are spurious because they have a decidedly local, Judeo-Christian and therefore un-Islamic tint.[32] On the whole, however, theologians in the early centuries of Islam do not seem to have doubted the earthly location of paradise.[33]

Muslim theologians also asked whether Adam and Eve's garden was identical with the garden to which the believers will return at the end of times. The Córdoban scholar Mundhir b. Saʿīd (d. 966) was one of the most outspoken deniers of the identity of the two abodes (he also denied that Adam's garden had been on earth).[34] His arguments centered around the idea that the eschatological paradise has a degree of perfection that the garden inhabited by Adam and Eve obviously lacks:[35] existence in the future paradise is never-ending and eternal, and there are no trials or tribulations in it. As the Qurʾān explicitly notes, there is not even idle talk in the future abode of the blessed (52:23). Therefore, the argument went, Satan is unable by definition to enter the eschatological garden; he could only whisper lies to Adam and Eve in 'their' garden. As some also pointed out, there is no sleep in the eternal garden.[36] Adam, however, slept in 'his' garden when Eve was created from his rib.

Ibn Qayyim, who may have held similar views, gives some room to these arguments, but in general those wishing to defend that the protological and the eschatological garden were the same appear to have been the majority.[37] The Qurʾān, they noted, makes no distinction between the two gardens, referring to both as '*the* Garden'.[38] In addition, some exegetes proposed that when, as the Qurʾān says, God 'chose' Adam after his banishment from the garden, this act of 'turning again unto him' (20:122) consisted in fact of the promise to resurrect and return Adam to the garden he had previously inhabited.[39] From Abū Mūsā l-Ashʿarī (d. *c.* 662) was related the following syllogism: The fruits in this world are fruits one also finds mentioned in the Qurʾānic descriptions

[31] Cf. G. Khoury, 'Wahb b. Munabbih', in *EI*[2], XI, pp. 34–36; G. Vajda, 'Isrāʾīliyyāt', in *EI*[2], IV, pp. 211–12.

[32] Ibn Qayyim, *Ḥādī* (n. 29 above), p. 26.

[33] Van Ess, *Theologie und Gesellschaft* (n. 1 above), IV, p. 550.

[34] Ibn Qayyim, *Ḥādī* (n. 29 above), p. 25. Cf. van Ess, *Theologie und Gesellschaft* (n. 1 above), IV, p. 553.

[35] Ibn Qayyim, *Ḥādī* (n. 29 above), pp. 31–4.

[36] When the Prophet was once asked whether the inhabitants of paradise sleep he answered that they do not, because 'sleep is the brother of death'. See Ibn Kharrāṭ (d. 1185), *K. al-ʿĀqiba*, Ṭanṭā, 1990, p. 325.

[37] Ibn Qayyim attributes this position to al-Ḥasan al-Baṣrī (d. 728) and the Muʿtazilites ʿAmr b. ʿUbayd (d. 761), Wāṣil b. ʿAṭā (d. 748), al-Jubbāʾī (d. *c.* 915) and al-Rumānī (d. 993). See Ibn Qayyim, *Ḥādī* (n. 29 above), p. 25. Cf. van Ess, *Theologie und Gesellschaft* (n. 1 above), IV, p. 553.

[38] Ibn Qayyim, *Ḥādī* (n. 29 above), pp. 29–30.

[39] Ibid., p. 30.

of the eschatological garden. Humankind owes the possession of these fruits and the knowledge required to cultivate them from Adam, who brought them with him when he was expelled from the garden. Therefore, Adam inhabited the eschatological garden.[40]

In sum, the mainstream position in medieval Muslim theology was to think that the eschatological and the protological garden were identical, and to situate this garden on earth. However, the tradition was flexible enough to preserve the memory of those who had argued for different positions on the matter. Not only was the discussion kept alive in this way, but every once in a while, it found a true *Sitz im Leben*. On various occasions in the history of Muslim societies, specific places were identified with paradise on earth, and the debate flared up again. The case of Wahb b. Munabbih's alleged attempt to link the earthly paradise to the Yemen has already been mentioned. However, more concrete instances of the 'edenification' of places can be summoned. In the following, a number of examples from the early centuries of Islamic history shall be discussed.

Paradise between Syria and the Ḥijāz

In 'The Virtues of the Holy City' (*Faḍā ʾil Bayt al-Muqaddas*) of al-Wāṣiṭī (d. *c.* 970), a preacher in Jerusalem's al-Aqṣā mosque,[41] one reads that after the conquest of the city in 637, one of the soldiers of the caliph ʿUmar b. al-Khaṭṭāb (r. 632–44), a man named Shurayk b. Khabāsha al-Numayrī, went to fetch water from a well on the Temple Mount. As al-Wāṣiṭī relates,

> suddenly the bucket fell from his [Shurayk's] hands, and so he descended [into the well] in search of it. A man appeared to him in the well, told him to follow him, took him by the hand and made him enter the Garden, and Shurayk took leaves [from a tree in the Garden]. Then the man brought him back, and he [Shurayk] exited [the well]. He went to his companions and told them about it. His story was brought before ʿUmar b. al-Khaṭṭāb, who said: 'Shall a man from this community enter the Garden while he is alive among you? Look at the leaves! If they have withered, they are not from the Garden. If they haven't withered, they are.' [...] And [indeed] the leaves had not withered.[42]

Shurayk is said to have kept a leaf from his subterranean visit to paradise in his personal copy of the Qurʾān for the rest of his life, and to have been buried with it, placed delicately between his chest and the burial shroud covering his corpse, when he was laid to rest in the Syrian village of al-Salāmiyya.[43]

In al-Wāṣiṭī's account of the story, popularly known as the 'Story of the Leaves' (*ḥadīth al-waraqāt*), the caliph ʿUmar at first inclines toward the transcendentalist view that this world and paradise belong to two radically different realms of existence. On the one hand,

[40] Ibid., p. 39.

[41] For the little biographical information available for al-Wāsiṭī, see al-ʿUlaymī (d. *c.* 1520), *al-Uns al-jalīl bi-tārīkh al-Quds wa-l-Khalīl*, Cairo, [1866 or 7], II, p. 482.

[42] Wāsiṭī (d. *c.* 971 or 2), *Faḍāʾil al-Bayt al-Muqaddas*, ed. I. Hasson, Jerusalem, 1979, pp. 93–4.

[43] Ibid., p. 94.

in his reaction to the news of Shurayk having returned from paradise, he invokes a Prophetic *ḥadīth* that raises the prospect that 'a man shall enter the Garden alive, while he is alive and walking on his two feet'. On the other hand, by asking for empirical evidence for Shurayk's otherworldly journey, ʿUmar's scepticism becomes evident. However, his intuition is proven wrong: the leaves Shurayk brings back from his tour of paradise, 'like the leaves of the Syrian peach tree, as big as the palm of a hand and with a pointed head',[44] do not wither. They partake of eternal life, thus demonstrating beyond doubt that Shurayk has really crossed the boundary between this world and the otherworld, and that humans, 'alive and walking on their feet', can indeed be transported to paradise and back again.

Jerusalem was held by many a medieval Muslim to mark the site of the future earthly paradise, a Jewish notion that was taken up in Islam.[45] However, as the 'Story of the Leaves' suggests, Jerusalem (or rather, the Temple Mount) was also perceived to *be* paradise, or in any case to be very close to it. Paradise was variously located above, on, or underneath the Temple Mount. Famously, the prophet Muḥammad would have ascended to heaven from the Rock.[46] In the several versions of the 'Ascension' (*miʿrāj*) of the Prophet, Muḥammad ascends a ladder until he reaches the roof of the world, where angels let him pass through a gate, emerging into the first heaven. In most versions of the *miʿrāj*, Mūḥammad catches glimpses of paradise and hell in the higher spheres of heaven, although in some versions he looks down to Jerusalem to behold the two abodes.[47] The east wall of the Temple Mount with the Gate of Mercy was sometimes equated by the exegetes with the barrier between paradise and hell 'with a door in it; inside it will be mercy, and outside it, in front of it, punishment' (57:13).[48] The notion that paradise has eight gates may also originate from a trend in early Islam to associate the Temple Mount, and more particularly the octagonal Dome of the Rock, with paradise.[49]

Finally, there are manifold traditions that place paradise *underneath* the Temple Mount, in the vein of the 'Story of the Leaves'. In Ibn ʿAbd Rabbih's (d. 940) 'The Unique Necklace' (*al-ʿIqd al-farīd*) one reads that the Rock sits on top of one of the gates of paradise.[50] In another tradition, the Rock itself is called the gate to paradise.[51] Al-Wāṣiṭī

[44] Ibid.

[45] Van Ess, *Theologie und Gesellschaft* (n. 1 above), IV, p. 395. Cf. van Ess, 'ʿAbd al-Malik and the Dome of the Rock: An Analysis of Some Texts', in *Bayt al-Maqdis: ʿAbd al-Malik's Jerusalem*, eds J. Raby and J. Johns, Oxford, 1992, I, pp. 89–103.

[46] A. Kaplony, *The Ḥaram of Jerusalem, 324–1099: Temple, Friday Mosque, Area of Spiritual Power*, Stuttgart, 2002, p. 359.

[47] The scholarly literature on the *miʿrāj* tradition is substantial. For a recent contribution that explores the gradual growth of the 'popular' *miʿrāj* narrative, see F. Colby, *Narrating Muḥammad's Night Journey: Tracing the Development of the Ibn ʿAbbās Ascension Discourse,* Albany NY, 2008.

[48] Ibn ʿAbd Rabbih (d. 940), *al-ʿIqd al-farīd*, eds M. S. al-ʿAryān et al, Cairo, 1953, VII, p. 256.

[49] C. Lange, 'The Eight Gates of Paradise-Tradition: A Genealogical and Structural Study', in *Roads to Paradise in Islam*, eds S. Günther and T. Lawson (Leiden, forthcoming).

[50] Ibn ʿAbd Rabbih, *al-ʿIqd al-farīd* (n. 48 above), VII, p. 256.

[51] Kaplony, *The Ḥaram* (n. 46 above), p. 511.

relates that the four rivers of paradise, Sayḥān, Jayḥān, Euphrates and Nile, spring from underneath the rock,[52] or that the Rock sits atop a date-palm, which has its roots in one of the rivers of paradise.[53]

The Temple Mount in Jerusalem was not the only earthly paradise known in early Islam. In the first century of the Islamic era, the sacred precinct of Thaqīf near Ṭāʾif in the Ḥijāz, in Wajj oasis, was said to have been the place where God had set His 'last step' (*ākhir waṭʾ*) on earth.[54] Hence some thought Wajj to be the earthly paradise.[55] It has been suggested that this notion was circulated in the Ḥijāz in order to oppose the claim of the Umayyad caliphate (r. 661–750) that the earthly paradise was located in Syria.[56] Wajj's career as paradise soon came to an end; but it may have set a precedent to think about earthly paradises in other local contexts. The many symbolic ways in which the Holy Sanctuary at Mecca connects its visitors with Eden and with the time before Adam and Eve's ejection from the Garden, has recently been the object of study.[57] In Syria, on the other hand, not only Jerusalem was in the race for Eden. Also Damascus, from early Islamic times, boasted a special connection with paradise. The famous Barada panels in the Umayyad mosque (completed in 714), which may have been inspired by the paradise mosaics in the south-eastern annexe floor in the Church of the Holy Cross (built in 559) of the northern Syrian city of Ruṣāfā/Sergiopolis,[58] are believed by many to depict paradise, while Islamic poets such as Ibn Qays al-Ruqayyāt (*fl.* 690) and Jarīr (d. *c.* 728) refer to a locality in the fertile valley (*ghūṭa*) of Damascus as *al-farādīs*.[59] These ideas reverberated in later centuries. In a eulogy composed after his visit to the city, the traveller Ibn Jubayr (d. 1217) wrote that Damascus is 'the paradise of the east (*jannat al-mashriq*)',

> garnished with the flowers of sweet-scented herbs, and bedecked by gardens [as if] in brocaded vestments. In the place of beauty she [Damascus] holds a sure position, and on her nuptial chair she is most richly adorned ..., [she lies at] a hill having meadows and springs[60] with deep shade and water of *salsabīl*, whose rivulets twist like serpents through every way, and the perfumed zephyrs of flower gardens breathe life to the soul ... To those who contemplate her she displays herself in a pure white dress calling to them: 'Come to the halting place of beauty, and take the midday repose'[61] ... By God, they spoke truth who said: 'If paradise is on the earth then Damascus

[52] Wāsiṭī, *Faḍāʾil* (n. 42 above), p. 68.

[53] Ibid., p. 67.

[54] *Musnad ʿUmar b. ʿAbd al-ʿAzīz*, ed. A. H. Harley, *Journal of Asiatic Society of Bengal, New Series*, 20, 1924, pp. 391–488 (421); Ibn al-Jawzī (d. 1201), *Dafʿ shubhat al-tashbīh wa-l-radd ʿalā l-mujassima,* Cairo, 1926, p. 56.

[55] See H. Lammens, 'La Cité arabe de Ṭāʾif', *Mélanges de l'Université Saint-Joseph*, 8, 1922, pp. 139–40; M. J. Kister, 'Some Reports Concerning al-Ṭāʾif', *Jerusalem Studies in Arabic and Islam*, 1, 1979, repr. in *Studies in Jāhiliyya and Early Islam*, London, 1980, XI, p. 18.

[56] See van Ess, *Theologie und Gesellschaft* (n. 1 above), IV, p. 396.

[57] Wheeler, *Mecca and Eden* (n. 28 above).

[58] Cf. T. Ulbert, *Die Basilika des heiligen Kreuzes in Resafe-Sergiopolis,* Mainz, 1986.

[59] Horovitz, 'Das koranische Paradies' (n. 25 above), pp. 61–2.

[60] The text here (*ilā rabwatin dhāti qarārin wa-maʿīn*) quotes the Qurʾān (23:50).

[61] Qurʾān 18:33.

without a doubt is in it. If it [paradise] is in the sky, then it [Damascus] vies with it and shares its glory.'[62]

Ibn Jubayr's description is a charming piece of Arabic rhetoric, ripe with quotations from, and allusions to, the Qurʾān's paradise passages. Damascus is likened to a bride embracing the visitor, a *houri* in the eternal garden awaiting the believer with outstretched arms. The smell of flowers and herbs, vegetation as lush and dark as brocade, the well of *salsabīl* and rivulets of water crisscrossing the landscape: all these are images traditionally associated with the Muslim paradise. There is undoubtedly a great deal of hyperbole in Ibn Jubayr's sentences, but it is not clear that his intended meaning is merely metaphorical. The quotation at the end of the passage suggests the possibility that Damascus is truly and fully 'paradise on earth' – even if some, as Ibn Jubayr does not fail to note, locate paradise 'in the sky'.

Madīnat al-Zahrāʾ: a paradise on earth?

Arguably the most sustained and systematic effort to create a paradise-city, however, is the palace complex known as Madīnat al-Zahrāʾ near Córdoba in Spain, built under the patronage of ʿAbd al-Raḥmān III (r. 912–61). It has become a standard scholarly tactic to question the facile assumption that Islamic art and architecture (including garden architecture) intends to create 'paradise on earth'. For example, the assertion that the Alhambra in Grenada was built to mirror the heavenly abode of the blessed rests on close to no archeological or textual evidence. Medieval Muslim writers very seldom offer symbolic interpretations of architecture, even in the case of buildings that appear ripe with eschatological connotations. Overall, Western interpreters of Islamic buildings and gardens seem to have been keener to speculate about such connotations than Muslim authors. Behind this Western propensity toward symbolic interpretation, as has been suggested, lies a reductionist agenda to de-contexualize works of Islamic art and posit instead a single dominant rationale for it, namely, a certain otherworldliness and total submission to the Islamic faith. Muslims, in this view, would be unable to appreciate art purely for esthetic reasons.[63]

If one wishes to write about the eschatological symbolism of Islamic architecture, one is well-advised to take such criticisms seriously. However, enough contextual evidence is available in the case of Madīnat al-Zahrāʾ to make a plausible case that its builder ʿAbd al-Raḥmān III intentionally sought to evoke images of paradise, albeit for reasons more political than spiritual.[64]

[62] Ibn Jubayr, *Riḥla*, ed. W. Wright, 2nd rev. edn by M. J. de Goeje, Leiden, 1907, pp. 260–61. I follow, with slight modifications, the translation by R. J. C. Broadhurst, *The Travels of Ibn Jubayr*, London, 1952, pp. 271–2.

[63] One of the most eloquent attacks on such reductionist interpretations of Islamic art has come from T. Allen, 'Imagining Paradise in Islamic Art', in *Five Essays on Islamic Art*, Occidental CA, 1988 (available at http://www.sonic.net/~tallen/palmtree/ip.html).

[64] M. Fierro, 'Madīnat al-Zahrāʾ, el Paraíso y los Fatimíes', *Al-Qanṭara*, 25/2, 2004, pp. 299–327.

Madīnat al-Zahrāʾ is located some seven kilometers west of Córdoba, in the foothills of the Sierra Morena overlooking the Guadalquivir valley to the south. It covers a rectangular area of roughly 1500m by 750m and is spread out over three large terraces. The caliphal palace and living quarters occupy the highest of the three terraces, while the second terrace is dominated by ʿAbd al-Raḥmān III's large audience hall, the Salón Rico. The urban hamlets of the city-complex are located on the third and lowest terrace. In 929, ʿAbd al-Raḥmān III, a descendant of the Umayyads, had declared himself caliph, God's representative on earth and both the religious and political leader of all Muslims. This declaration served to stake out his supreme authority in the face of two other dynasties claiming the caliphate, that of the venerable house of the ʿAbbāsids in Baghdad (r. 750–1258) and that of the Fāṭimids (r. 909–1171), the rising Shīʿī power in North Africa and later, Egypt. For the first ten years of his reign, ʿAbd al-Raḥmān III's imperial city had been Córdoba, then the largest city in Europe. Madīnat al-Zahrāʾ was built as a separate city, complete with its own palace, congregational mosque, jurisdiction and police. Yet it was not meant to supersede Córdoba, nor was its function secondary to it. In the words of its foremost interpreter, Madīnat al-Zahrāʾ 'was a whole new type of city, representing a new esthetics'.[65] Córdoba and Madīnat al-Zahrāʾ formed 'a single capital with a double pole'.[66]

Construction began in 940 shortly after the battle of Simancas, a military defeat that redirected ʿAbd al-Raḥmān III's energies from territorial expansion to symbolic representation of his empire. A significant moment in the building process was the addition of the Salón Rico between 953 and 957. The year 953 had marked the enthronement of a new Fāṭimid caliph, to which the construction of the Salón therefore appears to respond.[67] This new Fāṭimid caliph, al-Muʿizz (r. 953–75), explicitly styled himself as a messianic figure who would usher in the promised end of time. ʿAbd al-Raḥmān III's intention in building the splendid centrepiece of his new city may well have been to bring paradise, as it were, back home.

Madīnat al-Zahrāʾ, and the Salón in particular, evoke paradise in several ways: the Salón, like the eternal garden, had eight doors; in its centre was a large tank of mercury that produced dazzling light effects, reminiscent of the heavenly pool of al-Kawthar; and surrounding the building were the two pairs of the 'upper gardens' and the 'lower' gardens'. These two pairs of gardens seem to be inspired by the Qurʾān, which speaks of two pairs of paradise gardens sitting on top of each other (55:46–58). In the upper pair of gardens, as the Qurʾān says, are 'from each kind of fruit two varieties' (*min kulli fākiha zawjān*, 55:52), an expression which could be understood to refer to fantastic plants producing two different kinds of fruit at the same time. As has been suggested, it is

[65] D. Fairchild Ruggles, *Gardens, Landscapes, and Vision in the Palaces of Islamic Spain*, University Park PA, 2000, p. 53.

[66] M. Fierro, *ʿAbd al-Rahman III: The First Cordoban Caliph*, Oxford, 2005, p. 112.

[67] Fierro, 'Madīnat al-Zahrāʾ' (n. 64 above), p. 301.

precisely such asymmetrical, otherworldly plants that the panels covering the inside of the Salón aim to depict:

> [T]he decoration of its [the Salón's] walls … shows an astonishing proliferation of floral and vegetal motifs [which are] arranged alongside stems [and] asymmetrical, each side being different from its mirror counterpart… The pair of every fruit is understood by some interpreters of the Qurʾan as indicating two different kinds for each type of fruit, which will explain the asymmetry in the pairs of floral and vegetal motifs found in the decoration of the Hall.[68]

In sum, the ideological program of ʿAbd al-Raḥmān III in building Madīnat al-Zahrāʾ and the Salón in particular appears to have been to demonstrate that 'the caliph ensures salvation and therefore [that] it was as if paradise already existed in this world, in the town built by the caliph'.[69]

Contextual data confirm that the construction of Madīnat al-Zahrāʾ provoked associations with paradise. ʿAbd al-Raḥmān III's own passionate involvement in the construction went to such extremes that he failed to show up for Friday prayer in Córdoba's congregational mosque on a number of occasions. This promptly earned him a public rebuke from Córdoba's judge at the time, Mundhir b. Saʿīd.[70] Since 949, Mundhir had been one of the most high-ranking legal and religious authorities in al-Andalus. He also acted as a close advisor of ʿAbd al-Raḥmān III, which did not prevent him, however, from airing his criticism.[71] In his rebuke, 'full of allusions to the culpable behaviour of the monarch … and in particular to the criticisms addressed to him directly after he had built a pavilion covered with tiles gilded in gold and silver', Mundhir even accused ʿAbd al-Raḥmān III of 'allowing himself to be seduced by Satan to the point of acting like an infidel'.[72]

Mundhīr's most important disagreement may have been with the excessive luxury of Madīnat al-Zahrāʾ, but as Maribel Fierro has suggested, he may also have taken issue with ʿAbd al-Raḥmān III's hubris in desiring to build a paradise on earth.[73] Mundhir, as will be recalled, was among the theologians who most vociferously argued that the eschatological and protological paradise were not the same, and that the protological paradise inhabited by Adam and Eve was not on earth (see above). To see his monarch try to bring paradise down to earth may have provoked his theological sensibilities and confirmed him in his conviction that paradise was out of reach for all those who were 'still alive and walking on their two feet'.

[68] Fierro, *ʿAbd al-Rahman III* (n. 66 above), pp. 115–16. Cf. Fierro's detailed discussion of this point in 'Madīnat al-Zahrāʾ' (n. 64 above), pp. 307–8.

[69] Fierro, *ʿAbd al-Raḥmān III* (n. 66 above), p. 116.

[70] Maqqarī (d. 1632), *Nafḥ al-ṭīb*, ed. R. Dozy et al, 2 vols, Leiden, 1855–61, repr. Amsterdam, 1967, I, p. 375; idem, *Azhār al-riyāḍ fī akhbār ʿiyāḍ*, 2 vols, Cairo, 1939–40, II, p. 270. Cf. Fairchild Ruggles, *Gardens, Landscapes, and Vision* (n. 65 above), p. 60.

[71] C. Pellat, '(al-)Mundir b. Saʿīd', in *EI²*, VII, pp. 569–570.

[72] Ibid.

[73] Fierro, 'Madīnat al-Zahrāʾ' (n. 64 above), p. 325.

Memories of earlier Umayyad transgressions against this principle may have lingered in Córdoba. The Córdoban traditionalist and historian Aḥmad Ibn ʿAbd al-Barr (d. *c.* 948) wrote that the practice of planting trees in the courtyards of congregational mosques had been brought to al-Andalus in the eighth century from Syria,[74] a notion that may harken back to the Syrian Umayyads' efforts to transform such sacred sites as the Temple Mount in Jerusalem and the Umayyad Mosque in Damascus into earthly paradises. In the tenth century, under ʿAbd al-Raḥmān III and his successors, there were trees in the courtyard of the Great Mosque of Córdoba, and the famous mosaics of the prayer niche, commissioned by ʿAbd al-Ḥakam II (r. 961–76), were designed to emulate the paradise decoration of the Umayyad Mosque in Damascus.[75] However, the theologians and jurists increasingly seem to have viewed the use of such symbolism with unease, even in the context of mosques. In the eleventh century, a number of prominent jurists from al-Andalus, such as Ibn ʿAttāb (d. 1070) and Ibn Sahl (d. 1093), began to speak out against the practice of planting trees in mosque courtyards. If existing trees could not be cut, then at least ordinary believers must be prohibited from eating their fruit[76] – lest they feel they had achieved paradise already?

Conclusion

A nuanced periodization of conceptions of the earthly paradise (or indeed of the rejection of its existence), such as has been developed for the history of the European cosmological imagination,[77] remains a desideratum in the study of Islamic culture. Likewise, differences in opinion of the various theological and sectarian trends in Islam with regard to the earthly paradise deserve more careful attention. As noted above, the ʿIrāqī theologian Abū Ḥanīfa, whose followers, under the name of Ḥanafiyya-Māturīdiyya, formed the dominant school in Central Asia and Asia Minor in later centuries, located paradise in heaven rather than on earth. Theologians of the Muʿtazilite school, such as al-Jubbāʾī, who are known for their transcendentalist views, also held this opinion. Muslim mystics and philosophers tended to conceive of paradise and hell in purely spiritual terms; hence to them, the question of the physical location of paradise was of no interest at all.[78]

[74] Aḥmad Ibn ʿAbd al-Barr is quoted by Ibn Sahl (d. 1093), the leading Córdoban jurist of his time, in a *fatwā* prohibiting the planting of trees in the courtyards of mosques. See the text in *Wathāʾiq fī shuʾūn al-ʿumrān fī l-Andalus: al-masājid wa-l-dūr*, ed. M. ʿA. Khallāf, Cairo, 1983, pp. 49–50. I am grateful to Maribel Fierro for providing me with a copy of Ibn Sahl's *fatwā*.

[75] M. Fierro, 'En torno a la decoración con mosaicos de las mezquitas omeyas', in *Homenaje al Prof. Jacinto Bosch Vila,* Grenada, 1991, I, pp. 131–44.

[76] *Wathāʾiq fī shuʾūn al-ʿumrān* (n. 74 above), p. 50.

[77] A. Scafi, *Mapping Paradise: A History of Heaven on Earth*, London and Chicago IL, 2006.

[78] However, before the time of Ibn al-ʿArabī (d. 1240), the mystics of Islam only seldom questioned the physicality of the afterlife. See S. El-Saleh, *La Vie future selon le Coran*, Paris, 1971, pp. 89–120. For Ibn al-ʿArabī and his school, see W. Chittick, 'Death and the World of Imagination: Ibn al-ʿArabī's Eschatology', *Muslim World*, 78, 1988, pp. 51–82; R. Hajatpour, 'Auf der Suche nach dem verlorenen Paradies: Die Idee der Perfektibilität im Islam', in *Sehnsucht nach dem Paradies: Paradiesvorstellungen in Judentum, Christentum, Manichäismus und Islam*, eds J. Tubach, A. Drost-Abgarjan and S. Vashalomidze, Wiesbaden, 2010, pp. 159–70. For philosophical eschatology in Islam, see, for example, M. E. Marmura, 'Paradise in Islamic Philosophy', in *Roads to Paradise in Islam* (n. 49 above).

However, as I have argued in this essay, the notion that paradise overlaps with the sublunar world is pronounced in the Qurʾān, and the *ḥadīth* likewise offers ammunition for this view. Other *ḥadīth* narratives seem explicitly designed to sever the connection between this world and paradise; but these narratives could either be ignored or re-interpreted. Ibn Qayyim, for example, writes that one must distinguish between two types of entering the otherworld. One type is 'eternal entrance' (*dukhūl al-khuld*), while another type is 'accidental entrance' (*al-dukhūl al-ʿāriḍ*). 'Eternal entrance' is one-way and happens only once, that is, after resurrection on the Day of Judgement. 'Accidental entrance', however, is a return-trip that can take place anytime: Ibn Qayyim gives the example of the Prophet Muḥammad, who entered paradise during the *miʿrāj*.[79] Shurayk, the hero of the 'Story of the Leaves', would be another case in point. The tradition according to which very few if any enter paradise 'alive and walking on their two feet' thus could be interpreted as referring only to the 'eternal entrance' at the end of time, but not to the accidental visits to paradise of legendary figures and saints that one read about in pious stories or even in the *Arabian Nights*.[80]

One must also reckon with local traditions of sacred geography that developed around places like Jerusalem, Damascus and Mecca, connecting or in some cases even equating these places with the earthly paradise. By definition, as one might argue, such local traditions could not be indicated on 'universal' maps of the world. Sometimes, as in the case of Madīnat al-Zahrāʾ, a deliberate 'edenification' by the political authorities took place. Such political maneuvering, in turn, did not go unchallenged: the Córdoban judge and theologian Mundhir b. Saʿīd publicly questioned ʿAbd al-Raḥmān III's intention in building his new residence. In later centuries, the famous Ibn Taymiyya (d. 1328) waxed angrily about the 'edenification' of Jerusalem and popular traditions of hagiogeography such as are reflected in the 'Story of the Leaves'.[81]

The way in which a religious tradition imagines the hereafter has profound consequences for how it pictures this world, and the place of the human race within it. Although it may be true for certain strands within Islamic thought that paradise was believed to be entirely out of human reach, this cannot be said of Islamic religious history as a whole. Medieval Islamic maps and cosmographical charts may not show it, but as the examples adduced here suggest, in the Islamic religious imagination, paradise often appears near indeed, or even 'closer to you than the strap of your sandal'.

[79] Ibn Qayyim, *Ḥādī* (n. 29 above), p. 38.

[80] The hero of 'The Second Qalandar's Tale', for example, stumbles upon a subterranean cave in which he is served by a beautiful and ageless *houri*-princess who is kept prisoner there by an evil *jinnī*. When the hero, out of curiosity, provokes the *jinnī* to appear, he is expelled from the cave. Cf. R. Mottahedeh, '*ʿAjaʾib* in the Thousand and One Nights,' in *The Thousand and One Nights in Arabic Literature and Culture*, eds R. C. Hovannisian and G. Sabbagh, Cambridge, 1997, pp. 29–39.

[81] On Ibn Taymiyya's 'repugnance' against the exaggerated veneration of the *Ḥaram*, see I. Hasson, 'Introduction', in Wāsiṭī, *Faḍāʾil* (n. 42 above), pp. 17–18.

In Medieval Islamic Cosmography, where is Paradise?

Emilie Savage-Smith

In Arabic literature, by far the most common term for paradise is *jannah*, meaning 'garden'. It is used in two senses: (1) the primordial garden, the dwelling place of Adam and Eve, and (2) the after-world of the righteous. *Jannah* is not the only term for paradise, however, for the term *firdaws* – an Arabic form of an Old Iranian word meaning enclosure, that passed into Greek as παράδεισος from which we get our word paradise – was also used occasionally to designate paradise in general as well as a walled garden of trees within paradise.

Paradise, and its associated rewards, forms one of the major themes in the Qurʾān.[1] Subsequent Qurʾānic scholars interpreted the relevant verses in different ways, at times trying to reconcile varying or even conflicting interpretations of the same Qurʾānic passage, and at other times generating their own interpretations of selected traditions. For the purposes of such a paper as this, however, these complexities of interpretation have to be glossed over or drastically simplified.[2]

Because the terms for paradise (*jannah*, *firdaws*) meant 'garden', the representation of a garden is often interpreted by historians as a depiction of paradise. For example, two of the loveliest Islamic manuscript paintings of gardens in Islamic art, illustrating a Persian poetic anthology from Iran dated H 801 (AD 1398), are usually viewed by art historians as representing the Qurʾānic description of paradise.[3] They could, of course, just as easily be meant to evoke an idyllic landscape on earth, or simply a decorative design. Another example is the mosaic panel from the western ambulatory of the Umayyad Mosque in

The author wishes to thank Alessandro Scafi for inviting her to participate in the workshop on 'The Cosmography of Paradise' and the participants at the conference for making helpful suggestions. In particular, she wishes to thank Alasdair Watson for reading a draft and providing extremely useful suggestions, as well as Christian Lange, Yossef Rapoport, Colin Wakefield, Elvira Wakelnig, and M. B. Smith. The author also thanks Sheila Blair for providing a copy of her paper 'Ascending to Heaven: Fourteenth-century illustrations of the Prophet's Miʿrāj' prior to its publication. All misinterpretations, however, are solely the present author's responsibility.

Abbreviations used in this essay:

EI² = *The Encyclopaedia of Islam*, eds H. A. R. Gibb, B. Lewis, C. Pellat, C. Bosworth et al., 2nd edn, 11 vols, Leiden, 1960–2002.

EQ = *Encyclopaedia of the Qurʾān*, eds J. D. McAuliffe et al, 6 vols, Leiden, 2001–6.

[1] The word *jannah* occurs 77 times in the Qurʾān, while the term *firdaws* occurs only twice (Qurʾān 18:107 and 23:11), and in one of those occurrences the phrase is *jannat al-firdaws* ('the garden of paradise').

[2] In the following discussions of paradise and its attributes I am depending to a large extent upon A. M. Heinen, *Islamic Cosmology: A Study of as-Suyūṭī's* al-Hayʾa as-sanīya fī l-hayʾa as-sunnīya, with *critical edition, translation, and commentary*, Beirut, 1982; N. Rustomji, *The Garden and the Fire: Heaven and Hell in Islamic Culture*, New York, 2009; L. Gardet, 'Djanna', in *EI²*, II, pp. 447–52; and L. Kinberg, 'Paradise', in *EQ*, IV, pp. 12–19.

[3] İstanbul, Türk ve Islam Eserleri Müzesi, MS 1950, fol. 128v, dated Muḥarram 801 (September 1398), reproduced in T. W. Lentz and G. D. Lowry, *Timur and the Princely Vision: Persian Art and Culture in the Fifteenth Century*, Los Angeles CA, 1989, no. 14; some have suggested that these particular landscapes derive from Zoroastrian sources.

Damascus (AD 708–14), whose monumental trees and intricate buildings bordering a long stretch of water are often interpreted as representing paradise, even though there is no documentary evidence that a depiction of paradise was the intent of the producer of the mosaics.[4] If the trees are blooming while also in fruit (something that cannot occur on earth but is characteristic of trees in paradise), then paradise may be intended. But the state of the trees is unclear. The point I wish to make is that you will find people cataloguing all sorts of art – even rugs – as showing paradise.[5] In Islamic art, without documentation of the artist's intent, depictions of gardens become highly ambiguous. For that reason, for the purposes of this paper I will restrict myself only to images (and texts) that are specifically stated to concern paradise.

In Islam, the major source for information about the nature and structure of paradise was the first-hand experience of the Prophet Muḥammad himself. Through various visions in a waking state, and through dreams, Muḥammad was able to actually *observe*, it is believed, aspects of paradise and hell. This first-hand autopsy of paradise on the part of the Prophet, as elaborated in post-Qurʾānic narratives, was fundamental in establishing the Islamic conception of paradise.[6]

With regard to paradise, the most important event was his Night Journey and Ascension to Heaven (*al-israʾ wa-al-miʿrāj*). The event was a favourite topic for Safavid and Mughal painters, especially from the fifteenth century onwards, when illustrating copies of the classic Persian poetic anthology known as the *Khamsah* ('Quintet') by Niẓāmī (d. *c.* 613/1217) or *Yūsuf va Zulaykhah* of Jāmī (d. 898/1492), as well as devotional works about the Prophet's life and the class of Arabic treatises referred to as 'tales of the prophets' (*qiṣaṣ al-anbiyāʾ*).[7] There are also at least two illustrated treatises, called *Miʿrājnāma*, devoted entirely to the subject of the Prophet's ascension, with the earliest produced in the Timurid period (*c.* AD 1370–1500).[8]

[4] This particular mosaic on the Umayyad Mosque is known as the Barada panel after the river Barada that runs through Damascus. For details of portions of the mosaic, and a discussion of its association with paradise, see Rustomji, *The Garden and the Fire* (n. 2 above), pp. 132–5. See also *The Umayyads: The Rise of Islamic Art*, Amman and Vienna, 2000, p. 49; and R. Burns, *Damascus: A History*, Abingdon, 2005, pp. 116–17; F. B. Flood, *The Great Mosque of Damascus: Studies in the Making of an Umayyad Visual Culture*, Leiden, 2001, pp. 25–47.

[5] See, for example, S. S. Blair and J. M. Bloom, *Images of Paradise in Islamic Art*, Hanover NH, *c.* 1991, and *Hunt for Paradise: Court Arts of Safavid Iran, 1501–1576*, eds J. Thompson and S. R. Canby, New York and Milan, 2003.

[6] For Muḥammad's visions of paradise experienced while praying at the time of a solar eclipse, see Rustomji, *The Garden and the Fire* (n. 2 above), pp. 23–6 and, for those visions undertaken in a more mundane state, pp. 36–9. For the Night Journey and Ascension, see ibid., pp. 26–36, and *The Prophet's Ascension: Cross-Cultural Encounters with the Islamic Miʿrāj Tales*, eds C. Gruber and F. Colby, Bloomington IN, 2010.

[7] For illustrations of the Night Journey and for Ascension copies of the *Khamsa* by Niẓāmī, see Lentz and Lowry, *Timur and the Princely Vision* (n. 3 above), no. 140, and *Hunt for Paradise*, eds Thompson and Canby (n. 5 above), pl. 4.29. For an illustration of the ascent of the Prophet, preserved from a dispersed *Fālnāmeh* attributed to Jaʿfar al-Sādiq, see ibid., pl. 1.6, and M. Farhad with S. Bağcı, *Falnama: The Book of Omens*, Washington DC, 2009, item 22. For several illustrations from copies of Niẓāmī's *Khamsa* (Quintet) and Jāmī's *Yūsuf va Zulaykhah*, see *The Prophet's Ascension*, eds Gruber and Colby (n. 6 above).

[8] See C. Gruber, *The Timurid Book of Ascension (Miʿrajnama): A Study of Text and Image in a Pan-Asian Context*, Valencia, 2008, and her *The Ilkhanid Book of Ascension: A Persian-Sunni Devotional Tale*, London, 2009.

In the usual late-medieval version of the Night Journey, Muḥammad was guided from Mecca by the archangel Gabriel and transported to Bayt al-Maqdis (Jerusalem) on Burāq – a winged beast with human head, described as having a tail and feet like a camel's, a rump like that of a horse, and an emerald saddle, pearl reins, and turquoise stirrups. From Jerusalem, Muḥammad was transported to heaven. There he visited the seven levels of paradise and met some of its inhabitants and angels. He then returned to Mecca that same night. This Night Journey and Ascension of Muḥammad provided proof to believers of the afterworld. As a consequence, the Prophet's description of what he saw became an important source of information on paradise and the basis of subsequent commentaries and theories.

The manuscript paintings of the Night Journey and Ascension, however, tell us virtually nothing about the actual nature of paradise itself, but rather concentrate upon depicting the journey there and back. They usually show the Prophet (often with a nimbus of fire about his head) riding Burāq and accompanied by angels carrying vessels. It is only in the eighteenth century that we begin to get diagrams that attempt to visually or graphically represent paradise itself. Three of these relatively late manuscript illustrations I will use as convenient frameworks with which to understand the structure of paradise and its role within a traditional Islamic cosmology.

Fig. 1 is a Mughal manuscript illustration made possibly in Delhi around 1800.[9] No text accompanies the illustration, though it is paired with another showing the Night Journey and Ascension.[10] The bottom, blue-grey, band is occupied by the figure of Aaron (Hārūn) to the left of a central sketch of a building. In the small white space above, Hārūn is specified as the person depicted, while at the centre a red label reads 'the gate of the sixth heaven' (*bāb-i āsmān-i shishum*), applying either to the structure below or the doorway in the base of the tree above. The large yellow band has a label on the right-hand side stating that it is the 'sixth heaven' and is of pure gold (*az ṭilā*); in the middle is the base of a large tree, with the figure of Moses (Mūsá) to the left.[11] Above the gold band representing the 'sixth heaven' is a red-brown band labelled the 'seventh heaven' and said to be of ruby (*az yāqūt*); the figure of Abraham (Ibrāhīm) sits to the left of tree branches inhabited by birds.[12] All three figures (Aaron, Moses, and Abraham) are wearing face veils. At the base of the tree four grey offshoots arise, two on each side, which may represent rivers or, more likely, are intended simply as tree roots on which birds are perched. The notion of 'seven heavens' will be discussed below in greater detail.

[9] Washington DC, Freer Gallery of Art, Smithsonian Institution, inv. no. F1968.3, gift of Professor Alban G. Widgery of Winchester VA; the dimensions of the painting are 62.2 × 40.5 cm.

[10] Washington DC, Freer Gallery of Art, Smithsonian Institution, inv. no. F1968.2.

[11] For Moses being associated with the 'sixth heaven', see Gruber, *The Ilkhanid Book of Ascension* (n. 8 above), pp. 56–8.

[12] For Abraham being associated with the 'seventh heaven', see ibid., pp. 59–60.

From the trunk of the tree, which has a vertical label reading *maqām-i Jibrīl* ('the place of the archangel Gabriel'),[13] the tree opens upward and is espaliered in front of seven well-tended, tiered gardens, each with a name repeated four times on the four vertical bands of different colours that run through them. In ascending order, from the one adjacent to the 'seventh heaven' occupied by Abraham upward, the seven gardens are: 'the garden of the house of eternity' (*jannat dār al-khuld*), 'the garden of the house of splendour' (*jannat dār al-jalāl*), 'the garden of the house of peace' (*jannat dār al-salām*), 'the garden of Eden' (*jannat al-ʿadn*), 'the garden of rest' (*jannat al-maʾwá*), 'the garden of bliss' (*jannat al-naʿīm*), and, at the top, *firdaws* (*jannat al-firdaws*). A variety of birds inhabit the lower part of the tree below the first garden, along with two angels and a large pavilion. The six long drooping branches of the espaliered portion of the tree have angels carrying various baskets and containers, presumably harvesting fruit, and are draped here and there with strands of pearls and jewels as well as occasional scarves and ornamental clothing.[14]

At the top of the painting, above the espaliered tree and gardens, are five horizontal panels or bands, each with a white ground decorated with an overall geometric design drawn in gold. At the middle of the top panel the word *huwa* is written, meaning, in this context, God. To the left, in the same panel, the annotation *lā makān* (literally, 'no place') is written, meaning 'not a physical entity', that is, it is not a physical place as God is not located in a *locus*.[15] The same expression, *lā makān*, is written in the second band from the top, which is a completely empty stratum.

The Throne of God is depicted in the middle of the top five panels, with two angels on either side; it is labelled *maqām-i ʿarsh* ('the position of the throne'). Immediately below, in the fourth panel from the top, is the Footstool of God, labelled twice as *maqām-i kursī* ('the position of the footstool'). In the lowest of the top five panels there is a statement that the distance between the Footstool and paradise is the flight of a bird (*rafraf*) over a journey of thirty-six thousand years.[16]

To the left of God's Footstool we find a rooster or cock on whose side, written in gold, is the *shahādah* (the Muslim profession of faith), with a quotation from the Qurʾān written beneath in red ink.[17] Over the bird, there is a red label saying 'a prayer-calling angel (*firishta-i muʾazzin*) at the Lote Tree on the Boundary (*sidrat al-muntahá*), whose form is a white cockerel (*khuros*) suspended beneath the Throne'.

[13] For Gabriel as the guardian of paradise, see J. Pedersen, 'Djabrāʾīl', in *EI²*, II, pp. 362–4.

[14] In the texts relating the ascension it was said that when the Prophet viewed paradise he observed that 'the trees bear silken garments and whoever wears them never dies'; Gruber, *The Ilkhanid Book of Ascension* (n. 8 above), p. 70.

[15] The term *lā makān* is translated by Christiane Gruber as 'placeless realm', and it was used also for the space supporting the world; Gruber, *The Ilkhanid Book of Ascension* (n. 8 above), p. 62.

[16] Cosmic distances were usually measured in units of time. For the role of the Throne and Footstool of God (mentioned in the Qurʾān 9:129 and 2:253) in Islamic cosmology, see Heinen, *Islamic Cosmology* (n. 2 above), pp. 77–81, 130–34, 182–92, and 199–201.

[17] 'Everything will perish except His own face' (Qurʾān 28:88).

The appearance of such a creature might seem inexplicable, but a similar fowl appears in some illustrations of the Prophet's Night Journey. For example, an anonymous *Miʿrājnāma* probably made in Tabriz about AD 1317–35 has a painting of the Prophet Muḥammad, accompanied by a crowned angel (presumably, the archangel Gabriel), encountering a large celestial cockerel or rooster in the first heaven, witnessed by a group of angels raising their hands in prayer.[18] The celestial cockerel is included in many ascension texts, and is usually said to be positioned between the earth and God's Throne, while in this particular Ilkhanid illustration, the rooster stands on a polygonal dais. The cockerel is traditionally associated with the archangel Gabriel, who was responsible for counting the hours of the day and night in order to announce the hours of prayer and pronounce the *tasbīḥ* (a formulaic phrase beginning *subḥāna llāh* 'Praise to God!'), which was then repeated by roosters around the world. In the ascension text accompanying the Ilkhanid painting just described, the following account of this gigantic celestial cockerel is given:[19]

> I [the Prophet Muḥammad] saw a rooster standing on its feet. Its wings were whiter than snow. Its feet reached the earth, and its head breached through the skies. When it spread a wing, it stretched from the east to the west and beyond. It crowed and flapped its wings. It was uttering the prayer: 'Praise be to God the Great, the Exalted, there is no god but God, the Living, the Eternal'. When it crowed, the roosters from the east to the west crowed, and the crowing of those roosters was: 'Remember God, oh negligent ones!'
>
> I [the Prophet Muḥammad] asked, 'O Gabriel, what is this rooster?' He answered, 'Every night and day, this rooster utters a prayer twenty-four thousand times each hour. When they hear its crowing, all roosters crow. When it flaps it wings, they flap their wings. When it quiets down, the others quiet down'.

In this Indo-Persian depiction (Fig. 1), paradise is placed under the Throne of God (its most common placement by scholars), with two intervening strata, one containing the celestial rooster and God's Footstool. Paradise itself is clearly divided into seven levels – each indicated by one of the wooden (or brick) cross-pieces of the espaliered tree. There is a door at each place where a wooden tier (or brick wall) crosses the main tree trunk, and on the lintel of each square door frame is written the Muslim profession of faith (*lā ilāha illā llāh wa-Muḥammad rasūl Allāh*, 'there is no god but God, and Muḥammad is His messenger').[20]

[18] İstanbul, Topkapı Sarayı Kütüphanesi, MS H. 2154, fol. 61ᵛ; reproduced and discussed by C. Gruber, 'The Ilkhanid *Miʿrājnāma* as an Illustrated Sunni Prayer Manual', in *The Prophet's Ascension*, eds Gruber and Colby (n. 6 above), pp. 27–49, esp. pp. 35–8, fig. 2.3, and pl. 8. The painting is also reproduced in Gruber, *The Ilkhanid Book of Ascension* (n. 8 above), pl. 6.

[19] Gruber, *The Ilkhanid Book of Ascension* (n. 8 above), p. 46; the translation is hers.

[20] According to the traditionalist al-Bukhārī (d. 256/870), the key to opening the eight principle gates of paradise is to (1) proclaim the divine unity (*tawḥīd*), (2) obedience to God, and (3) abstention from all unlawful deeds; see Gardet, 'Djanna' (n. 2 above), p. 448.

Topographical strata and features

In his Night Journey, Muḥammad stressed that paradise was stratified both ethically (in terms of piousness) and topographically. Seven levels or tiers formed the most common division of paradise. The term *firdaws* sometimes designated only the seventh, highest space. For some commentators, the paradise of Adam and Eve, the Garden of Eden, was placed in the seventh level. During the Night Journey, the archangel Gabriel took Muḥammad to visit the various levels of paradise, and at approximately the seventh level, Muḥammad meets Abraham (Ibrāhīm) and proceeds as far as the 'Lote Tree on the Boundary' (*sidrat al-muntahá*), mentioned in the Qur'ān 53:14, beyond which no one can pass; there Muḥammad gazes upon God and bows down to worship Him.[21] The Lote Tree is not illustrated in Fig. 1, but much was written about it in the medieval texts: it had fruits of indescribable colours, the size of earthenware jars, and leaves as large as an elephant's ears. Some said a single leaf from the Lote Tree on the Boundary could shade the entire community of the Faithful.[22] Others stated that *al-jannah* (paradise) consists of one hundred levels, of which *firdaws* is the best, while it was also said by some that *al-firdaws* itself consists of four gardens, two of gold and two of silver.[23]

Another common division of paradise was into two gardens. This accorded with the statement in the Qur'ān (55:46), saying that two gardens await those who fear God. Commentators suggested various interpretations, including one garden for those who obey God, another for those who refrain from sin; or, one for the *jinn* (shape-shifting demons mentioned in the Qur'ān) and a second for humans; or one for those who worship God secretly and one for those who worship openly.[24]

All the gardens – whatever the number or stratification – have wonderful fruit trees, with no thorns; they bear fruit born at the same time as they are flowering (an earthly impossibility). The trees are of a wondrous size: for example, it would take a horse 100 years at a gallop to emerge from the shade of a banana-tree.[25]

The gardens were irrigated by underground rivers. Rivers of Paradise (*anhār al-jannah*) are mentioned forty-six times in the Qur'ān.[26] They are described variously as water always gushing and flowing, milk whose taste never alters, wine which does not intoxicate, or pure honey. According to one tradition, there are four rivers, two are internal to paradise and two external (the Nile and the Euphrates), while other traditions say all four are external (two in Iraq: the Nile, the Euphrates, and two in Anatolia: the Sayḥān and Jayḥān). According to others, the four Rivers of Paradise flow from under the Lote Tree of the Boundary. While the artist of the painting in Fig. 1 has not labelled the Rivers of

[21] A. Rippin, 'Sidrat al-Muntahā', in *EI*², IX, p. 550.

[22] Gardet, 'Djanna' (n. 2 above), pp. 448–9.

[23] Kinberg, 'Paradise' (n. 2 above), p. 13.

[24] Ibid., p. 15

[25] Gardet, 'Djanna' (n. 2 above), p. 448, citing al-Bukhārī; Gruber, *The Ilkhanid Book of Ascension* (n. 8 above), p. 70.

[26] See Amira El-Zein, 'Water of Paradise', in *EI*², V, pp. 466–7.

Paradise, it is likely that the four vertical bands of different colours, two on either side of the tree, running through the gardens are depicting them, with their different colours representing water, milk, wine, and honey; they appear to be traversing the seven gardens in stone conduits from which smaller horizontal channels irrigate the gardens. Given the fact that there is not a single river in all of Arabia, it is not surprising that there was so much interest shown in irrigated gardens.

Four mountains were assigned by many to paradise (Uḥud, Sinai, Lebanon, Ḥaṣīb) and large valleys are described. Inhabitants were said to include horses and camels 'of dazzling whiteness' and also sheep and goats. According to some, *al-jannah* is the abode of Gabriel and the angels, or of green birds containing the souls of martyrs, without specifying a location or a garden; such notions are perhaps reflected in the painting of Fig. 1.

Ethical or moral stratification

The inhabitants of paradise could be grouped in various ways according to their level of piety. Some placed together those who lived prior to Muḥammad, then those who contributed to Islam in its first stages, and finally the righteous who came later. The angels and Jesus (ʿĪsá) are usually given special places, as are Adam and Eve who were nearly always assigned particular locations.

Paradise (in all its variations) in the medieval Islamic tradition is an eternal abode with physical attributes; its inhabitants are envisioned as living beings with human sensations – not souls divorced from bodies. In fact, one tradition says that 'each of the elect will have the same stature as Adam (i.e., 60 cubits by 7) and the same age as Jesus (i.e., 33 years)'.[27]

The chosen will rejoice in the company of those of their parents, wives and children who are faithful and righteous and hence eligible for paradise.[28] There is the assumption that the goods of paradise will be perfected replicas of goods on earth. Robes and adornments will be marvellous. There was also the notion that unimagined and unimaginable delights were awaiting the blessed.[29]

A detailed description of the rewards bestowed upon its inhabitants is provided in the Qurʾān,[30] where it states that in the Garden of Bliss (*jannat al-naʿīm*) there will be a number of people reclining on couches or thrones encrusted (with gold and precious stones). Youths of perpetual youthfulness will serve them, using goblets and beakers and

[27] Gardet, 'Djanna' (n. 2 above), p. 449.

[28] Qurʾān 13:23 and 36:56.

[29] For example the Qurʾān states: 'Now no person knows what delights of the eye are kept hidden (in reserve) for them, as a reward for their (good) deeds' (32:17). Traditionalists such as Muslim ibn al-Ḥajjāj (d. 261/875) recorded a tradition that the Prophet said 'I have prepared for my pious servants that which no eye has ever seen, and no ear has ever heard, and no human heart has ever imagined' (Muslim ibn al-Ḥajjāj, *al-Jāmiʿ al-ṣaḥīḥ*, *Sahih Muslim,* English transl. ʿAbd al-Hamīd Siddīqī, Lahore, 1971–5, XL, no. 6780).

[30] Qurʾān 56:12–26.

cups filled from clear-flowing fountains. They will suffer no discomfort, and they will not become intoxicated. They can have any fruits they wish and eat of any fowl they wish. There will be companions with beautiful, large, lustrous eyes – eyes like pearls – who will be a reward for the deeds of their past lives.

The Qurʾānic description of these companions, the *houris* (*ḥūr*, plural of *ḥawrāʾ*), whom neither man nor *jinn* will have touched (55:56, 55:74), prompted endless commentaries. The later traditions described many physical and sensual joys that the elect – presumably *male* elect – would derive, as well as adding other details: their eyebrows are black lines; their forehead is a crescent moon, and their shining faces reflect the divine light. Their flesh is so delicate that the marrow of their shank bones can be seen, even through seventy silken garments. They dwell in pavilions or castles hollowed out of a single enormous pearl and furnished with seventy sumptuous beds and seventy couches of ruby covered with seventy rugs, and they are surrounded by 70,000 maids-of-honour. Their age equals that of their husband in paradise (i.e., 33, the age of Jesus). Their virginity is perpetually renewed. Some elect, such as Jesus, will have a hundred *houris* at their disposal and will be able to have intercourse with each of them as many times as he has fasted during the days of Ramaḍān and performed good works. Being purified, the *houris* know neither menstruation, nor human needs, nor pains of childbirth, for they bear no children.[31]

Such are the pleasures afforded men in paradise, but what of women? Women who have been admitted to paradise through their good deeds will rank 70,000 times greater than the *houris* in the eye of God. While there are no sensual pleasures comparable to those provided by the *houris* for men, women in paradise will know none of the cares of earthly life and will enjoy numerous pleasures. Women who had several husbands will choose the one they prefer (the assumption being that all the husbands will be in paradise), while polygamous husbands will be allowed to keep all their earthly wives (who qualify for paradise).[32]

Depicting Paradise

If we wish to graphically depict paradise within a larger cosmological framework, then we must turn to a popular late Ottoman Turkish treatise titled *Maʿrifetnāme* ('The Book of Gnosis') written in AD 1756–7 (H 1170) by the mystic-scholar Ibrāhīm Ḥaqqī (İbrahim Hakkı).[33] They include two cosmographical diagrams reproduced in Figs 2 and 3 from a copy made in the city of Erzerum in eastern Turkey and completed in July to August of 1820 (Shawwāl, H 1235). Both diagrams show the entire universe enveloped

[31] Gardet, 'Djanna' (n. 2 above), p. 449; A. J. Wensinck and C. Pellat, 'Ḥūr' in *EI*², III, pp. 581–2.

[32] Wensinck and Pellat, ibid., p. 582.

[33] London, British Library, MS Or. 12964, fols 23ᵛ and 24ʳ, copied in Erzerum in Shawwāl 1235 (July-August AD 1820). The illustrations were published (without the labels or detailed discussion) in A. T. Karamustafa, 'Cosmological Diagrams', in *The History of Cartography,* II/1*: Cartography in the Traditional Islamic and South Asian Societies*, eds J. B. Harley and D. Woodward (Chicago IL and London, 1992), pp. 71–89, fig. 3.18 and pl. 3.

on three sides by six rectangular bands (see Fig. 4). These are labelled, beginning with the topmost:

1. 'The realm of the Godhead' (*ʿālam-i lāhūt*)
2. 'The Throne of God' (*ʿarsh aʿẓam*)
3. 'The realm of divine omnipotence' (*ʿālam al-jabarūt* or *ʿālam-i jabarūt*)
4. 'The Footstool (*kursī*) [of God]'
5. 'The realm of the souls and the kingdom' (*ʿālam al-arwāḥ wa-al-malakūt*)
6. 'The trumpet of Isrāfīl' (*ṣūr-i Isrāfīl*)

Isrāfīl, named in the innermost band, was an archangel of astounding size.[34] His head was said to support the divine throne, while his feet extended under the 'seven earths' (a concept to be discussed below). He is the angel who reads out the divine decisions from the 'guarded tablet' and transmits them to the archangel Gabriel. Three times a day, and three times at night, he looks into hell and is so grieved by what he sees that he cries tears that produce the rainfall on earth. It was said that he continually holds a trumpet to his mouth in order to be ready to blow it, as soon as God commands, in order to arouse humans from their graves, and some accounts specified the length of this trumpet to be forty thousand years, with a width of twenty thousand years.[35]

In addition to these six enclosing bands, both diagrams also share other features (Fig. 4). At the top of each diagram are eight graduated tiers. The base of the tiers bears the numeral '1' and has eight small colourful doors in it, with the label 'Gates of Paradise' (*abwāb al-jinān*) written beneath. Over the seven tiers of paradise itself, as well as the base with paradisiacal gates, there hang branches from the heavenly tree called Ṭūbā (*shajarah-i Ṭūbá*). To the upper left of the tiered paradise are the 'guarded tablet' (*lawḥ maḥfūẓ*) and the pen or stylus (*qalam*), both of which are important concepts of Islamic revelation though usually of less importance in cosmological speculation.[36] Above the right-hand corner of the second tier of paradise on both diagrams there is a label *jabal al-ḥamd* ('the mountain of praise [of God]'). Above the latter enigmatic inscription, on fol. 23^{v} only (see Figs 2 and 5), there is a small banner on a standard, labelled *liwāʾ al-ḥamd* ('the banner of praise [of God]'); this banner occurs on fol. 24^{r} in a different position and in a larger size (see Figs 3 and 6).

Both diagrams also have an identical lower design depicting hell (Fig. 4). The top of hell is demarcated by an arched bridge, called 'the bridge (*al-ṣirāṭ*) over the top of hell (*al-ṣirāṭ ʿalá matn jahannam*)' and 'the body of the bridge over the top of hell (*jism-i ṣirāṭ ʿalá matn jahannam*)', with seven gates opening downward from it. The word *al-ṣirāṭ* was a

[34] A. J. Wensinck, 'Isrāfīl', in *EI*^{2}, V, p. 211.

[35] For these cosmic dimensions of the trumpet, given as usual in terms of time, see Gruber, *The Ilkhanid Book of Ascension* (n. 8 above), pp. 65–6.

[36] For the role of the guarded tablet and the stylus in theological and cosmological thought, see Heinen, *Islamic Cosmology* (n. 2 above), pp. 81–5, 135–78, and 192–6.

specific name for the bridge, 'narrower than a hair and sharper than a sword', that extended over the pit of hell; believers will cross it with the speed of lightning, while the wicked will fall into the fire.[37]

Immediately beneath the middle of the bridge is a rectangle filled with the tree of hell, Zaqqūm (*shajarah-i Zaqqūm*)[38] – a tree mentioned three times in the Qurʾān as growing in hell, with bitter fruit which the damned are condemned to eat. Beneath are two sets of red tiered platforms, numbered 1 to 7, so that hell consists of seven levels comparable to those of paradise above.

While paradise and hell, as well as the surrounding six frames, are identical on these two diagrams, the middle portions differ.

The diagram reproduced in Fig. 2 (and with explanatory labels in Fig. 5) portrays the earth and its relationship to the firmament, with paradise above and hell below, according to traditional Qurʾānic cosmology. Immediately beneath the Gates of Paradise (see Fig. 5) are three rectangles. The one on the right has three layers of blue, the top labelled 'a boiling-over sea' (*baḥr masjūr*),[39] the middle 'an expansive sea' (*baḥr manshūr*), and the lower 'a blue ruthless sea' (*baḥr zaraq maghsūm*); immediately beneath is a suspended disc labelled 'the substance of day' (*jawhar-i nahār*). The left-hand rectangle is also in three blue layers, but without labels; the suspended disc beneath reads 'the substance of night' (*jawhar-i layl*). The middle grey-brown rectangle also has no internal labels, but written beneath is 'the partition of the angels' (*ḥajb al-malāʾikah*).

In the centre of the diagram (see Fig. 5), representing earth, is the Kaʿba in Mecca (labelled *Kaʿbah-i Mukarramah*) and prominently painted in black, orange, and grey-brown. Above it are three small coloured lozenges, each one labelled. The brown, lower lozenge reads 'clouds and vaporous clouds' (*ghamām wa-saḥāʾib*), the middle, light-brown one 'mountains of snow and hailstones' (*jibāl-i salj* [=*thalj*] *wa-barad*), while the top grey lozenge has the puzzling label 'a secluded [isolated] sea' (*baḥr maʿkūf*).[40]

The Kaʿba and the three small lozenges are enclosed by a pink arching band which contains the sun and moon and lots of red dots (presumably representing stars); the bottom portion of this pink band has two labels: 'stars' (*kawākib*) and 'a surrounding sea' (*baḥr muḥīṭ*). These two labels representing the sun, moon, and stars is somewhat confusing and suggests that the artist failed to depict the sea that was thought to surround the inhabited land on earth and simply wrote the label for it wherever there was space.

This configuration then rests on top of a seven-tiered structure with seven blue arches rising over them and connecting to either end of each tier. The seven blue arches are

[37] For *al-ṣirāṭ* in this specific sense, see G. Monnot, 'Ṣirāṭ', in *EI*², IX, p. 670.

[38] C. E. Bosworth, 'Zaqqūm', in *EI*², XI, p. 425.

[39] The 'boiling-over' or 'swollen' sea (*baḥr masjūr*) is mentioned in the Ilkhanid *Miʿrājnāma* and illustrated with a painting of the Prophet Muḥammad being carried on the shoulders of Gabriel while flying over this swollen sea that lies beyond the seventh heaven; see Gruber, *The Ilkhanid Book of Ascension* (n. 8 above), fol. 42ᵛ.

[40] For the notion of 'an enclosed sea' in the sky, under the Throne, see Heinen, *Islamic Cosmology* (n. 2 above), p. 117.

labelled 'the seven heavens' (*al-samawāt al-sab'ah*), while the tiers are numbered 1 to 7, beginning with the lowest and largest tier. At the base of the seven tiers is the label: 'the location of the angel who supports the seven earths' (*makān malak al-ḥāmil | al-araḍīn al-sab'ah*). At each of the outside corners of the base is the label 'the mountain of Qāf' (*jabal-i Qāf*), and the same label occurs on either side of the base of the Ka'ba.

The legendary Mountain of Qāf was said to encompass the sea surrounding the inhabited world.[41] It was the source of all mountains on the earth and made of green topaz or emerald, causing the greenish colour of the sky which is reflected onto the green sea.[42] The Mountain of Qāf is occasionally illustrated on Islamic world maps, a particularly outstanding example being included in a copy made in 1622 (H 1032) of the *Marvels of Things Created and Miraculous Aspects of Things Existing* (*Kitāb 'Ajā'ib al-makhlūqāt wa-gharā'ib al-mawjūdāt*) composed by the cosmographer al-Qazwīnī (d. 682/1283).[43] That map shows the inhabited earth surrounded by a sea (a blue circle); the sea, in turn, is enclosed by a ring of impassable land. Beyond the ring (looking like mountains of fire) is the legendary Mountain of Qāf. Through a break in the ring representing the Mountain of Qāf and the ring of uncrossable land, the Fountain of Life (*'ayn al-ḥayāh*) pours into the Surrounding Sea.

The notion of seven heavens and seven earths reflects a cosmology derived from the Qur'ān as well as from a pre-Islamic tradition.[44] According to one version, God's throne was said to be above eight mountain goats, standing on an ocean, and beneath this ocean are seven heavens, and below these were seven earths, at distances of 500-year units. A variant tradition puts the celestial ocean between earth and the lowest of the seven heavens and omits the mountain goats. In the painting illustrated in Figs 2 and 5 there are no mountain goats, but there are a number of oceans or seas between paradise and the outermost heaven, as well as 'an isolated sea' between the innermost heaven and earth (as represented by the Ka'ba). The seven heavens were sometimes associated with different substances, such as (in ascending order), water, white marble, iron, copper, silver, gold, and ruby.[45]

The notion of 'seven earths' is even more difficult for a modern reader than seven heavens. There were numerous interpretations. Most commonly they were interpreted as the seven stages of hell, each with a storehouse for a torment: the first (in Figs 2 and 5 at the top of the pyramid, labelled number 7) was earth itself; second, the sterile

[41] M. Streck and A. Miquel, 'Qāf', in *EI*², IV, pp. 400–402; this conception of a mythical mountain range at the edge of, and surrounding, the terrestrial world reflects an ancient Iranian tradition.

[42] Heinen, *Islamic Cosmology* (n. 2 above), p. 118.

[43] Gotha, Forschungs-Bibliothek, MS Orient. A. 1507, fols 95ᵛ–96ʳ, copied in 1032/1622. For a colour reproduction, see *History of Cartography,* II/1, eds Harley and Woodward (n. 33 above), pl. 9.

[44] Qur'ān 65:12. For the concept of seven heavens and seven earths, see Heinen, *Islamic Cosmology* (n. 2 above), pp. 85–8, 138–45, and 196–209.

[45] Heinen, *Islamic Cosmology* (n. 2 above), p. 86, gives a variant association, in which the first heaven is called *raqī'a*, a green smaragh; the second *araqlūn*, white silver; the third *qaydūm*, a ruby; the fourth *mā'ūnā*, pearl; the fifth *dī'ā*, red gold; and the sixth *daquā*, green hyacinth; while the seventh is simple *'arībā*, light.

wind; third, the stones of hell; fourth, scorpions; fifth, vipers; sixth, brimstones; and seventh, the devil. Whether İbrahim Hakkı thought of these seven earths as stages of hell is somewhat doubtful, since he depicts hell separately below, at the bottom.

It was also sometimes said that the seven earths rested upon a foundation of water, or on a fish, or on a green rock that was held accountable for the green colour of the ocean.[46] In Figs 2 and 5 we have 'the green rock' (*al-ṣakhrah al-khaḍrāʾ*) and 'the fish in the sea' (*al-ḥūt fī al-baḥr*), but the item between is an unexplained red band labelled 'the red light' (*nūr al-aḥmar*) or possible the red fire. An entirely different tradition, separate from that positing 'seven earths', states that a cosmic fish supports the earth; this cosmic fish is in water over stones that are held by an angel who in turn stands on rocks, supported by winds.[47]

Turning to the second illustration from the *Maʿrifetnāme* composed by İbrahim Hakkı (Figs 3 and 6), the area between paradise and hell is a schematised presentation of the Day of Judgement. Immediately beneath paradise there is what appears to be a large reservoir of water and cups hung in a circle around; one side is labelled 'the cups' (*al-kāʾsāt*) and the lower side 'the basin of *al-kawthar*' (*ḥawḍ al-kawthar*); the word *al-kawthar* occurs only once in the Qurʾān, and later commentators generally identified it with one of the four Rivers of Paradise.[48] This particular illustration seems to reflect an interpretation of *al-kawthar* found in the Persian *Miʿrājnāma* preserved in an Ilkhanid manuscript of about 1317–35, in which the text states that Gabriel replied to a query from the Prophet Muḥammad by saying:

> This is what God has promised you: *To you have We granted abundance* (*Kawthar*) [Qurʾān 108:1]. These four rivers flow into your basin. When you arrive there on the Resurrection, the basin will have been filled.'[49]

In other words, by this interpretation, *al-kawthar* refers not to one of the four Rivers of Paradise, but to the confluence of all of them, which will fill a basin that will be waiting in paradise on the Day of Judgement. The artist preparing the diagram in Fig. 6 has drawn a large circular pool or basin, surrounded by small cups as if they were suspended from its rim.

To the right of the 'basin of *al-kawthar*' is a rectangular, colourful grid with a four-line label reading: 'the form of the heights'[50] and 'the location of the insane and

[46] An alternative cosmology has (in descending order) the earths, oceans, rock, angel, and fish; Heinen, ibid., pp. 88 and 115. See also Streck and Miquel, 'Qāf' (n. 41 above), p. 401, for other variants regarding supports for the earth.

[47] Heinen, ibid., p. 110.

[48] Qurʾān 108:1; the other three names of Qurʾānic origin given the Rivers of Paradise are Kāfūr, Tasnīm, and Salsabīl; see El-Zein, 'Water of Paradise' (n. 26 above), p. 466; see also Gardet, 'Djanna' (n. 2 above), p. 448; J. Horovitz and L. Gardet, 'Kawthwar [*sic*]', in *EI*², IV, pp. 805–6; and I. K. Poonawala, 'Āb, Water in Muslim Iranian Culture', in *Encyclopaedia Iranica*, I, London and Boston MA, 1985, pp. 27–8.

[49] Gruber, *The Ilkhanid Book of Ascension* (n. 8 above), p. 64.

[50] The term for 'heights' or 'elevated places', *al-aʿrāf*, is also the name of the seventh sūrah of the Qurʾān; in 7:46 and 48 those who are on 'the heights' appear neither in paradise nor in hell, but in an intermediate place or condition.

children of the polytheists' (*ṣūrat al-aʿrāf | makān al-majānīn | wa-aṭfāl | al-mushrikīn*). At the centre is an oval enclosed by a band with the label inside it reading 'The fire surrounding the people [on the day] of judgement (*al-nār muḥīṭ bi-ahl al-maḥshar*)'. Enclosed by the oval ring are a number of items: to the left are three *minbārs*, or pulpits, for the prophets (labelled *manābir al-anbiyāʾ*), with the label 'the place of judgment' (*arḍ-i maḥshir*) written under them. Immediately to the right of the three pulpits is a 'banner of praise [to God]' (*liwāʾ al-ḥamd*), and beneath the banner are footstools for the religious scholars (*kursī | al-ʿulamāʾ*), who presumably oversee the process of judgement. The circular decoration in the centre is labelled 'a praiseworthy place' (*maqām maḥmūd*), that is, a place promised to the Prophet Muḥammad. To the left is a balance for weighing human deeds, appropriately labelled 'balance of the deeds' (*mīzān al-aʿmāl*). At the bottom, next to the base of the standard holding the banner, the phrase *dafātir al-aʿmāl* ('the records of deeds') is inscribed.

In this diagram of the Day of Judgment, the only exit from the encircling oval is at the lower left, labelled 'Path to hell' (*ṭarīq-i jahannam*). From there, a person must cross the bridge over the top of hell, with its seven gates leading downward into hell, before one can proceed upward along the path on the right-hand side of the diagram which is labelled 'Path to paradise' (*ṭarīq al-jinan*).

These two diagrams are unique in trying to illustrate the relationship between paradise (and hell) and other elements of Islamic cosmology. They incorporate nearly all the concepts one finds discussed in writings on Islamic cosmology. They also emphasize the great distance that paradise is from the mundane world, and that it lies well above, or past, the 'Seven Heavens', and far distant indeed from the sphere of the fixed stars and the sun and moon and other planets.

Occasionally we do find illustrations of other aspects of paradise. For example, an Ottoman devotional miscellany copied by Ḥājj Muḥammad Rāsim (Hacı Mehmed Rasim) and illuminated by al-Ḥājj Aḥmad Ayasofyah in 1877 (H 1294) provides an image of the Tree of Paradise and an illustration of the mansions of the inhabitants of paradise, which are shown to be domed structures set on a gold ground with a blue hemispherical heaven above.[51]

From all these illustrations and textual sources, it is evident that while paradise is distant from the world we know, it is furnished with features recognisable from our earthly existence. Moreover, paradise is an *afterworld* and not so much an *afterlife*. That is, the emphasis is upon knowing and describing the physical details of the afterworld, rather than dwelling upon the afterlife of the soul.

[51] Ghassan I. Shaker coll., MS 69, fols 208^{v}–209^{r}, reproduced in N. F. Safwat, *Golden Pages: Qurʾans and other manuscripts from the collection of Ghassan I. Shaker*, London and Oxford, 2000, item 69, pp. 268–75 (275). For an illustration of the Garden of Paradise and a painting of the Day of Judgement from a *Fālnāmeh*, see Farhad with Bağcı, *Falnama* (n. 7 above), items 57 and 55.

Locating Paradise

If, however, paradise is so much of this world in its nature (although enhanced), then why is it not indicated on Islamic maps? This is a topic that has intrigued me. The depiction of paradise is commonly found on Latin and Greek maps, and occasionally on Syriac and Hebrew ones.[52] Why are there not similar representations in Arabic, Persian, or [Ottoman] Turkish maps?

To illustrate the tradition of placing paradise on non-Islamic maps, let us look at two examples – both drawn from the southern rim of the Mediterranean and in areas soon to come (if not already) under Muslim rule and hence produced in a locality potentially able to influence Islamic thinkers and scholars. The first example is a Greek world map illustrating the *Christian Topography* composed in Alexandria *c.* AD 547–9 by Cosmas Indicopleustes, a Christian merchant turned monk.[53] At the centre of the map the inhabited world (*oikoumene*) is depicted as a rectangle surrounded by a great ocean that humans could not cross. To the east (toward the right-hand side of the map), beyond the ocean, is paradise, separated from the inhabited world not only by the Surrounding Sea but by additional land – land which, in turn, encompasses the entire Surrounding Sea that encircles the inhabited world. From this paradise, four rivers are drawn flowing *underground* to become the Tigris, Euphrates, Nile (called Gihon or Gheon), and Pison or Phison (variously identified as the Danube, or Indus, or Ganges).

There is of course an equivalent in the Islamic world to this notion of land enclosing the ocean that surrounds the inhabited world, for the Mountain of Qāf (discussed above in connection with Figs 2 and 5) played a similar role in traditional Islamic cosmography. Various theories arose about what might lie on or beyond the Mountain of Qāf: was it the land inhabited by angels, or a land of the *jinn*, or a land of gold, silver and musk? Most commonly, the land beyond the Mountain of Qāf was said to be the haunt of the fabulous bird Sīmurgh – but never was the Mountain of Qāf or the land beyond it associated with paradise.[54]

The second example of medieval cartographic depictions of paradise is provided by the world maps associated with copies of *Commentarius in Apocalypsin*. This is a

[52] For a detailed survey and interpretation of such depictions, see A. Scafi, *Mapping Paradise: A History of Heaven on Earth*, London and Chicago IL, 2006. David G. K. Taylor, of the Oriental Institute, Oxford, is currently exploring the topic of Syriac maps showing paradise.

[53] For a reproduction of Vatican City, Biblioteca Apostolica Vaticana, MS Vat. Gr. 699, fol. 40^{v}, see Scafi, *Mapping Paradise* (n. 52 above), p. 161; for an analysis of another copy, now in Florence, see M. Kominko, 'New Perspectives on Paradise – The Levels of Reality in Byzantine and Latin Medieval Maps', in *Cartography in Antiquity and the Middle Ages: Fresh Perspectives, New Methods*, eds R. J. A. Talbert and R. W. Unger (Technology and Change in History, 10), Leiden, 2008, pp. 139–53. For the role of paradise in Jewish cartographic traditions, see J. M. Scott, 'On Earth as in Heaven: The Apocalyptic Vision of World Geography from *Urzeit* to *Endzeit* according to the *Book of Jubilees*', in *Geography and Ethnography: Perceptions of the World in Pre-Modern Societies*, eds K. A. Raaflaub and R. J. A. Talbert, Oxford and Chichester, 2010, pp. 182–96.

[54] Streck and Miquel, 'Qāf' (n. 41 above), p. 402. The land behind the Mountain of Qāf was, however, on occasion said to be the location of hell; see C. Lange, *Justice, Punishment, and the Medieval Imagination*, Cambridge, 2008, p. 121 n. 123.

commentary on the Apocalypse of Saint John, composed in northern Spain in AD 776 by Beatus of Liébana. On these Latin maps paradise is not a separate island, but a dwelling place located somewhere in the east, where a vignette is placed showing Adam and Eve in the primordial garden, immediately after their act of disobedience.[55]

In the Islamic world, however, not even Adam and Eve appear on maps, though there are full-page Islamic paintings of Adam and Eve in paradise. The stories behind some of these elaborate paintings help to explain the absence of paradise on Islamic world maps.

Angels bowing before Adam and Eve in paradise are the subject of a painting made in Qazvin (Iran) of the mid-1550s or early 1560s from a dispersed *Fālnāmeh*, or illustrated treatise on bibliomancy, attributed to Jaʿfar al-Sādiq.[56] According to the Qurʾān,[57] God taught Adam the names of all things, and he in turn taught them to the angels. Subsequently God ordered the angels to bow down before Adam, and this they did with the exception of Iblīs (the Muslim equivalent of Satan), who said he was of higher rank because he was created of fire, whereas Adam was made of clay.[58] For being disobedient, Iblīs was expelled from the garden (*jannah*), but Adam was left to live there, with the restriction that he must not eat the fruit of one particular tree. This story was elaborated by compilers of tales about the prophets. Details such as elaborate garments, the introduction of Eve (who, according to the Qurʾān, was not yet fashioned by God when the events in this story took place), and the description of the dark-skinned Iblīs, were added to the accounts and incorporated into later Safavid paintings.

Another painting associated with this same dispersed *Fālnāmeh*, depicts Adam and Eve apparently outside of paradise (*jannah,* which, as noted above, could mean either garden or paradise), watched by two rows – one atop and one below the painting respectively – of seven angels.[59] In the centre, Eve is shown riding a peacock, who was considered the gatekeeper of paradise, while Adam is depicted riding a huge serpent. The dark-skinned Iblīs appears in the lower left of the central panel. The painting appears to be illustrating a later elaboration in which Iblīs was intent on re-entering the Garden of Paradise to foil Adam and Eve.[60] In order to achieve this, Iblīs appeals to the peacock's

[55] For a reproduction of the world map from a copy made in 1109 in the Monastery of Santo Domingo de Silos (London, British Library, Add. MS 11695, fols 39^{v}–40^{r}), see Scafi, *Mapping Paradise* (n. 52 above), pp. 112–13, fig. 5.14 and pl. 3b; and, for related copies, see pp. 105–6 and J. Williams, *The Illustrated Beatus: A Corpus of the Illustrations in the Commentary on the Apocalypse,* 5 vols, London, 1994–2003.

[56] Washington DC, Sackler Gallery, Smithsonian Institution, inv. no. S1986.254; for colour reproductions, see *Hunt for Paradise*, eds Thompson and Canby (n. 5 above), pl. 1.4, and Farhad with Bağcı, *Falnama*, (n. 7 above), item 12.

[57] Qurʾān 2:30–8; 7:11–25; 15:26–39.

[58] J. Pedersen, 'Ādam', in *EI*^{2}, I, pp. 176–8; C. Schöck, *Adam im Islam*, Berlin, 1993. For Iblīs, see A. J. Wensinck and L. Gardet, 'Iblīs', in *EI*^{2}, III, pp. 668–9.

[59] Washington DC, Sackler Gallery, Smithsonian Institution, inv. no. S1986.251a; for colour reproductions, see *Hunt for Paradise,* eds Thompson and Canby (n. 5 above), pl. 4.35, and Farhad with Bağcı, *Falnama* (n. 7 above), item13. For another painting of a similar scene, from a copy of the *Stories of the Prophets* (*Qiṣaṣ al-anbīyāḏ*) by Abū Isḥāq al-Nīshabūrī, see Farhad and Bağci, item 65.

[60] Pedersen, 'Ādam' (n. 58 above), p. 177, citing the accounts given by al-Thaʿlabī (d. 427/1035) and al-Kisāʾī (uncertain dates, before 13th century AD).

vanity by telling it that all creatures shall die but that he (Iblīs) can show the peacock the tree of eternity. The peacock then tells the serpent, who goes to Iblīs (who is not in the garden/paradise at this time) and offers to carry Iblīs into paradise (the garden) so that he can be shown this tree of immortality. Iblīs then rushes into the serpent's mouth and seats himself between the serpent's fangs (or in his belly, according to one version). In this manner Iblis re-entered the garden and ultimately, by speaking through the serpent, tricked Eve into eating the fruit of the forbidden tree. For this transgression, they were all (including the serpent) thrown out of paradise.

This literal interpretation of the fall from grace of Adam and Eve as an expulsion downward is, I think, fundamental to our understanding why paradise does not occur on Islamic maps, as opposed to Greek, Latin, and Syriac maps. While there are no Islamic maps depicting Adam and Eve, or an earthly paradise, there are a few Islamic maps on which the location of the place where Adam fell to earth is indicated. The clearest example is a world map produced in Baghdad by al-Kāshgharī between AD 1072 and 1077 to illustrate a comparative dictionary of the Turkish languages, *Compendium of the Languages of the Turks* (*Dīwān lughāt al-Turk*); the map is preserved today in a copy made in Damascus in 1266 (H 666).[61] The circular world map indicates the place where Adam fell to earth with a label reading 'the landing place of Adam, on him be peace' (*mahbiṭ-i ādam ʿalayhi al-salām*). It is in the east, between the labels for Sri Lanka (Sarandīb) and the 'mountain of Sri Lanka' (jabal-i Sarandīb) and the label indicating the wall built by Alexander the Great to contain Gog and Magog.[62]

Similar information is found in the text accompanying a unique map of the Indian Ocean preserved in a late twelfth-century copy of an anonymous Arabic treatise titled *Kitāb Gharāʾib al-funūn wa-mulaḥ al-ʿuyūn* ('The Book of Curiosities of the Sciences and Marvels for the Eyes') composed in Egypt between AD 1020 and 1050.[63] On the Indian Ocean map, the island of Sri Lanka (*Sarandīb*) is a large circle in the middle of the map with several red large dots on it. The accompanying text has the following description:[64]

[61] İstanbul, Millet Kütüphanesi, Ali Emiri coll., MS 4189, fols 22v–23r; see A. Kaplony, 'Comparing al-Kāshgharī's Map to his Text: On the Visual Language, Purpose, and Transmission of Arabic-Islamic Maps', in *The Journey of Maps and Images on the Silk Road*, eds P. Forêt and A. Kaplony, Leiden, 2008, pp. 137–53, colour figs 7.1 and 7.2, and 'Appendix: List of Geographical Nomenclature in al-Kāshgharī's Text and Map', pp. 209–5. See also A. Hermann, 'Die älteste türkische Weltkarte (1076 n. Chr.)', *Imago Mundi*, I, 1935, pp. 21–8.

[62] For Gog and Magog in Islamic sources, see E. van Donzal and A. Schmidt, *Gog and Magog in Early Eastern Christian and Islamic Sources*: *Sallam's Quest for Alexander's Wall*, Leiden, 2010.

[63] Oxford, Bodleian Library, MS Arab. c. 90; the map of the Indian Ocean is on fols 29v–30r. For a reproduction of the map, see Y. Rapoport, 'The *Book of Curiosities*: A Medieval Islamic View of the East', in *The Journey of Maps*, eds Forêt and Kaplony (n. 61 above), pp. 155–71; and Y. Rapoport and E. Savage-Smith, *An Eleventh-Century Egyptian Guide to the Universe: The Book of Curiosities, edited and translated*, Leiden, 2013, pp. 154–7 and 443–6.

[64] Oxford, Bodleian Library, MS Arab. c. 90, fol. 37v, lines 8–20.

> The island of Sarandīb [Sri Lanka]: a great country on the equator, with several great cities, located in the Bay of Bengal. It is ruled by two kings and is inhabited by members of every nation. There is the Mountain of al-Rahūn, which is the place where Adam, may the blessings of God be upon him, fell [from heaven]. The trace of his foot is in the rock, but it has now been submerged by water, so anyone wishing to observe it needs to dive in order to see it. Fish as red as blood surround [the trace], and whoever eats this fish dies instantly. In Sri Lanka there are plantations of aloes-wood of unparalleled quality, and mines of gold, as well as of red, yellow and blue corundum, mines of diamonds, and corundum-like stones ... No other country on the face of the earth equals the wealth of Sri Lanka.

These examples reflect the common tradition that Adam was expelled from paradise and after his descent landed in Sri Lanka. The writers of stories regarding the prophets elaborated upon this theme, and said that Adam alighted in *Sarandīb* (Sri Lanka) or in India,[65] Eve in Jedda, Iblīs in Basra or another Iraqi village, and the serpent in Isfahān (or, according to some, the desert). Later on Adam and Eve met in ʿArafah (or ʿArafāt, a plain east of Mecca) and begat children.

The relevance of these legends to the depiction of paradise is that because Adam was ordered to 'go down' (*habaṭa*) to earth – clearly leaving paradise – paradise could not by definition be anywhere on earth. The striking feature of all accounts of paradise in Islam is its great distance and its remoteness from earth and from humans.[66] While its structure and contents were described in materialistic or worldly terms, it was not 'worldly' in the sense of being part of a sub-lunar or sub-stellar world. Because of this extreme distance, there was no concept at all of an earthly paradise.[67] Consequently, there was no room for it to be illustrated on Islamic maps of the world, even though it was a common cartographic convention in nearby societies with whom Islam shared many of its cosmological theories.

[65] Pedersen, 'Ādam' (n. 58 above), p. 177, citing al-Ṭabarī (d. 310/923) as saying that neither Muslim, Christian, nor Jewish scholars have refuted the tradition that Adam was placed in India (*al-Hind*); the most common tradition, however, was that Adam landed in Sri Lanka (also known as Ceylon).

[66] There was a later tradition that considered the garden (Eden) of Adam to be an earthly garden distinct from the heavenly paradise, but this was not widely ascribed to; see Gardet, 'Djanna' (n. 2 above), pp. 449–50.

[67] Paradise was also said to be of extreme dimensions. A paradise whose width is equal the width of heaven and earth is mentioned in the Qurʾan (57:21).

Valhalla and Heaven: Scandinavian Images of Paradise in a Period of Religious Change

Anders Hultgård

Introduction

Two Old Norse poems of the tenth century, the *Eiríksmál* and the *Hákonarmál*, recount the welcoming of a king to Valhalla and are the first and only explicit witnesses to the idea of a glorious afterlife in the hall of the god Odin.[1] Strangely enough the poems were composed in honour of two Norwegian rulers whom tradition regarded as having been Christian. We are thus faced with a paradox that immediately brings us to the core of our subject matter: images of paradise in a period of religious change.

The word 'paradise' refers to an eschatological concept common to Jewish, Christian and Islamic traditions, and also to Zoroastrianism from where the term itself is ultimately derived.[2] Paradise can be defined as an otherwordly place usually depicted as a beautiful garden which is the abode of divine beings and where righteous men may dwell in the afterlife. Using that concept in the study of other religious traditions is not without problems, because 'paradise' implies specific connotations that may not apply to traditions having a different world-view. Preconceived notions may thus influence the scholar's interpretation. Keeping this in mind it can nevertheless be justified, I believe, to look for ideas about the afterlife in ancient Scandinavian religion that broadly correspond to the term 'paradise'.

The period with which we are concerned here, the ninth to the twelfth centuries, is marked by a cultural transformation of utmost significance. Christianity, a new religion with a different world-view is slowly paving its way, exerting vague influences in the ninth century but attaining complete power in the twelfth. The end of the process meant a pervasive change of religion and the prohibition of all pagan cults, a clear parallel with what happened in the Roman Empire at the end of the fourth century. Information on the afterlife ideas of the ancient Scandinavians is scarce,

Abbreviations used in this essay:
GNH = *Gamal Norsk Homiliebok*
IHB = *The Icelandic Homily Book*

[1] Cf. G. Neckel, *Walhall. Studien über germanischen Jenseitsglauben*, Dortmund, 1913, p. 53; F.-X. Dillmann, 'Valhǫll', in *Reallexikon der Germanischen Altertumskunde*, eds H. Beck, D. Guenich and H. Steuer, Berlin and New York, 35, 2007, pp. 341–66 (342). Some stanzas in eddic poems also mention Valhalla: foremost 'Grímnismál' 8–10, 'Vafþrúðnismál' 41 and 'Vǫluspá' 33.

[2] For the Iranian paradise and the history of the term and the concept, see A. Hultgård, 'Das Paradies vom Park des Perserkönigs zum Ort der Seligen', in *La Cité de Dieu/Die Stadt Gottes*, eds M. Hengel, S. Mittmann, A.-M. Schwemer, Tübingen, 2000, pp.1–43; J. N. Bremmer, *The Rise and Fall of the Afterlife*, London, 2002, pp. 109–19; B. Lincoln, 'À la recherche du paradis perdu', *History of religions*, 43, 2004, pp. 139–54; and the essay by Antonio C. D. Panaino in this volume.

The Cosmography of Paradise: The Other World from Ancient Mesopotamia to Medieval Europe, Warburg Institute Colloquia 27, 2016

whereas Christian medieval eschatology is well documented in Old Norse vernacular texts from the twelfth and thirteenth centuries although descriptions of the end of the world are less common than those pertaining to heaven and hell. In native Scandinavian tradition cosmic eschatology is represented by the myth of the Ragnarok which also includes the renewal of the world.

In spite of the few texts on individual eschatology that have come down to us from a pre-Christian milieu, Old Norse sources and archaeological evidence hint at a wide variety of afterlife beliefs ranging from coming to the realm of Hel below, survival after death in mounds and hills, to living with the gods in their dwelling places.[3] In what follows I shall first discuss the character of the otherwordly places in ancient Scandinavian tradition and relate them to the general notion of paradise. Particular attention will be given to the cosmographical context as far as it can be elucidated. In the second part an enquiry will be made into the way in which the new paradise concept was presented to the first generations of Scandinavian Christians. Thirdly, the questions of continuity and influences will be addressed.

Ancient Scandinavian afterlife places

One of the connotations inherent in the 'paradise' concept of Judaism, Christianity and Islam is that it also refers to the existence of another place which is its opposite. To denote this altogether different place a pre-Christian word common to all Germanic languages was taken up by the church. This word (Gothic *halja*, Old English *hell*, Old High German *hella*, Old Saxon *hellia*; Old Scandinavian *hel*) originally referred to a particular place in the afterlife. Unlike its Christian counterpart the Germanic belief in the nether world was not associated with pain and punishment, but neither had it the bright colours of a paradise.[4] To this place men as well as women could come, irrespective of the way they died.[5] Even gods could end up in Hel as the Baldr-myth shows. Hel was also the name of

[3] The following description of afterlife places partly summarizes my article 'Óðinn, Valhǫll, and the Einherjar. Eschatological Myth and Ideology in the Late Viking Period', in *Ideology and Power in the Viking and Middle Ages*, eds G. Steinsland, J. Viðar Sigurðsson, J. E. Rekland and I. Beuermann, Leiden and Boston MA, 2011, pp. 297–328. For an overview, see H. Ellis Davidson, *The Road to Hel. A Study of the Conception of the Dead in Old Norse Literature,* Cambridge, 1943, pp. 65–99; A. Hultgård, 'Mytische Stätten, Tod und Jenseits', in *Reallexikon der Germanischen Altertumskunde* (n. 1 above), 20, 2002, pp. 472–7; A. Nordberg, *Krigarna i Odins sal*, 2nd edn, Stockholm, 2004, pp. 69–73; R. Simek, 'Totenreiche', in *Reallexikon der Germanischen Altertumskunde* (n. 1 above), 31, 2006, pp. 89–92.

[4] In the medieval cosmography of Snorri there is a tendency to depict Hel in dark and inhospitable colours, e.g. *Snorri Sturluson. Edda. Gylfaginning og prosafortellingene av Skáldskaparmál*, eds A. Holtsmark and J. Helgason, 2nd edn, Copenhagen, Oslo, and Stockholm, 1968, ch. 34 (on this passage, see *L'Edda. Récits de mythologie nordique*, transl. from the Old Irish, intr. and annotated by F.-X. Dillmann, 10th edn, Paris, 2008, pp. 170–71). According to some scholars this passage reflects Christian influence, e.g. J. de Vries, *Altgermanische Religionsgeschichte, I–II*, Berlin, 1956–57, § 581; R. Simek, *Religion und Mythologie der Germanen*, Darmstadt, 2003, p. 209.

[5] Cf. E. F. Halvorsen, 'Hel', in *Kulturhistorisk leksikon for nordisk middelalder*, 6, 1961, pp. 304–5. Snorri reports that men who died of sickness and old age ended up in Hel: *Snorri Sturluson. Edda* (n. 4 above), ch. 34.

the female sovereign of the nether world.[6] Reminiscences of Hel (and Valhalla) as a general afterlife place are attested in the thirteenth-century saga literature by the concept and custom of *binda helskó*: that is to bind a particular pair of shoes on the feet of the dead with which to walk to the other world.[7]

By contrast, another Germanic idea, that of a *post mortem* existence in the green meadows of the other world can be inferred from the use of the Old English term *neorxnawang* for the Christian paradise (the first element is unexplained) where the term *wang* ('field, meadow') corresponds with Gothic *waggs* that renders the Greek παράδεισος in the Bible.[8] One of the names of heaven in Old Saxon is *heƀanwang* ('the heavenly meadow') recalling its character of a paradise.[9] Most probably the term *wang* was chosen in all these cases because it served to denote an otherwordly place imagined as a heavenly meadow in the pre-Christian culture. In Old Norse mythology the goddess Freyja is thought to live in Fólkvangr[10] and the abode of the god Thor is in one branch of tradition called Þrúðvangar (see below). The second element of the name (Old Norse *vangr* 'field, meadow', plural *vangar*) indicates the nature of their dwelling places. By this terminology we may trace the contours of an ancient Germanic 'paradise' of which, however, we do not know the details.

The most well-known afterlife place, in the general mind at least, is Valhalla where Odin welcomes fallen heroes. As already mentioned the two main textual sources are memorial lays in honour of two Christian kings.[11] Yet their poets promised them an

[6] The problem of this personification of *hel* is discussed by R. Simek, *Dictionary of Northern Mythology*, transl. A. Hall, Cambridge, 1996, repr. 2008, p. 137 and by H. Beck, 'Hel', in *Reallexikon der Germanischen Altertumskunde* (n. 1 above), 14, 1999, pp. 257–60.

[7] On this idea and custom, see D. Strömbäck, 'Att binda helskor: Anteckningar till Gisle Surssons saga', in *Kungl. Humanistiska Vetenskaps-Samfundet i Uppsala. Årsbok*, Uppsala, 1952, pp. 139–48, and J. Wellendorf, *Kristelig visionslitteratur i norrøn tradition*, Oslo, 2009, pp. 104–10. According to the short redaction of *Gísla saga Súrssonar* (M redaction), ed. A. Loth, Copenhagen, 1956, p. 14, Þorgrímr says to Gísli: 'þat er tizka ... at binda mǫnnum helskó, þá er þeir skulu ganga til Valhallar' ('It is customary... to tie Hel shoes on men when they shall walk to Valhalla'). In the longer redaction (S) we find a somewhat different, though more genuine, version without the mention of Valhǫll (cf. Dillmann, 'Valhǫll' [n. 1], p. 349): 'þat er nú siðr ... at binda helskó at fótum mǫnnum áðr í haug eru lagðir því at þat var þá mælt at þeir skyldi til heljar fara ...' (normalized after the text of *Gísla saga Súrssonar. Membrana regia deperdita* (S redaction), ed. A. Loth, Copenhagen, 1960, p. 29): 'It is now tradition to tie Hel shoes on the feet of men before they are brought into the mound because it was then said that they should go to Hel'.

[8] II Corinthians 2:14. The other passages in the Greek Bible where παράδεισος occurs have been lost in the Gothic version. *Neorxnawang* and the adjective *neorxnawangile* appear in the Old English translation of Gregory's *Dialogues* and his *Pastoral Care*, and in Ælfric as well as in some other texts.

[9] *Heliand und Genesis*, ed. O. Behaghel, 10th rev. edn by B. Taeger, Tübingen, 1996, p. 275.

[10] 'Grímnismál' 14, in *Snorri Sturluson. Edda* (n. 4 above), ch. 24.

[11] It has been suggested that the iconography of the Gotlandic picture stones from the late 8th century includes the Valhalla motif: so, for example, L. Buisson, *Der Bildstein Ardre VIII auf Gotland. Göttermythen, Heldensagen und Jenseitsglaube der Germanen im 8. Jahrhundert n. Chr.*, Göttingen, 1976, pp. 88–109; A. Andrén, 'Doors to the Other Worlds: Scandinavian Death Rituals in Gotlandic Perspectives', *Journal of European Archaeology*, 1, 1992, pp. 33–56 (41); D. Ellmers, 'Valhalla and the Gotland Stones', in *The Ship as Symbol in Prehistoric and Medieval Scandinavia*, eds O. Crumlin-Pedersen and B. Munch Thye, Copenhagen, 1995, pp. 165–71. I am not convinced of this interpretation; cf. the reservations made by S. Lindquist, *Gotlands Bildsteine*, I, Stockholm, 1941, pp. 86–101, and in particular by Dillmann, 'Valhǫll' (n. 1 above), pp. 359–63.

afterlife in the 'paradise' of the most prominent non-Christian deity. The reasons for this are not quite clear and we are faced with a problem involving politics, mentality and religious change. I have tried to elucidate this problem elsewhere.[12] Since the idea of Valhalla (hereafter the Old Norse form *Valhǫll* will be used) has been treated repeatedly in previous research, it is not my purpose to reconsider it in detail.[13] Here I will address the questions of the character and topography of Valhǫll and for whom this 'paradise' was created.

While the poems 'Eiríksmál' and 'Hákonarmál' transmit the myth of Odin welcoming fallen heroes in Valhǫll, the eddic lay of 'Grímnismál' describes the appearance and the site of Valhǫll itself.[14] According to stanza 8 it is located in a place or territory called Glaðsheimr, 'the radiant land'.[15] As is indicated by the second name element in Valhǫll, Old Norse *hǫll* 'hall', and by the characterization of the building as 'broad' (*víð*) and 'shining with gold' (*gullbjǫrt*), Valhǫll was thought of as a large magnificent building, an otherworldly reflection of the impressive halls that the Scandinavian aristocracy began to erect already in the fourth century AD.[16] The huge measures of Valhǫll are further underlined in stanza 23: the building has no less than 640 doors. The hall's interior has spear-shafts for rafters, it is thatched with shields and on the benches are strewn mail-coats (stanza 9). The description of 'Grímnismál' strengthens the impression that the mythic dwelling-place of Odin is modelled on the aristocratic hall.[17]

Close to the hall there is a tree called Læraðr and on the roof of the hall a goat named Heiðrún and a stag called Eikþyrnir are browsing from the branches of the tree ('Grímnismál' 25–6).[18] Both these animals fulfil specific functions. One, the goat, provides the inhabitants of Valhǫll with a beverage, the bright shining mead, which is constantly flowing. The other, the stag, drips water through his horns into a source called Hvergelmir from which all the rivers of the world spring. Belonging to Valhǫll are also

[12] See Hultgård, 'Óðinn, Valhǫll, and the Einherjar' (n. 3 above).

[13] Among the many contributions the following have a particular significance: Neckel, *Walhall* (n. 1 above); H. Ellis Davidson, *The Road to Hel* (n. 3 above); Nordberg, *Krigarna i Odins sal* (n. 3 above); and Dillmann, 'Valhǫll' (n. 1 above).

[14] The dating of eddic poems is a matter of controversy. For a discussion of this question, see B. Fidjestøl, *The Dating of Eddic Poetry*, ed. O. E. Haugen, Copenhagen, 1999. The poems 'Grímnismál' and 'Vafþrúðnismál' are usually thought to have been composed in the 10th century; for a different view, see R. Simek, 'The use and abuse of Old Norse religion', in *Old Norse Religion in Long-Term Perspectives*, eds A. Andrén, K. Jennbert and C. Raudvere, Lund, 2006, pp. 377–80. The precise dating of eddic mythological poems is of less significance for their use as sources of ancient Scandinavian religion since they have preserved a pre-Christian mythic tradition.

[15] For the meaning of *Glað-* in *Glaðsheimr*, see Dillmann, *L'Edda* (n. 4 above), p. 155. Snorri makes *Glaðsheimr* both a building (*hús*) and a temple (*hof*), see *Snorri Sturluson. Edda* (n. 4 above), ch. 14.

[16] For these halls, see below.

[17] Cf. A. Nordberg, 'En varg hänger väster om salen, en örn sänker sig ner från ovan. Om Odins sal i Grímnismál 9 och 10', *Arkiv för Nordisk Filologi*, 2007, pp. 115–36 (123 and 132–3). Nordberg also suggests (126–7) that the phrase 'a wolf (*vargr*) hangs to the west of the doors' alludes to the motif of 'the young hanged aristocratic revenger' (*vargr*).

[18] For the dual role of these animals in the mythology, see E. Mundal, 'Heiðrun: Den mjødmjølkande geita på Valhalls tak', in *Eyvindarbók*, eds F. Hødnebø et al., Oslo, 1992, pp. 240–47.

other animals: in the first place those which are especially associated with Odin, the two ravens, Huginn and Muninn, and the two wolves Geri and Freki.[19] Quoting three verse lines of a lost poem Snorri informs us that 'Glasir', a grove with golden leaves, stands outside Valhǫll, the most beautiful grove to be found among gods and men: [20]

Glasir stendr	'Glasir stands
með gullnu laufi	with golden leaves
firir Sigtýs sǫlum	in front of Sigtýr's halls'

The early poetic texts unanimously regard Valhǫll as the pre-eminent dwelling place of Odin[21] although the god has other places, too, in which to stay.[22] In Valhǫll belong the Valkyries and the retinue of Odin, the Einheriar.[23] The Valkyries, also called 'Odin's maidens', are in the earliest sources described as divine women of a friendly and radiant appearance.[24] They live in the 'paradise' of Odin and were believed to manifest themselves for those who were going to die in combat. The company of otherworldly elite warriors, the Einheriar, is formed by those who fall in battle. They have received a new status in the afterlife and have one great task. In the future they will fight together with the gods at Ragnarok against their enemies. In the open grounds (*tún* in the plural) outside Valhǫll they perform their daily martial exercise ('Vafþrúðnismál' 43). Referring to the 'Hákonarmál' commentators state that Valhǫll was also thought of as a gloomy place where one feared to come after death. As I have pointed out this is a misinterpretation of the poem, and the image of Valhǫll is primarily that of a glorious place to which it was a great honour to be welcomed and where one would take part in a banquet with the gods.[25] To conclude, Valhǫll qualifies for being a paradise in several respects: as the dwelling place of a god (or gods), as the place where men are believed to join him (or them) after death, as predominantly a place of joy and happiness and, finally, Valhǫll is associated with trees and animals.

The question has to be raised whether Valhǫll was considered the dwelling place of Odin exclusively or whether it was the home of the other gods too.[26] In fact, one of the two poems

[19] Other animals are the cock Gullinkambi ('Vǫluspá' 43) perhaps the same as Salgofnir ('Helgakviða Hundingsbana' II, 49) and the boar Sæhrímnir; cf. Dillmann, 'Valhǫll' (n. 1 above), p. 353.

[20] *Snorri Sturluson. Edda* (n. 4 above), ch. 34. The prose runs: 'in Asgard there is a grove (*lundr*) in front of the doors of Valhalla which is called Glasir and its leaves are all of red gold as it is said here'; the verse quotation is followed by the remark: 'that is the most beautiful wood (*viðr*) among gods and men.'

[21] Cf. Dillmann, 'Valhǫll' (n. 1 above), p. 345.

[22] 'Grímnismál' 7 mentions Søkkvabekkr which he shares with the goddess Sága who probably is identical with Frigg. Stanza 6 describes a mansion (*bǿr*) called Valaskiálf which the Prose-Edda of Snorri (*Snorri Sturluson. Edda* [n. 4 above], ch. 17) attributes to Odin and where the high seat of the god is placed.

[23] On the meaning of *einheriar*, see the discussion in Dillmann, 'Valhǫll' (n. 1 above), pp. 352–4.

[24] This is emphasized by Dillmann, ibid., p. 346, referring to 'Haraldskvæði' (end of the 9th century) and 'Hákonarmál' (10th century).

[25] See the discussion in Hultgård, 'Odin, Valhǫll, and the Einherjar' (n. 3 above).

[26] Cf. Dillman, 'Valhǫll' (n. 1 above), p. 353.

describing how Odin receives a king in his hall also mentions that 'all the gods' welcome him, and would share the ale with him.[27] On the other hand the 'Grímnismál' clearly states that the other deities had their own abodes. Mythic and poetic thinking follows paths of its own, however, and we may assume that the poet imagined the divine assembly coming together in Odin's hall on specific occasions like that of the welcoming of a king in Valhǫll.

Who were the persons chosen to dwell in this 'paradise'? The place has in all probability its name from the Old Norse *valr* which denotes those who die on the battle field, and the 'Grímnismál' (stanza 8) explicitly states that Odin in Valhǫll takes to himself 'men killed in battle' (*vápndauðir verar*).[28] Several features in the description of Valhǫll in the'Grímnismál' point in the same direction. The poet says that it is easy to recognize the hall of Odin because of its martial aspect (see above) and other marks. On the western side of the building a wolf is hanging and over him an eagle hovers. The fence that surrounds Valhǫll has an entrance called Valgrind, the gate through which the *valr*, the fallen ones, pass ('Grímnismál' 22). As mentioned, the poems 'Eiríksmál' and 'Hákonarmál' describe the passage into the afterlife of two kings killed in battle.

The evidence thus strongly suggests that the myth of Odin welcoming his heroes in Valhǫll was part of a male warrior ideology associated with the higher strata of society.[29] We are not, I believe, entitled to suppose that all other dead, women as well as men, came to Hel. Allusions to the destiny of women beyond death are extremely rare in the literary sources. Yet in view of the richly furnished female graves from the Viking period, such as the ship burials of Oseberg and Tuna in Badelunda, we have to ask: what were the afterlife expectations of aristocratic women in the Viking age?[30] They must have hoped for something more than a shady existence in a gloomy nether world.

Hel, which is not a 'paradise', and Valhǫll are the only afterlife places of which details have been preserved in Old Norse literature. This should not lead us to the assumption that they were the only ones. It is suggested here that the world-view of the ancient Scandinavians included not one but several 'paradises', some of them also afterlife places for humans.

The idea of Valhǫll, and to a certain extent that of Hel, shows a particular type of afterlife concept: the expectation to join a particular deity in his or her dwelling place. Allusions in eddic poems and saga texts, together with archaeological evidence, indicate that this concept was not restricted to Odin in Valhǫll or Hel in the realm bearing her

[27] 'Hákonarmál', stanzas 16 and 18.

[28] For the significance of *val* in Valhǫll, see Neckel, *Walhall* (n. 1 above), pp. 2–14; Dillmann, 'Valhǫll' (n. 1 above), pp. 341–2.

[29] Cf. Hultgård, 'Óðinn, Valhǫll, and the Einherjar' (n. 3 above), p. 325, and J. P. Schjødt, 'The Warrior in Old Norse Religion', in *Ideology and Power in the Viking and Middle Ages*, eds G. Steinsland, J. Viðar Sigurðsson, J. E. Rekland and I. Beuermann, Leiden and Boston MA, 2011, pp. 269–95 (279).

[30] For the Oseberg ship burial, see A. E. Christensen, A. S. Ingstad and B. Myhre, *Osebergdronningens grav. Vår arkeologiske nasjonalskatt i nytt lys*, Oslo, 1992; for Tuna, see E. Nylén and B. Schönbäck, *Tuna i Badelunda: guld, kvinnor, båtar*, Västerås, 1994.

name. For the goddess Freyja we have the stanza 14 of the 'Grímnismál' stating that she lives in Fólkvangr. She decides who shall sit where in her hall (*í sal*) and she also chooses half of the fallen men (*hálfan val*) each day, whereas Odin owns the other half. The name Fólkvangr alludes, as we have seen, to the green fields in the other world and the term *fólk* seems here to denote a 'host of warriors'. *Egils saga Skalla-Grímssonar* may hint at a belief of women joining Freyja in the afterlife. Having been informed of the death of her brother and of Egil's intention not to live any longer Þorgerðr refuses to take her evening meal, saying, she will not have it until she joins Freyja.[31]

The place where the god Thor lives is, according to 'Grímnismál' stanza 4, called *Þrúðheimr* 'home of strength'; in Snorri it appears as *Þrúðvangar* ('fields of strength').[32] His hall bears the name *Bilskirnir* (or *Bilskírnir*) which perhaps alludes to its brightness.[33] The large size is its most prominent feature.[34] An eddic poem, the 'Hárbarðsljóð', states in stanza 24 that Odin owns the *jarlar* who fall in battle but Thor owns the slaves (*þræla kyn*). This is a contrastive device characteristic of the poem's genre and, on the part of Thor, certainly biased since his worship was particularly associated with the free farmers, the *karlar*. Anyway the passage hints at the belief that some men were to join this god in the afterlife.

Excavations in the Mälar region of eastern Sweden have yielded a surprisingly large number of burials containing iron rings to which small pendants in the form of hammers were attached. They can be dated to a period from the eighth to the tenth centuries and were in all probability worn as a necklace. The rings with the pendants, symbols of the god, had been carefully placed in the centre of the graves in or around the urn which contained the ashes.[35] More than half of all burials were female graves. The hammer as a symbol of the god Thor can convey different meanings, but a small Viking age amulet from southeastern Sweden now confirms the protective nature of the miniature hammers as symbols of Thor. It most probably belonged to a grave and carries a runic inscription with a prayer formula 'may Thor protect him with the hammer that ...' (the sense of the following words is uncertain).[36] The archaeological evidence thus suggests a commitment to the god which was most probably thought to continue beyond death.

[31] Ch. 78: 'engan hefi ek náttverð haft, ok engan mun ek, fyrr en at Freyju ... vil ek ekki lifa eptir fǫður minn ok bróður'.

[32] *Snorri Sturluson. Edda* (n. 4 above), ch. 21. The Uppsala codex has the word in the singular.

[33] Cf. Dillmann, *L'Edda* (n. 4 above), p. 163.

[34] Bilskirnir is explicitly attributed to Thor only in *Snorri Sturluson. Edda* (n. 4 above), ch. 21. The 'Grímnismál' 24 states that it has 640 'rooms' or 'boards' (*gólf*) in all (*með bugom*).

[35] For this type of material, see K. Ström, 'Thorshammerringe und andere Gegenstände des heidnischen Kultus', in *Birka II/1 Systematische Analyse der Gräberfunde*, ed. G. Arwidsson, Stockholm, 1984, pp. 127–40.

[36] The Kvinneby amulet from Öland: 'Þōrr gæti hans meR þeim hamri...' For editions and interpretations, see S. Lindqvist, *Religiösa runtexter III. Kvinnebyamuletten*, Lund, 1987; B. Westlund, 'Kvinneby-– en runinskrift med hittills okända gudanamn?', *Studia anthroponymica Scandinavica*, 7, 1989, pp. 25–52; and O. Grønvik, 'En hedensk bønn. Runeinnskriften på en liten kobberplate fra Kvinneby på Öland', in *Eyvindarbók. Festskrift til Eyvind Fjeld Halvorsen*, eds F. Hødnebø, J. G. Jørgensen, E. Mundal, and V. Ólason, Oslo, 1992, pp. 71–85.

The nature of the god Thor is usually described on the basis of the Old Icelandic mythological texts that predominantly show his martial side and convey the impression of a somewhat rough character. It cannot be assumed, however, that these texts give a full picture of the nature and qualities of Thor as he was worshipped among all Scandinavians. The notice on the sacrificial worship of the Svear at Uppsala given by Adam of Bremen in his *Gesta Hammaburgensis* conveys a somewhat different image. He reports that Thor had the place of honour in the cult hall (*triclinium*) being the mightiest (*potentissimus*) of the three principal gods. He appears as a sky god: 'Thor, they say, is in the sky. He rules over thunder and lightning, over winds and rains, clear weather and (good) crops'.[37]

According to Adam, people sacrificed to him when disease and famine were impending.[38] There is also evidence in the Old Norse literature to suggest that Thor was the chief personal deity in the Viking period. The relation of a man to his god was often expressed by the concept of *fulltrúi*: 'somebody in whom one has trust or confidence'. The reports of Thor as the *fulltrúi* of men could be interpreted as medieval constructs.[39] On the other hand oral tradition may well have preserved memories of personal religion in a period which was not after all very distant from the time of the saga authors and the *fulltrúi* concept may therefore reflect pre-Christian ideas.[40] It is reasonable to assume that the personal relationship with the deity would not end with death. To conclude: the belief in an afterlife for both men and women in the dwelling place of the sky god Thor may not appear as strange as it would seem at first.

The cosmographic context

Where were Valhǫll and the abodes of the other deities situated in the mythical geography? Our knowledge of ancient Scandinavian cosmography is incomplete, being based on some eddic poems only and on scattered allusions in other tenth-century poetry.[41] To supplement this fragmentary picture with information drawn from the early thirteenth-century Prose-Edda of Snorri is a risky endeavour, but can be justified in cases where it seems probable that Snorri draws upon now lost poetic sources. The problem of Snorri's cosmography and its relationship with pre-Christian ideas of the universe has

[37] Adam of Bremen, *Gesta Hammaburgensis Ecclesiae Pontificum*, ed. B. Schmeidler, 3rd edn, Hannover and Leipzig, 1917, IV, 26: 'Thor, inquiunt, presidet in aere, qui tonitrus et fulmina, ventos ymbresque, serena et fruges gubernat.'

[38] *Gesta Hammaburgensis* (n. 37 above), IV, 27: 'Si pestis et fames imminet, Thor ydolo lybatur.'

[39] This is the interpretation that seems to prevail today: see B. Maier, *Die Religion der Germanen. Götter, Mythen, Weltbild*, Munich, 2003, pp. 32–40, and J. Zernack , 'Fulltrúi', in *Reallexikon der Germanischen Altertumskunde* (n. 1 above), 10, 1996, pp. 243–5.

[40] Cf. A. Hultgård, 'The Religion of the Vikings', in *The Viking World*, ed. S. Brink, London, 2008, pp. 212–18 (213).

[41] Among eddic poems foremost the 'Grímnismál' and some stanzas of the 'Vafþrúðnismál': 16, on the river that separates the land of the gods from that of the giants; 40–41, on the open grounds (*tún* in the plur.) where the *einherjar* fight; 43, on the nine worlds and the *Niflhel neðan*.

been much discussed, but is beyond the scope of the present contribution since my analysis of the cosmographic context is mainly based on the early poetic sources.

The 'Grímnismál' appears to be the most reliable source and, in the first part of the poem (stanzas 4 to 26), the dwelling places of the main gods and goddesses are named and briefly described. The divine world is a holy territory. Grímnir, the masked Odin, begins his teaching of mythic knowledge by stating: 'the land is sacred which I see lying near the gods and the elves'.[42] It has valleys, hills, rivers, woods and grasslands like the natural environment of men and in this sacred land the deities have their homes and halls. We have to imagine a particular territory (*heimr* or *land*) for each deity in which they have their own residence denoted with the terms *hǫll* or *salr*, both of which can be translated with 'hall'.[43] Thus we find *Þrúðheimr* for Thor, *Glaðsheimr* for Odin, *Álfheimr* for Freyr, *Þrymheimr* for Skaði, but *land* for Viðarr. In other cases the abode of a god or goddess is mentioned by using names that indicate the natural environment. Valleys of ewe (*Ýdalir*) for Ullr, skyhills (i.e. 'land of high hills', *Himinbjǫrg*) for Heimdallr, open enclosed places for ships (*Nóatún*) for Njǫrðr, field for warrior hosts (*Fólkvangr*) for Freyja, land with wide views (*Breiðablik*) for Baldr.[44] The god's name is most probably included in the territorial designation of the divine abode *Válaskjálf* which may be translated as 'Váli's rock shelf'.[45] In the case of the god Forseti only his hall *Glitnir* ('the glittering one') is mentioned.[46] The world of the gods was thought of as a land flourishing with plants and trees. The 'Hákonarmál' (13) reports that the two Valkyries shall ride 'to the green lands of the gods' (*grǿna heima goða*), to tell Odin that a mighty ruler is on his way to Valhǫll.

The localization of the land of the gods within pre-Christian Scandinavian cosmology is not quite clear since a coherent cosmography does not appear until the thirteenth century in the Prose-Edda of Snorri. The land is also called *Ásgarðr* ('the home of the gods'), and is without doubt an ancient term.[47] It is a parallel formation to *Miðgarðr*, the ancient Germanic name for the inhabited world (Gothic *midjungards* translating the Greek οἰκουμένη in the Gothic bible, Old Saxon *middilgard*, Old English *middangeard*). The location of the land of the gods, as well as the name *Ásgarðr,* is not explicitly

[42] 'Grímnismál' 4: 'Land er heilakt, er ek liggia sé ásom ok alfom nær'.

[43] For these terms, see S. Brink, 'Political and social structures in Early Scandinavia: a settlement-historical pre-study of the central place', *Tor. Journal of Archaeology*, 28, 1996, pp. 235–81 (251–8).

[44] *Breiðablik* in 'Grímnismál' 12 should be understood as referring to the land of the god; it is mentioned in the same stanza with the words *á því landi* ('in that land').

[45] The interpretation of the name is debated: cf. de Vries, *Altgermanische Religionsgeschichte* (n. 4 above), § 515; Dillmann, *L'Edda* (n. 4 above), p. 160. It can also be read as Valaskiálf with a short 'a' in the first syllable. The meaning of the second element – *skjálf* – is not quite clear but I take it as referring to a natural vantage point in the landscape, a high rocky shelf; cf. Old English *scylf* ('crag, ledge, shelf'), see J. R. Clark Hall, *A Concise Anglo-Saxon Dictionary*, 4th edn, Cambridge, 1960, p. 299; see also the discussion in E. Elgqvist, *Skälv och skilfingar*, Lund, 1944

[46] 'Grímnismál' 15.

[47] Cf. P. Vikstrand, 'Ásgarðr, Miðgarðr, and Útgarðr. A Linguistic Approach to a Classical Problem', in *Old Norse Religion in Long-Term Perspectives*, eds A. Andrén, K. Jennbert and C. Raudvere, Lund, 2006, pp. 354–7.

mentioned in the 'Grímnismál'. However, allusions within the poem suggest somewhere in the world tree, the ash of Yggdrasill, but above the homes of mankind and giants. For these groups of beings the poem tells that the world tree has three roots: Hel lives underneath one, the second covers the land of the frost giants (*hrímþursar*) and below the third lies the home of men (*mennzkir menn*; 'Grímnismál' 31).

Finally there is the much debated mention of a place called *Gimlé* in the eddic poem 'Vǫluspá', stanza 64.[48] The prophetess sees a golden hall fairer than the sun where trustful men will live and always enjoy happiness. This sounds like a description of the Christian paradise and the stanza has also been taken by many to reflect Christian ideas; but the image is so vague that it is open for different interpretations.[49] It should be emphasized, however, that Gimlé is not quite an ordinary paradise where people come after death. Gimlé does not yet exist and will not be realized until another world appears. Its description bears the stamp of ancient Scandinavian world-view. The stanza must be interpreted as an integrated part of the description of the new earth that will rise from the sea after the conflagration of the world at the Ragnarok. Seen in this perspective the assumption of a Christian influence loses much of its weight.

To conclude: it has been shown that Odin's abode, the *Glaðsheimr*, where Valhǫll is situated, clearly appears as an eschatological paradise. For the abodes of the other deities the evidence suggests that the *Fólkvangr* of the goddess Freyja and the dwelling place of the god Thor, *Þrúðheimr* or *Þrúðvangr*, also were considered as paradises in their own right where people could join the deity in the afterlife.

The coming of the Christian Paradise

With the penetration of Christianity into Scandinavia a different and more uniform afterlife concept appeared which, during some centuries, competed with native eschatological ideas. Christian efforts of conversion were in the first place directed to the political elite and missionary preaching concentrated on a few chosen points thought to impress the 'pagans', in particular the doctrine of heaven and hell. Our knowledge of the way in which the Christian paradise was presented to the Scandinavians in the transition period, roughly from the tenth to the twelfth centuries, is essentially derived from three types of sources.[50] These are: first, runic inscriptions mainly from the eleventh century;

[48] For the interpretation of the name, see F.-X. Dillmann, 'Gimlé', in *Reallexikon der Germanischen Altertumskunde* (n. 1 above), 12, 1998, pp. 111–14.

[49] S. Bugge, *Studier over de nordiske Gude- og Hektesagns oprindelse*, 2 vols, Oslo, 1881–89, II, pp. 416–17, suggests that Gimlé reflected the New Jerusalem of Christian eschatology. So also P. Pétursson, 'Manifest and Latent Biblical Themes in *Vǫluspá*', in *The Nordic Apocalypse. Approaches to Vǫluspá and Nordic Days of Judgement*, eds T. Gunnell and A. Lassen, Turnhout, 2013, pp. 185–201 (186). A. Olrik, *Ragnarök. Die Sagen vom Weltuntergang*, Berlin, 1922, pp. 123–4, concludes that 'Gimlé zwischen Christentum und Heidentum schwebt'. In *The Poetic Edda*, II, *The mythological poems*, ed. and transl. U. Dronke, Oxford, 1997, pp. 60, 104, Ursula Dronke interprets the idea of Gimlé in the light of ancient Scandinavian world-view.

[50] See further A. Hultgård, 'Old Scandinavian and Christian eschatology', in *Old Norse and Finnish Religions and Cultic Place-Names*, ed. T. Ahlbäck, Åbo, 1990, pp. 344–57.

second, poetry in praise of Christian rulers and saints; and third, didactic and homiletic literature in the vernacular.

Runic inscriptions

Runic inscriptions constitute direct sources from the late Viking period which convey important information on the christianization process in Scandinavia. Most of the inscriptions belong to the eleventh century and are found on raised stones, the great majority occurring in the Mälar region of eastern Sweden. The runes are usually carved within the lines of an animal, a serpent or a dragon (see Fig. 1) and, on the stones that were erected by Christian families, a large cross is depicted. The inscriptions are usually short and formulaic in character stating who had the stone raised in memory of whom. Besides the conventional formula many stones with a Christian iconography add a short prayer with an eschatological connotation. The formula varies but runs basically as follows: 'May God (Christ, God's mother or St Michael) help his soul (or spirit)'.[51] The formula alludes to the passage of the soul into the other world which could be difficult or dangerous. The main inspiration of this formula seems to have come from Christian ritual, more precisely the *commendatio animae*, the funeral mass of the Roman church, and to have reached Scandinavia through the early mission.[52]

Sometimes the prayer adds a reference to the place where one hopes the dead will arrive in the afterlife. The Christian paradise is certainly meant, although the terms to denote it may vary. In runic inscriptions we find *himinrīki* ('heaven', literally 'heaven's realm'), *himinríkis vist* ('heaven's dwelling place') and even *paradīs(i)*. For heaven we have the prayer on a rune stone from Västergötland, Sweden (Vg 186): 'God and God's mother help his soul, holy Christ in heaven.'[53] Two rune stones from Uppland in Sweden and one from the island of Bornholm in Denmark use the phrase 'light and paradise'.[54] The Risbyle stone (U 160) says:

> Ulfketill and Gyi and Unni had this stone raised after Ulf, their father, a good man. He lived in Skulhambr. God and God's mother help his spirit and soul, grant him light and paradise.[55]

The inscription of the Bornholm stone ends with a similar formula: 'Christ and Saint Michael help the souls of Ødbjørn and Gunnhild into light and paradise.'[56] The third

[51] For the variation of the formula, its distribution and frequency in the Mälar region, see F. Herschend, *The Recasting of a Symbolic Value--Three Case Studies on Rune Stones*, Uppsala, 1994, pp. 29–61.

[52] Cf. P. Beskow, 'Runor och liturgi', in *Nordens kristnande i europeiskt perspektiv*, eds P. Beskow and R. Staats, Skara, 1994, pp. 16–36.

[53] *Västergötlands runinskrifter*, IV, ed. E. Svärdström, Uppsala,1958. In eastern Viking age Scandinavian: 'Guð hialpi sālu hans ok Guðs mōðir, hæilagʀ Kristr ī himinrīki'.

[54] For these stones, see also S. B. F. Jansson, *Runes in Sweden*, Stockholm, 1987, pp. 113–14.

[55] *Upplands runinskrifter*, III/1, eds E. Wessén and S. B. F. Jansson, Uppsala, 1949. The concluding prayer runs in eastern Viking age Scandinavian: 'Guð hialpi hans and ok sālu ok Guðs mōðir, lē hanum liūs ok paradīs'.

[56] Klemensker stone 1. In eastern Viking age Scandinavian: 'Kristr hialpi siālu Auðbjarnar ok Gunnhildar, ok Santa Mikel ī liūs ok paradīs'.

runic inscription that refers to the paradise is the Folsberga stone (U 719): 'May Christ let Tumme's soul come into light and paradise and into the world best for Christians' (Fig. 1).[57]

The last phrase could be seen as an implicit reference to the fact that people who were not Christian believed in other afterlife places. The use of the word *hæimr* is also significant since the term may allude to the *hæimar*, the 'worlds' or 'abodes' of the Scandinavian gods. In eleventh-century Sweden, particularly the provinces around Lake Mälaren, large groups of the common people and some families of the aristocracy were still adherents of the traditional Scandinavian religion.

Early Christian Eulogistic Poetry

This poetry is, by its form and style, a continuation of an established pre-Christian genre but the religious world-view is a new one. The poems expressing the Valhǫll idea were dedicated to rulers who in one way or another became involved in the confrontation of Christianity with ancient Scandinavian traditions that were still vital in the tenth century. By the early eleventh century the afterlife hoped for by many of the ruling elite seems to have changed and now clearly conveyed Christian eschatological ideas. Valhalla was replaced by heaven and Odin by Christ. The first testimony of the changing afterlife concept for rulers is found within the genre of *erfidrápa* ('poem recited at a funeral feast'). The last stanza of the 'Erfidrápa', composed by Hallfrøðr Óttarson vandræðaskáld in the first years of the eleventh century and dedicated to Óláfr Tryggvason, expresses the hope that the pure Christ may have the soul of the wise king in heaven:[58]

kǿns hafi Kristr inn hreini	'May the pure Christ have
konungs ǫnd ofarlǫndum	the wise king's soul in the land above'

The 'Glælognskviða' of Þórarinn loftunga, which in all probability dates from the late thirties of the eleventh century, focuses on the sainthood of Óláfr Haraldsson[59] but the poem also mentions that the king 'went to heaven' (*hvarf til himinríkis*, stanza 3) and that he had struggled to arrive there (stanza 4) :

hafði sér	'Harald's son
harðla ráðit	won himself
Haraldz sonr	Heaven through
til himinríkis	hardship'

[57] Folsberga stone: 'Kristr lāti koma and Tumma ī liūs ok paradīsi ok ī þann hæim bæzta kristnum'.

[58] F. Jónsson, *Den norsk-islandske skjaldedigtning*, I, A: Tekst efter håndskrifterne [= text after/from the MSS]; B: rettet tekst [= corrected text], Copenhagen, 1912, repr. 1967 and 1973, p. 166.

[59] Ibid., pp. 324–7. The poem is also edited and commented upon by W. Lange, *Studien zur christlichen Dichtung der Nordgermanen 1000–1200*, Göttingen, 1958, pp. 113–20.

Sigvatr's 'Erfidrápa Óláfs helga' is likewise more concerned with the holiness and miracles of Óláfr than with his being welcomed by God in heaven. Only a short reference in stanza 24, *hann sótti guð* ('he went to God'), alludes to this theme. In his *erfidrápa* for Haraldr harðráði, composed around 1067, Arnórr Þórðarson jarlaskáld states briefly that the king is now in heaven (stanza 18): *hefr afreka it øfra … heilǫg fold* ('The holy land above (now) has the forceful man'). A fragment of another *erfidrápa* for the same king composed by Stúfr enn blindi takes up the wording of Hallfrøðr Óttarson's poem: [60]

Hafi ríks þars vel líkar	'May the eternal abode have
vist of aldr með Kristi	the powerful Harald's soul with Christ
Haralds ǫnd ofar lǫndum	in the land above where it is good to be'

The eleventh-century laudations appear allusive in character when addressing the afterlife destiny of the rulers, just as are the contemporary runic inscriptions that honour Christian chieftains and farmers in eastern Scandinavia. In general both groups of sources limit themselves to mentioning the place where the dead come in the afterlife. The early examples denote this place by using, on one hand, the traditional concept of 'heaven' (*himinríki*); on the other by presenting terms that could evoke associations with pre-Christian notions, 'the land above' (*ofarlǫnd*), and the phrase 'the holy ground above' (*heilǫg fold … it øvra*) which is reminiscent of the *heilakt land* in 'Grímnismál' stanza 4 to denote the world of the gods. The use of *heimr* to denote paradise or heaven, the Christian afterlife place, is also significant.

By the twelfth century such poems can be seen to be a little more explicit, adding a note of divine welcome and by giving some details of paradise. The traditional skaldic diction is also reinforced. The theme of the dead ruler being welcomed in heaven by God appears clearly in the *Geisli* ('Beam of light'), composed by Einarr Skúlason in praise of St Olaf.[61] The poem was commissioned by the Norwegian king Eysteinn and was recited in 1153 by the poet himself 'in the very confines of the Christ Church' in Trondheim.[62] God had heaven opened up for the brave king: 'lét himinríki opnask fyr gram snjǫllum' (stanza 16 in prose syntax). Taking up the motif of Christ as 'the light of the world' (stanza 2) the poet calls him 'the light of the sun' which was extinguished on the cross, but announced to men 'a better light in the holy abode of another sun's wide seat', i.e. 'heaven': 'fyr betra (ljósi) heilags setrs ómjós annars rǫðuls' (stanza 3 in prose syntax) where *ómjós annars rǫðuls* is a kenning for God.[63]

[60] For the text, see Jónsson, *Den norsk-islandske skjaldedigtning* (n. 58 above), I, A, p. 404; Lange, *Studien* (n. 59 above), p. 30, and *Poetry from the Kings' Sagas*, ed. K. E. Gade, Turnhout, 2009, p. 353.

[61] Used here is M. Chase, *Geisli. A Critical Edition*, Toronto, 2005.

[62] *Morkinskinna*, ch. 97; cf. also Lange, *Studien* (n. 59 above), p. 121 and K. Attwood, 'Christian Poetry', in *A Companion to Old Norse-Icelandic Literature and Culture*, ed. R. McTurk, Oxford, 2005, pp. 43–63 (51); Chase, *Geisli* (n. 61 above), pp. 9–10.

[63] Cf. Chase, *Geisli* (n. 61 above), p. 128.

The use of Old Scandinavian ruler epithets and images is revived. God is called 'the highest chieftain' (*hæstr skjǫldungr*), who invites men to 'heaven's [?] abode' (*til himins vistar*, stanza 6). St Olaf is called 'prince' (*gramr*, *stillir*, e.g. stanzas 13, 15, 16, 20) and 'the ray from God's hall' (*geisli goðs hallar*, stanza 7), and he lives in 'Christ's highest hall' (*Krists lífir hann í hæstri hall*, stanza 11).

Bergsbók , a medieval manuscript of *Óláfs saga Helga*, has preserved a poem (*drápa*) in honour of Óláfr Tryggvason which is attributed to Hallfrøðr vandrædaskáld but generally considered to be a composition of the twelfth century.[64] The poem reviews Óláfr's forced christianization of the countries in the north and describes the battle at Svolder. The concluding stanza 28 is a prayer to Christ that he might give (*veita*) the dead king 'a place of light and the highest protection' (*stað bjartan ok hæsta hlíf*).

The Rekstefja, composed by Hallar-Steinn, is likewise a twelfth-century poem dedicated to Óláfr Tryggvason much in the same vein as the anonymous *drápa* to the same king.[65] It is said in stanza 33:

Þengill þróttar strǫngum	'Christ, heaven's king
þeim bauð Kristr af heimi	invited the powerful man from the world
byrtjalds beztu heilli;	to the best of happiness;
bragningi Guð fagni;	may God welcome the prince;
ygglaust alla þiggi	may the brave Olaf
eljunfimr á himnum	fearless receive in the heavens
Óláfr ózta sælu	all the highest bliss of the glorious abode
ítrbóls með gram sólar	together with the sun's king'

Paradise is here localized in the heavens (*á himnum*) and depicted as 'the glorious abode' (*ítrból*), which gives the righteous happiness (*heill*) and bliss (*sæla*) of an unsurpassable kind. The wording of the praise poems strikes a tone of ancient skaldic style but expresses an eschatological wish which is in full accordance with medieval Christian instruction as presented in the early vernacular texts.

Laudatory poems were also composed for Christian saints and apostles. Four stanzas of a poem called 'Jóansdrápa' by Gamli the cleric (latter half of the twelfth century), dedicated to the apostle Jón, are preserved in *Jóns saga postola*. In one of them (stanza 3), according to the introductory line of the saga, the presence of the Heavenly Jerusalem in Jón's soul is described:[66]

Systrungr hefr nú alt þats æstir	'The kinsman (Jón) of the chieftain of men (Christ)
ýta hilmis giptu flýti	now has all for which he prays the Giver of happiness:
sólar vangs með sjǫlvum deili	free from sins (he dwells) in full bliss
synda lauss í fullu yndi.	with the Ruler of paradise himself.'

[64] Jónsson, *Den norsk-islandske skjaldedigtning* (n. 58 above), I, A, pp. 573–8; I, B, pp. 567–74. Jónsson states: 'et digt fra det 12. årh.', I, A, p. 573; F. Paasche, *Kristendom og Kvad*, Kristiania [Oslo], 1914, p. 83.

[65] Jónsson, *Den norsk-islandske skjaldedigtning* (n. 58 above), I, A, pp. 543–52; I, B, pp. 525–34.

[66] For the text, see ibid., I, A, p. 561 and Lange, *Studien* (n. 59 above), p. 82.

The stanza concludes with the statement that 'the apostle Jón got all glory (*dyrð*) from the Prince of the angels (*engla stillir*)'. It is noteworthy that paradise is here denoted with the expression 'the sun's meadow' (*sólar vangr*). The use of *vangr* may be fortuitous here but the term could also have been chosen to evoke pre-Christian notions of paradise.[67]

Didactic and homiletic texts in the vernacular

The eschatology of the Christian runic inscriptions and the early Christian poetry of praise is allusive in character. I suggest, however, that these sources presuppose the type of fuller descriptions that we find in the early didactic and edifying texts translated into Old Norse mainly during the twelfth century, but occasionally already in the latter part of the eleventh century. These texts are for the most part based on Latin models[68] but the translations range from literal renderings to free adaptations. Passages that from the outset were composed in the vernacular are sometimes added. *The Icelandic Homily Book* (IHB) and the *Old Norwegian Homily Book* (*Gamal Norsk Homiliebok*, GNH) are the two most important collections of didactic and edifying texts in the vernacular. These texts were studied in monasteries and priest schools and, if not read as such, they provided models for the sermons recited to the common people on Sundays and Church holidays.[69] The pieces that make up these collections are of varied origin, the model texts can in many cases be identified, but in others they are unknown. Also useful is the *Elucidarius* written by Honorius Augustodunensis around the year 1100 which was very influential in spreading Christian world-view and theology to broader circles of the population in medieval Western Europe.[70] It was translated into Old Norse already in the twelfth century and soon became popular also in the north.[71] In a wide sense early sagas of saints, translated and adapted to a Scandinavian milieu, can be regarded as edifying texts directed to a wider audience. *Placidus saga* and *Clemens saga* are works of this genre that were produced in the latter half of the twelfth century.[72] Some specimens of the visionary literature in Old Norse,

[67] Compare Fólkvangr, the 'paradise' of Freyja, and the variant *Þrúðvangar* for the otherworldly abode of Thor, cf. above. On the other hand, the early Christian literature in the vernacular rarely denotes the paradise with *vangr*, either alone or in compounds; it is not attested, for example, in the *Old Icelandic Homily Book* and the *Lexikon Poeticum* gives only the above mentioned passage.

[68] Old English models are likely to have played some part too. For the difficulty of finding the precise models used by the Old Norse translators, see Wellendorf, *Kristelig visionslitteratur* (n. 7 above), pp. 30–35.

[69] Cf. S. Óskarsdóttir, 'Prose of Christian Instruction', in *A Companion to Old Norse-Icelandic Literature and Culture*, ed. McTurk (n. 62 above), pp. 338–53 (341). It seems that the custom of giving a sermon at every Sunday mass was not fully established until the 13th century; cf. A. Nedkvitne, *Lay Belief in Norse Society 1000–1350*, Copenhagen, 2009, p. 52.

[70] Cf. *Elucidarius*, in *Old Norse Translation*, eds E. S. Firchow and K.Grimstad, Reykjavík, 1989, pp. xvii and xxii.

[71] See *Elucidarius* (n. 70 above), pp. xxiii and xxviii.

[72] For the latter, see *Clemens saga. The Life of St Clement of Rome*, ed. and transl. H. Carron, London, 2005.

in particular those found in the *Dialogues* of Gregory the Great, can also be dated to the twelfth century; but the majority of the Old Norse visionary texts belong to the thirteenth and fourteenth centuries.[73]

The manuscripts in which this literature has come down to us belong to the oldest ones that have survived. They can be dated to the decades around the year 1200 and, in some respects, they show an earlier linguistic stage than the manuscripts and texts of the thirteenth and fourteenth centuries.[74] The texts have been studied in the first place for linguistic, philological and palaeographical reasons, but less so for their religious and cultural interest. Yet they form the most important group of sources for the study of the type of Christian teachings that reached Viking age and early medieval Scandinavia, although no specimens of the missionary preaching itself have survived.[75]

The Christian afterlife place of the righteous is usually denoted with terms based on the word *himinn* ('heaven') either in the singular or the plural. The most common term is *himinríki* ('heaven', literally 'the realm of heaven') or *himnaríki* (literally 'the realm of the heavens').[76] Often *himinn* alone stands for the afterlife place.[77] Common also is the expression *himinríkis vist* ('heaven's abode').[78] The word *paradís* or *paradísus* in its eschatological sense occurs rather frequently and is sometimes combined with *vist* ('abode').[79] The expression *eilíf vist* ('the eternal abode') is also used.[80] Another image is that of a city in heaven called *himneskr Ierusalem* ('the Heavenly Jerusalem') or *himnesk borg* ('the heavenly city').[81] Sometimes heaven is described as a splendid hall as in the

[73] For the visionary pieces in the *Dialogues* an early date is apparent: cf. Wellendorf, *Kristelig visionslitteratur* (n. 7 above), p. 157. *Páls leizla* has been supposed to belong to the 12th century, but the arguments for that dating can be seriously questioned: see Wellendorf, ibid., pp. 142–4.

[74] The oldest Norse manuscript known is a fragment usually dated to *c.* 1150 containing a homily. The manuscripts of the above mentioned collections (Cod. AM 619 4° and Perg. 15 4° of the Royal Library in Stockholm) are dated to around the year 1200: *GNH, Cod. AM 619 4°*, ed. G. Indrebø, Oslo, 1931, pp. *38–*39, and *IHB Perg. 15 4° in the Royal Library, Stockholm*, ed. A. de Leeuw van Weenen, Reykjavík, 1993, Introduction, pp. 3–4. The oldest manuscript of *Elucidarius* can be dated approximately to about 1200: see the discussion in *Elucidarius* (n. 70 above), pp. xxviii–xxxi. Linguistic features pointing at an earlier language stage are, for example, *es* and *vas* for later *er* and *var*, and the reflexive form ending in – *sk* of verbs for later – *st* and – *z*.

[75] The question of the missionary preaching is addressed by Lange, *Studien* (n. 59 above), pp. 253–64 and by L. von Padberg, *Die Inszenierung religiöser Konfrontationen. Theorie und Praxis der Missionspredigt im frühen Mittelalter*, Stuttgart, 2003. Lange is, in my view, too sceptic about the use of the Old Norse homily collections as sources for getting an idea of the preaching of the missionaries.

[76] *GNH* (n. 74 above), 54, 3, 33; 65, 13; 66, 4 (*bis*); 71, 25; 88, 28 (*bis*); 154, 27; 158, 21, 28, 32 and 34; 159, 8. *IHB* (n. 74 above), 1r15, 17; 7v12, 32; 9v17.

[77] *GNH* (n. 74 above), 32, 16 (plur); 33, 34 (plur); 35, 11 (plur); 99, 34 (sing); 134, 11–12 (*bis* sing).

[78] Ibid., 37, 33; 101, 12; 149, 34. *IHB* (n. 74 above), 62v21; 97v16. *Clemens saga* (n. 72 above) 36, 13.

[79] *GNH*, (n. 74 above), 71, 23; 142, 13–14 (AM 237 a; Michael chieftain of paradise); 165, 5–6. *IHB* (n. 72 above), 11v7; 12v1; 22r8; 23r25; 29r5 (for all attestations, see A. de Leeuw van Weenen, *Lemmataized Index to The Icelandic Homily Book*, Reykjavík, 2004, p. 126); *paradísar vist*: *IHB* (n. 74 above), 20v15; 27r18.

[80] *IHB* (n. 74 above), 7r22.

[81] *GNH* (n. 74 above), 54, 31–32; 89, 25–26. *IHB* (n. 74 above), 7v30; 8r5; 85r20; 86v1. *Elucidarius* (n. 70 above), p. 136 (Latin text III, 74: *caelestis Ierusalem*).

Elucidarius: 'God almighty made for himself a splendid hall (*albjǫrt hǫll*) which he called the realm of heavens (*ríki himna*)'.[82] The term *konungshǫll* ('the King's hall') also occurs.[83] Other instances may indicate the same usage but here the 'hall' could also be interpreted as a building of its own within the realm of paradise.[84]

To give some idea of how paradise was presented to Scandinavians in the early Christian period, i.e. the eleventh and twelfth centuries, three passages from sermons in Old Norse language are here cited. The first two descriptions are characterized by the use of an emphatic *þár* ('there'). The following passage is taken from the 'Sermo necessaria' of the Old Norwegian collection:

> Heaven (*himinríki*) is so good that no man can imagine it or tell it to anybody else.
> There everlasting life and eternal light are found.
> There delight and joy and all sorts of beauty (*prýði*), glory and happiness without end (shall be),
> there a person will not fall ill nor will he grow old (*ofgamall*),
> there no hunger and thirst shall be,
> there no anxiety and sorrow.
> There every man always loves his fellow man as himself.
> There all good men (*góðir menn*) shall see God for ever and ever and live with him and his angels in all eternity.
> Blessed (*sælir*) are those men who shall experience such things.[85]

The second one comes from a sermon in the *Icelandic Homily Book* introduced by the words *Frá upphafi heims* ... 'From the beginning of the world ...':[86]

> This seems now beautiful to us, gold and precious objects, fine halls and houses (*hallir góðar ok bøir góðir*), and other things we own (*eiga ór ǫnnur*). Yet let us consider how (much more) beautiful are the halls where all seats shine with gold and precious stones (*í gulli ok gimsteinum*).
> There the blossom of the herbs (*blómi grasa*) does not fade,
> there the fruits of the woods (*aldin skóga*) do not decay,
> there streams of honey fall on evergreen fields (*á ei grønum vǫllum*),
> there all herbs smell like the finest incense (*sem it dýrsta reykelsi*),
> there light is shining without shadow, the air is bright without mist and (it is) eternal day without darkness.[87]

[82] The Latin model has: 'Primo igitur Deus, ut praepotens rex, constituit sibi praeclarum palatium, quod dicitur regnum caelorum', *Elucidarius* (n. 70 above), p. 25.

[83] Ibid., p. 30. Another metaphor for heaven is 'the day's hall' (*dags hǫll*), 'Leiðarvísan' 45.

[84] 'Leiðarvísan' 38.

[85] *GNH* (n. 74 above), 88, 27–35.

[86] Two other versions of this sermon are known in Old Norse: see A. Loth, 'Til Sebastianus saga', in *Opuscula*, 5 (Bibliotheca Arnamagnæana 31), Copenhagen, 1975, pp. 103–22, and S. Karlsson, 'Om himmel og helvede på gammelnorsk, AM 238 XXVIII fol.', in *Festskrift til Ludvig Holm-Olsen*, eds B. Fidjestøl et al., Øvre Eirvik, 1984, pp. 185–96. The contents derive partly from Pseudo-Ambrose's *Acta S. Sebastiani martyris*: see *IHB* (n. 74 above), Introduction, p. 13.

[87] *IHB* (n. 74 above), 66r7–13.

The third passage occurs in the sermon 'Postolamál' of the *Icelandic Homily Book* and is a description of the heavenly city inspired by Revelation Chapter 21:[88]

> These are the twelve precious stones above the twelve gates of the Heavenly Jerusalem that St John the Evangelist (*Johannes postoli*) saw in his heavenly vision. Jerusalem means 'vision of peace' (*friðar sýn*) and signifies (*merkir*) the heavenly peace which God gives his friends (to be) with himself in the heavens *(á himnum*). The twelve gates of this city (*borg*) are the teachings (*kenningar*) of the twelve apostles. They are the ones who open up for us the entrance to heaven (*inngǫngu himinríkis*) and to the universal faith (*almennilig trúa*). This city has properly four corners (*es makliga ferskeytt*) and is equally large (*jafnmikil*) on all sides because one is the universal faith that is preached among all the generations of the world and one the glory of heaven that is offered to all those who go or come to her (the city) with just (*jǫfnun*) merits. The twelve precious stones above the gates are the twelve apostles who adorned the whole of God's Christendom in their life and teachings. We should be aware of the fact that John saw the names of the apostles written above the gates of the Heavenly Jerusalem. Why would the names of the apostles be written above the gates of the heavenly city if it were not for some token (*tákn*) attached to their names that should show us the way to heaven's glory? To that end it is necessary that we render in our own language (*þýðem til várs máls*) their names. Then we shall seek out each meaning (*skynsemi*) we find in the names. Peter means (*þýðisk*) stone, Andrew brave (*drengiligr*) ... (there follow the other apostles' names rendered in the vernacular).[89]

The relationship of the names with the afterlife is then explained. Everyone who wants to have a share in heaven with God's apostles must be a Peter, an Andrew etc.: 'he must become Peter, that is a living stone in the temple of God.'[90] Similar allegorical interpretations are then given of the names of the other apostles.

The emphasis of the passage is put on the symbolic explanation of the heavenly city although its real existence is in no way denied. The allegoric connection between the form of the city and the merits of the righteous is underlined by a play on the word *jafn* in its sense of both 'equal' and 'just, right'.

The three passages cited convey different aspects of paradise. The sermon in the Old Norwegian collection indicates several recurrent topics of Christian paradise descriptions whereas the Old Icelandic sermon takes up a less prominent aspect of these descriptions, viz. the green meadows and groves, and the splendid halls. In general, emphasis is placed on the light (*ljós*) and the glory (*dýrð*) of paradise.[91] Frequently the terms *fagnaðr* ('joy') and *sǽla* ('bliss') are used to sum up the excellence

[88] On other sources for this passage, see ibid., Introduction, p. 8.

[89] Ibid., 7v28–8r9.

[90] Ibid., 8r13–14.

[91] As to the light see, for example, *GNH* (n. 74 above), 32, 13; 38, 1; 88, 30; 158, 7; *IHB* (n. 74 above), 24r22. The glory of paradise is mentioned, e.g., in *GNH* (n. 74 above), 31, 8; 32, 16, 19; 54, 26; 57, 9; *IHB* (n. 74 above), 5r4; 9v21; 13r19; 17r21; 22r6; 29r4; *Elucidarius* (n. 70 above), p. 126 (rendering the Latin III, 19: *gloria*).

of paradise.[92] The main activity of the righteous is to behold God and to live with him, Christ, the angels and the saints.[93]

Paradise also involved a theological problem. According to early Christian theology it is the event of resurrection and judgement that decides the place for each individual in the afterlife, be it paradise or hell. Meanwhile the righteous were thought to await the resurrection in some place of the nether world which, in Western Christendom from the twelfth to the thirteenth centuries, eventually developed into purgatory.[94] However, already in the beginnings of Christianity, the belief emerged that martyrs and even all Christians went straight to paradise after death. Two types of eschatology thus competed in the early and medieval Church.[95]

Conclusion

The introduction of a new paradise concept among the Scandinavian peoples can first be seen in the eleventh-century poems of praise and in contemporary runic inscriptions. Both types of sources hint at the new idea of paradise by using similar expressions – sometimes with connotations to pre-Christian culture. Generally it is a European medieval eschatology transposed into the vernacular but, in style and expression, adapted to a Scandinavian context.[96]

The early Scandinavian attestations are allusive in character but I suggest that the underlying eschatology has to be reconstructed from the fuller descriptions of heaven and hell as they are found in the Old Norse religious texts from the twelfth century.[97] In fact vocabulary and main points of emphasis support this hypothesis. The terms 'paradise' (*paradīs*) and 'heaven' (*himinrīki*) already occur on the earliest Christian rune stones, as well as the 'light' (*liūs*) to characterize its excellence. The poet of the 'Glælognskviða' uses *himinríki* for the afterlife dwelling-place of king Olaf.

The twelfth-century poetry adds more evidence for the influence of the contemporary Christian teaching in the vernacular. The use of the expressions 'heaven's abode' (*himinríkis vist*) and 'highest bliss' (*ózta sǽla*) are good examples. A sepulchral stone with runes incised, belonging to the ancient cathedral of Oslo (Hallvardskirken) and dated to the end of the twelfth century, also shows that influence. The inscription ends with the

[92] For the term *fagnaðr* (or *fǫgnuðr*), see, for example, *GNH* (n. 74 above), 32, 16; 81, 30; 88, 31; 94, 34; 108, 28; 111, 20; 143, 23; 158, 7; 170, 1; *IHB* (n. 74 above), 13r24; 15r22; 32r2–3; 33v9; 34r7, 11; *Elucidarius* (n. 70 above), p. 137 (III, 76: *gaudium*), 142 (*fagnaþir* renders III, 87: *gaudia*). 'Bliss' *sǽla* e.g. *IHB* (n. 74 above), 15r22.

[93] See *GNH* (n. 74 above), 31, 7–8; 32, 10–11; 33, 34; 35, 11; 88, 33–5; 95, 1; 99, 34–5; 134, 22–3; 143, 23; 158, 7; *IHB* (n. 74 above), 22r5; 23r30; 29v8; 32r4.

[94] For the development of this idea and of purgatory, see Bremmer, *The Rise and Fall of the Afterlife* (n. 2 above), pp. 56–70; J. Le Goff, *La Naissance du purgatoire*, Paris, 1981 (English version: *The Birth of Purgatory*, Aldershot, 1990); and C. Hill, *Regnum Caelorum: Patterns of Future Hope in Early Christianity*, Oxford, 1992.

[95] See Hill, *Regnum Caelorum* (n. 94 above), pp. 13–18.

[96] Cf. Paasche, *Kristendom og kvad* (n. 64 above), pp. 35–40.

[97] Cf. also Hultgård, 'Old Scandinavian and Christian eschatology' (n. 50 above), pp. 347–9.

following exhortation: 'May everyone who knows to read (*rædr*) this piece of (runic) art (*list*) be granted heaven's abode (*himinríkis vist*)'.[98]

In Denmark and Sweden the eschatological allusions on the Christian rune stones from the eleventh, and the end of the tenth, centuries belong to a period before the Church had been fully established in the north and when literacy was not widespread. The references to paradise or heaven on these stones must reflect oral teaching in different settings, such as gatherings of common people at markets and assemblies (*þing*), at celebrations of the Christian mass, and in private instruction.

Medieval reception of ancient Scandinavian world-view

The reception of ancient eschatological myth and its partial adaptation to a Christian world-view appears most clearly in texts of the thirteenth century: in the first place the Prose Edda of Snorri Sturluson, but this is mainly a literary and antiquarian phenomenon which goes beyond the scope of the present study.[99] For the eleventh and twelfth centuries, with which we are concerned here, the evidence for a continuity of living faith is, however, much more limited. In this respect the Valhǫll idea as it appears in the poem known as 'Krákumál' appears important. The poem shows clear links with the Orkney Islands and was composed in praise of the legendary king Ragnar Loðbrok in the latter half of the twelfth century.[100] Lines 2 to 6 of the final stanza (29) express Ragnar's hope for an afterlife with Odin:

heim bjóða mér dísir	'The divine maidens invite me home
þærs frá Herjans hǫllu	those whom Odin has sent to me
hefr Óðinn mér sendar,	from the hall of Herjan,
glaðr skal ek ǫl með ásum	glad shall I drink the ale with the gods
í ǫndvegi drekka	in front of the high-seat'

This twelfth-century mention of the Valhalla idea anticipates the description in the Prose-Edda of Snorri and one has to ask for the reason why the poet took it up. The Orkney Islands were long since converted to Christianity and it is difficult to consider the contents of the stanza as a pagan survival.[101] It could either represent a deliberate revival in contrast to Church teaching or a literary device to render the afterlife hope of a pre-Christian chieftain.

Paradise descriptions compared

From the outset the comparison is complicated owing to the imbalance of the source materials. As already stated ancient Scandinavian ideas on paradise have to be extracted

[98] M. Olsen, with collaboration of A. Liestøl, *Norges Innskrifter med de Yngre Runer*, 5, Oslo, 1960, No. 549.

[99] For that type of adaptation, see A. Hultgård, 'Fornskandinavisk hinsidestro i Snorre Sturlusons spegling', in *Religion och samhälle i det förkristna Norden*, eds U. Drobin et al., Odense, 1999, pp. 109–23.

[100] See H.-P. Naumann, 'Krákumál', in *Reallexikon der Germanischen Altertumskunde* (n. 1 above), 17, 2001, pp. 299–302.

[101] Cf. Dillmann, 'Valhǫll' (n. 1 above), p. 356.

from a very limited number of sources whereas Christian texts on the afterlife, even in the vernacular, are much more numerous. Bearing this in mind it is nevertheless appropriate to make some remarks on continuity and influences.

In some respects ancient Scandinavian and Christian medieval images of paradise present close affinities. The green lands of the gods with the splendid halls where they live recall the evergreen fields and shining halls to which God welcomes the righteous. These similarities are not to be interpreted as an influence from pre-Christian tradition. Flowering fields, fruitful gardens and magnificent buildings belong with the set of images that make up paradise descriptions in Western as well as Eastern Christian traditions of the Antiquity and early Middle Ages. The relevance of the similarities lies elsewhere. In the mind of newly converted Scandinavians Christian teaching of the green meadows of paradise would have represented an image with which they were already familiar.

Along the same line we may interpret another correspondence. In death the Scandinavian heroes were thought to join Odin in his hall and, as we saw, these Einheriar, these otherworldly elite warriors, formed the retinue of the god. In Valhǫll they prepared themselves for the last battle at Ragnarok. Similarly there was an idea in medieval Christianity of a heavenly host of soldiers (*milites)*, composed of angels but also of men who by their martyrdom had qualified for that position. They formed, as it were, the retinue of God or Christ. According to the *Elucidarius* their number was fixed and could not be exceeded. After having described how God made heaven as a splendid hall the Old Norse translation continues: 'For His hall (*til hallar sinnar*) He established a certain number of chosen warriors (*ríðarar*) which had to be filled but was not permitted to be exceeded'.[102]

One of the chosen heroes to become such a warrior was Stephanus. The Old Norse sermons read to his honour call him 'Christ's soldier' (*Krists ríðari*) and he receives the attribute *sigrsæll* ('victorious'). Like one of the Einheriar he went in death to the divine hall where he joined the heavenly host.[103]

More important for the question of acculturation are the terms chosen to translate or adapt the Latin descriptions to Old Norse language and culture. For the first generations of Scandinavian Christians some of these terms still carried meaningful connotations to their recent pagan past. The use of the word 'hall' (Old Norse *hǫll*) to denote the Christian paradise or its buildings is illuminating. In the cases where a Latin model could be identified we find *palatium* (so *Elucidarius*) or *domus* (so in the *Acta Sancti Sebastiani martyris*).[104] The fact that many instances of the metaphor 'hall' for heaven or paradise occur in texts and passages composed directly in the vernacular indicates that this

[102] The Latin text runs: 'Ad quod palatium praedestinavit quemdam certum numerum electorum militum, quem nec liceret excedi et quem necesse esset compleri.' *Elucidarius* (n. 70 above), pp. 25–6.

[103] For this passage *GNH* (n. 74 above), 43, 32; 44, 1 and *IHB* (n. 74 above), 81r4–5, present the same wording: 'en í dag fór sigrsæll ríðari til himins hallar.'

[104] The Latin models of these texts are found in *Elucidarius* (see n. 68 above), and in Loth, 'Til Sebastianus saga' (n. 86 above).

metaphor was deliberately chosen to evoke associations with the halls of Viking Age Scandinavia. These halls were wooden, boat-shaped longhouses, often of considerable proportions, the residence of chieftains and well-to-do farmers.[105] The Old Norse word to denote the 'hall', also in its mythic sense, was *salr* or *hǫll*.[106] The hall was more than just an impressive building: it was a multifunctional building expressing the culture and the values of the Scandinavian aristocracy.[107] The hall was a place for religious, social and political ceremonies and could also serve as a dwelling-place. Several passages from Beowulf give a glimpse of the function of the hall which can shed light also on the Scandinavian halls.[108] In the course of the eleventh and twelfth centuries it seems that this type of cultural institution gradually lost its importance although wooden banqueting halls (*veizluhallir*) continued to be built in the thirteenth century, especially in Norway but now in a Christian cultural setting.[109] The hall as a pre-Christian phenomenon was replaced by other types of buildings and institutions: manors for the ruling nobility, preferably built in stone or brick, and churches for religious purposes.[110] The religious function was transferred to new specific buildings. The hall as a multifunctional building characteristic of ancient Scandinavian culture thus disappeared. Yet the concept of paradise or heaven as a great hall, promulgated in the Old Norse religious texts of the twelfth century, did still carry the association with the pagan past and its splendid halls.

[105] The characteristics of the Scandinavian hall as a building are indicated in F. Herschend, 'Halle', in *Reallexikon der Germanischen Altertumskunde* (n. 1 above), 13, 1999, pp. 414–25; J.-H. Fallgren, 'Farm and village in the Viking Age', in *The Viking World*, ed. S. Brink, London and New York, 2008, pp. 67–76.

[106] For the discussion of these terms, see Brink, 'Political and Social structures in Early Scandinavia' (n. 43 above), pp. 251–8.

[107] Examples of multifunctional chieftain halls include Lejre and Tissø in Denmark: T. Christensen, 'Lejre beyond legend: the archaeological evidence', *Journal of Danish Archaeology*, 10, 1991, pp. 163–85; L. Jørgensen, 'Manor, Cult and Market at Lake Tissø', in *The Viking World*, ed. Brink (n. 105 above), pp. 77–82; Borg and Skiringssal in Kaupang, both in Norway: G. Stamsø et al., *Borg in Lofoten. A Chieftain's Farm in North Norway*, Trondheim, 2003; *Kaupang in Skiringssal*, ed. D. Skre, Aarhus, 2007; Old Uppsala, Lunda and Slöinge in Sweden: *Arkeologi och miljögeologi i Gamla Uppsala*, ed. W. Duczko, Uppsala, 1993; J. Callmer, 'Slöinge', in *Reallexikon der Germanischen Altertumskunde* (n. 1 above), 29, 2005, pp. 112–14; and Hofstaðir in Iceland: *Hofstaðir 2002. Framvinduskýrslur/Interim Report*, ed. G. Lucas, Reykjavík, 2003; G. Lucas and T. McGovern, 'Bloody Slaughter: Ritual Decapitation and Display at the Viking Settlement of Hofstaðir', *European Journal of Archaeology*, 10, 2007, pp. 7–30.

[108] *Beowulf*, ed. C. L. Wrenn, fully rev. W. F. Bolton, London, 1973: pp. 64–85, 491–8 and 607–41 are illuminating.

[109] Cf. Brink, 'Political and Social structures in Early Scandinavia' (n. 43 above), pp. 242–3.

[110] For these changes cf. also Orri Vésteinsson, 'Archaeology of Economy and Society', in *A Companion to Old Norse-Icelandic Literature and Culture*, ed. McTurk (n. 62 above), pp. 7–26 (22).

Figures

Fig. 1. Nativity, Monastery of Hosios Loukas. Greece, 11th century (photo: Jean Housen).

Fig. 2. Theophany, Monastery of Hosios Loukas. Greece, 11th century (photo: Hans Rosbach).

Fig. 3. Anastasis, Monastery of Hosios Loukas. Greece, 11th century (photo: Wikimedia).

Fig. 4. Nativity, Perivleptos Church, Mystra, Peloponnese, 14th century (photo: Directmedia).

Fig. 5. Anastasis, Chora Church. Istanbul, 1316–1321 (photo: José Luiz Bernardes Ribeiro).

Fig. 6. Dome of Genesis, St. Mark's Basilica, Venice, 1220–1240 ca. (photo: Wikimedia).

Fig. 7. Theophanes the Greek, Icon of the Transfiguration, Cathedral of the Transfiguration in Pereslav, 1403 (photo: Wikimedia).

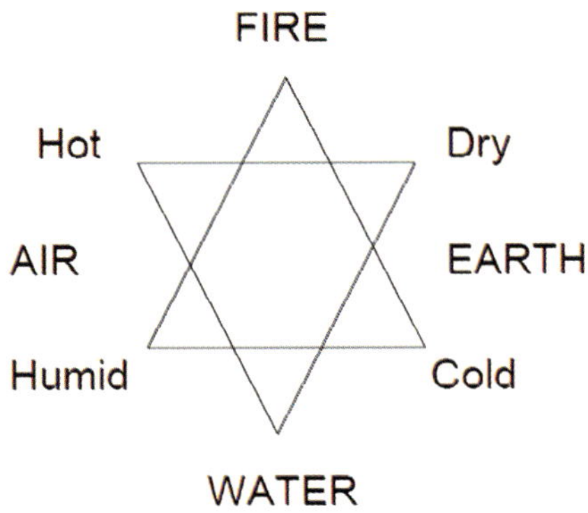

Fig. 8. Solomon's seal after M. Chelli, *Manuale dei simboli nell'arte: l'era paleocristiana e bizantina*, Rome, 2008 (design by the author).

Fig. 9. Illumination from Four Gospels ms., with prophets Elijah and Ezekiel proclaiming their visions of God presented as the youthful Christ Emmanuel framed by the symbols of the four evangelists. Constantinople, mid-12th century, Venice z540=557 (Biblioteca Nazionale Marciana).

Fig. 10. Mosaic of the Majestas Domini on the central dome of Saint Mark's basilica, Venice, 13th century (photo by the author).

Fig. 11. Ascension, Istituto e Museo Ellenico, Venice, second half 15th century (photo: Istituto Ellenico e Museo Ellenico di Venezia).

Fig. 12. Cosmas Indicopleustes's tabernacle-shaped cosmos, Sin. Gr. 1186, fol. 69r (by permission of Saint Catherine's Monastery, Sinai, Egypt).

Fig. 13. Christ Pantocrator surrounded by the signs of the zodiac and the angelic orders. Holy Monastery of Docheiariou, Mount Athos, 1568 (photo: Fr. Apollò Docheiarite).

Fig. 14. Transfiguration from the *katholikon* of the Holy Monastery of Docheiariou, Mount Athos, 1568 (photo: Fr. Apollò Docheiarite).

Fig. 15. Polyeleos, Theoskepastou Monastery, Sochos, Greece (photo: Fr. Apollò Docheiarite).

Fig. 1. Paradise in a Mughal manuscript illustration.
Washington DC, Freer Gallery of Art, Smithsonian Institution, inv. no. F1968.3, gift of Professor Alban G. Widgery of Winchester VA, *c.* 1800.

Fig. 2. Cosmographical diagram from Ibrāhīm Ḥaqqī (İbrahim Hakkı), *Maʿrifetnāme* ('The Book of Gnosis'), AD 1756–7 (H 1170). London, British Library, MS Or. 12964, fol. 23 v, copied in Erzerum, AD 1820 (H 1235).

Fig. 3. Cosmographical diagram from Ibrāhīm Ḥaqqī (İbrahim Hakkı), *Maʿrifetnāme* ('The Book of Gnosis'), AD 1756–7 (H 1170). London, British Library, MS Or. 12964, fol. 24 r, copied in Erzerum, AD 1820 (H 1235).

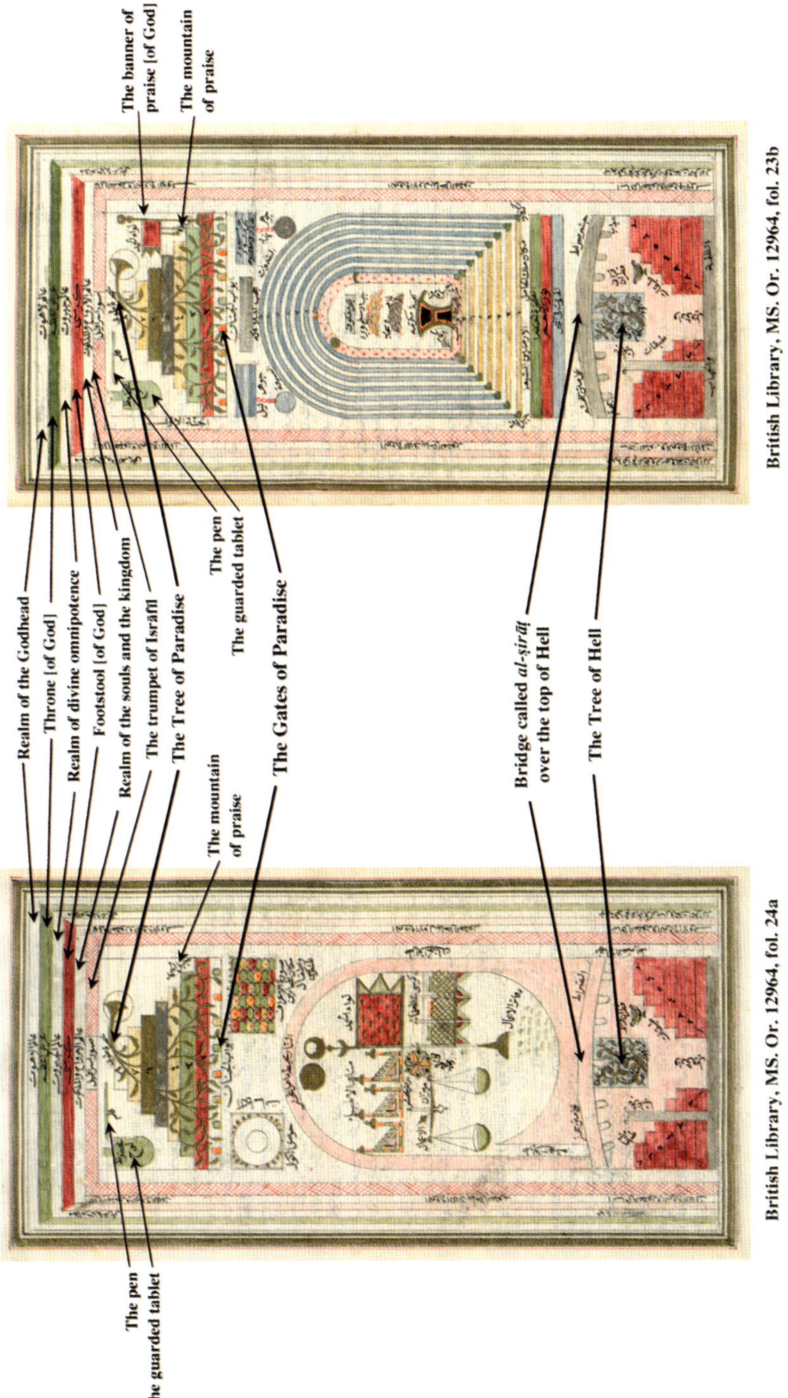

Fig. 4. Explanation of Figs 2 and 3.

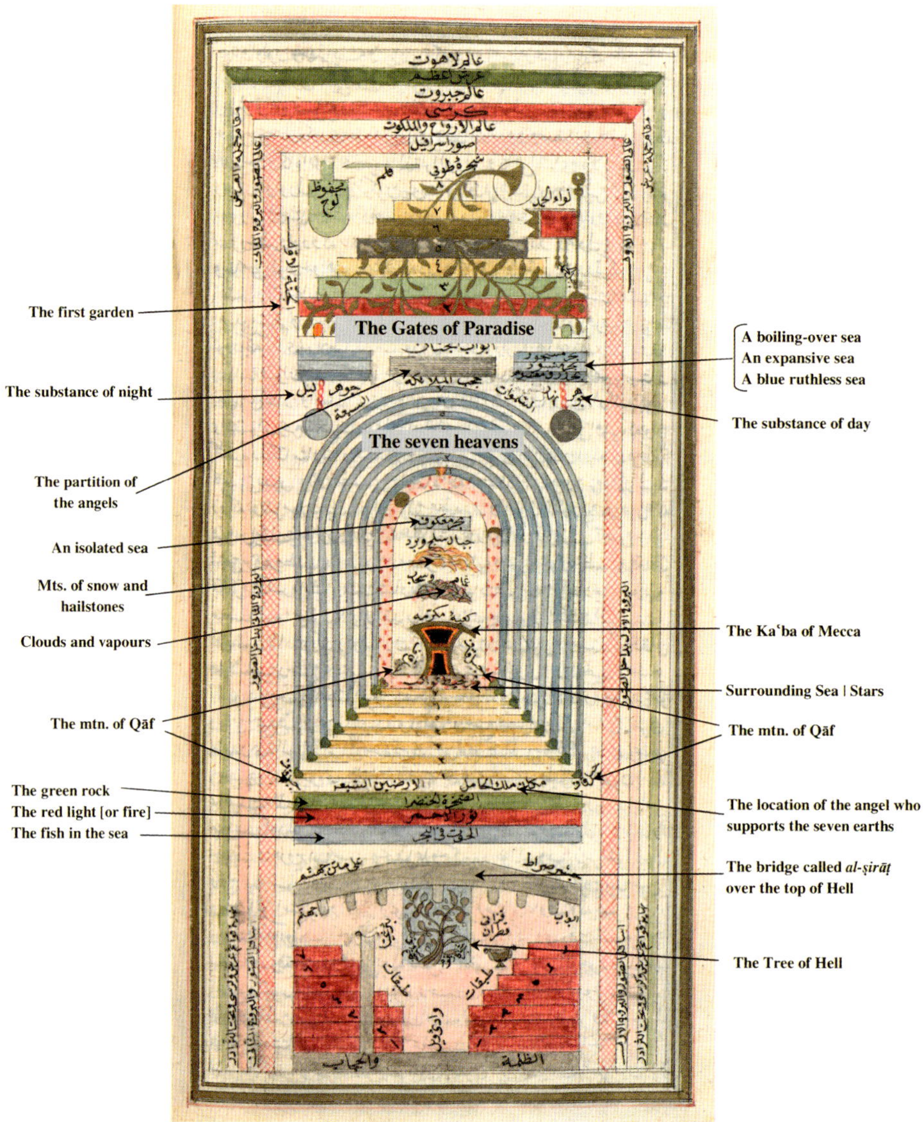

Fig. 5. Explanation of Fig. 2.

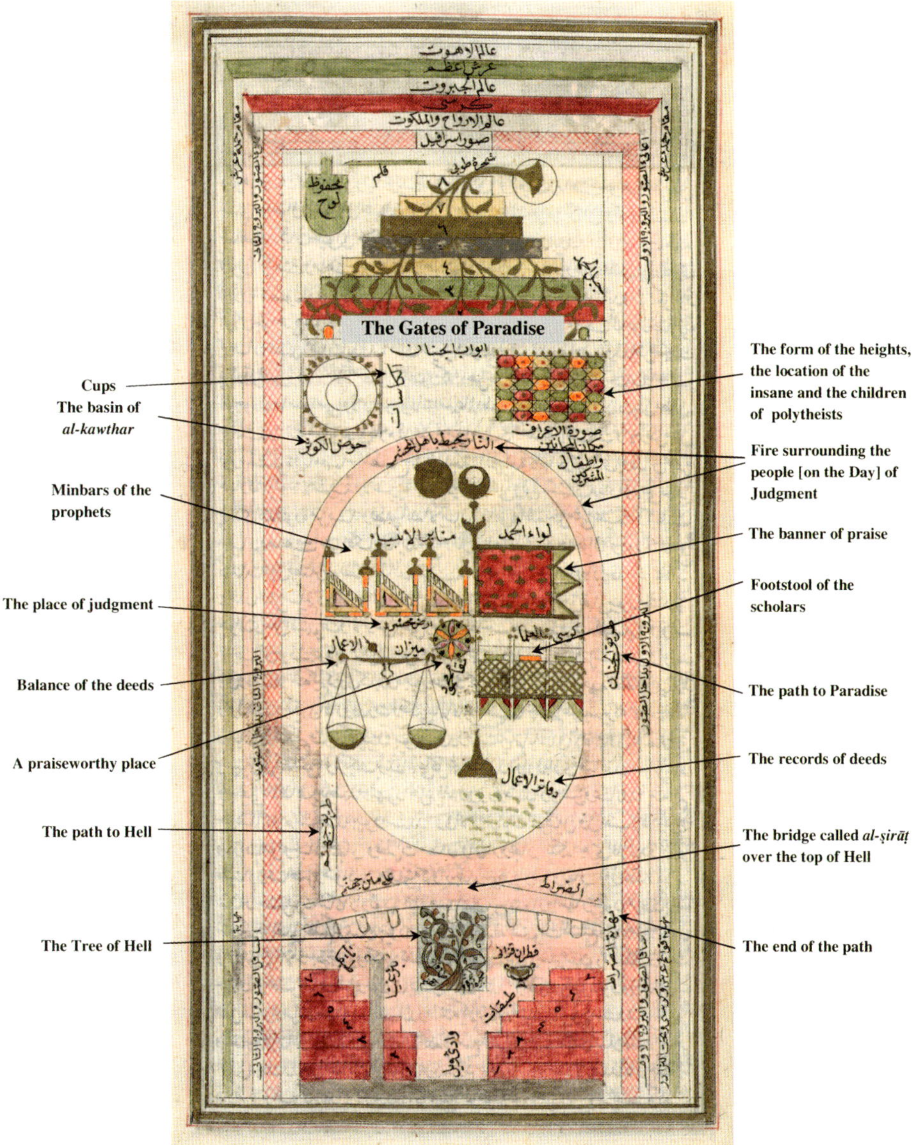

Fig. 6. Explanation of Fig. 3.

Fig. 1. The Folsberga rune stone in Uppland (U 719), Sweden, 11th century (photo: Ulla-Maj Hultgård).

Index of Names